Essential Oil-Bearing Grasses

The genus *Cymbopogon*

Medicinal and Aromatic Plants — Industrial Profiles

Individual volumes in this series provide both industry and academia with in-depth coverage of one major genus of industrial importance.

Series Edited by Dr. Roland Hardman

Essential Oil-Bearing Grasses

The genus *Cymbopogon*

Edited by **Anand Akhila**

Medicinal and Aromatic Plants — Industrial Profiles

CRC Press
Taylor & Francis Group
Boca Raton London New York

CRC Press is an imprint of the
Taylor & Francis Group, an **informa** business

CRC Press
Taylor & Francis Group
6000 Broken Sound Parkway NW, Suite 300
Boca Raton, FL 33487-2742

ISBN 13: 978-0-367-44603-1 (pbk)
ISBN 13: 978-0-8493-7857-7 (hbk)

Library of Congress Cataloging-in-Publication Data

Essential oil-bearing grasses : the genus Cymbopogon / editor: Anand Akhila.
     p. cm. -- (Medicinal and aromatic plants--industrial profiles ; v. 46)
  Includes bibliographical references and index.
  ISBN 978-0-8493-7857-7 (hardcover : alk. paper)
    1. ˜Cymbopogon. 2. Cymbopogon--Industrial applications. 3. Cymbopogon--Therapeutic use. ˜I
Akhila, Anand. II. Title. III. Series: Medicinal and aromatic plants--industrial profiles ; v. 46.

QK495.G74E745 2010
661'.806--dc22
                                                    2009024407

Visit the Taylor & Francis Web site at
http://www.taylorandfrancis.com

and the CRC Press Web site at
http://www.crcpress.com

Contents

Preface to the Series

There is increasing interest in industry, academia, and the health sciences in medicinal and aromatic plants. In passing from plant production to the eventual product used by the public, many sciences are involved. This series brings together information that is currently scattered through an ever-increasing number of journals. Each volume gives an in-depth look at one plant genus about which an area specialist has assembled information ranging from the production of the plant to market trends and quality control.

Many industries are involved, such as forestry, agriculture, chemical, food, flavor, beverage, pharmaceutical, cosmetic, and fragrance. The plant raw materials are roots, rhizomes, bulbs, leaves, stems, barks, wood, flowers, fruits, and seeds. These yield gums, resins, essential (volatile) oils, fixed oils, waxes, juices, extracts, and spices for medicinal and aromatic purposes. All these commodities are traded worldwide. A dealer's market report for an item may say "drought in the country of origin has forced up prices."

Natural products do not mean safe products, and account of this has to be taken by the above industries, which are subject to regulation. For example, a number of plants that are approved for use in medicine must not be used in cosmetic products.

The assessment of "safe to use" starts with the harvested plant material, which has to comply with an official monograph. This may require absence of, or prescribed limits of, radioactive material, heavy metals, aflatoxin, pesticide residue, as well as the required level of active principle. This analytical control is costly and tends to exclude small batches of plant material. Large-scale, contracted, mechanized cultivation with designated seed or plantlets is now preferable.

Today, plant selection is not only for the yield of active principle, but for the plant's ability to overcome disease, climatic stress, and the hazards caused by mankind. Such methods as in vitro fertilization, meristem cultures, and somatic embryogenesis are used. The transfer of sections of DNA is giving rise to controversy in the case of some end uses of the plant material.

Some suppliers of plant raw material are now able to certify that they are supplying organically farmed medicinal plants, herbs, and spices. The Economic Union directive CVO/EU No. 2092/91 details the specifications for the obligatory quality controls to be carried out at all stages of production and processing of organic products.

Fascinating plant folklore and ethnopharmacology lead to medicinal potential. Examples are the muscle relaxants based on the arrow poison curare from species of *Chondrodendron*, and the antimalarials derived from species of *Cinchona* and *Artemisia*. The methods of detection of pharmacological activity have become increasingly reliable and specific, frequently involving enzymes in bioassays and avoiding the use of laboratory animals. By using bioassay-linked fractionation of crude plant juices or extracts, compounds can be specifically targeted which, for example, inhibit blood platelet aggregation, or have antitumor, antiviral, or any other required activity. With the assistance of robotic devices, all the members of a genus may be readily screened. However, the plant material must be fully authenticated by a specialist.

The medicinal traditions of ancient civilizations such as those of China and India have a large armamentarium of plants in their pharmacopoeias that are used throughout Southeast Asia. A similar situation exists in Africa and South America. Thus, a very high percentage of the world's population relies on medicinal and aromatic plants for their medicine. Western medicine is also responding. Already in Germany all medical practitioners have to pass an examination in phytotherapy before being allowed to practice. It is noticeable that medical, pharmacy, and health-related schools throughout Europe and the United States are increasingly offering training in phytotherapy.

Multinational pharmaceutical companies have become less enamored of the single compound, magic-bullet cure. The high costs of such ventures and the endless competition from "me-too" compounds from rival companies often discourage the attempt. Independent phytomedicine companies have been very strong in Germany. However, by the end of 1995, 11 (almost all) had been acquired by the multinational pharmaceutical firms, acknowledging the lay public's growing demand for phytomedicines in the Western world.

The business of dietary supplements in the Western world has expanded from the health store to the pharmacy. Alternative medicine includes plant-based products. Appropriate measures to ensure their quality, safety, and efficacy either already exist or are being answered by greater legislative control by such bodies as the U.S. Food and Drug Administration and the recently created European Agency for the Evaluation of Medicinal Products based in London.

In the United States, the Dietary Supplement and Health Education Act of 1994 recognized the class of phytotherapeutic agents derived from medicinal and aromatic plants. Furthermore, under public pressure, the U.S. Congress set up an Office of Alternative Medicine, which in 1994 assisted the filing of several Investigational New Drug (IND) applications required for clinical trials of some Chinese herbal preparations. The significance of these applications was that each Chinese preparation involved several plants and yet was handled as a *single* IND. A demonstration of the contribution to efficacy of each ingredient of each plant was not required. This was a major step toward more sensible regulations in regard to phytomedicines.

I always look forward to the journal *HerbalGram* (HG), of the American Botanical Council, which was founded by Mark Blumenthal in 1988. He continues as the Executive Director and as the Editor of HG. In it he regularly justifies the status of a medicinal plant and challenges official bodies when necessary. In HG Number 80 (2008), he tells the U.S. Food and Drug Administration to rescind the 1991 IMPORT ALERT on the herb stevia (Vol. 19 in this series) because the United Nations and the World Health Organization have concluded that stevia extract, containing 95% stevia glycosides, is safe for human use as a sweetening agent, in the range 4 mg/kg body weight per day. This has paved the way for regulatory approval around the world for the use of this low-cost noncaloric material and notably for those obese persons facing cardiovascular disease and diabetes.

For this volume, I thank its editor, Dr Anand Akhila, for his dedicated work and the chapter contributors for their authoritative information. My thanks are also due to Barbara Norwitz of CRC Press and her staff for their unfailing help.

Roland Hardman, BPharm, BSc (Chem), PhD (London), FR Pharm S.

Preface

The essential oils of the grasses of species of *Cymbopogon* have an industrial profile; they are used in beverages, foodstuffs, fragrances, household products, personal care products, pharmaceuticals, and in tobacco.

These oils are sourced from around the world by, for example, Fuerest Day Lawson (FDL) Ltd. It is fascinating that this company is on the same site where plant raw material has arrived for the past 400 years—close to the Tower of London and the River Thames. No longer in a warehouse but in a modern tower block, FDL has all the equipment to test the efficacy of an essential oil for a particular purpose. For the producer of the oil, FDL will advise on the development of a modern production process (in the country of origin if required), even to the stage of the final saleable product.

FDL's technical director, David A. Moyler, regularly attends meetings of the European Union in Brussels concerned with the regulations covering the commercial production and sale of such products, and he has contributed a relevant chapter to this book.

The genus *Cymbopogon* (family Gramineae) has many species of grasses that grow in tropical and subtropical regions around the world from mountains to grasslands to arid zones. These plants produce essential oils with pleasant aromas in their leaves.

Five species yield the three oils of main commercial importance: lemongrass from *C. citratus* of Malaysian origin (West Indian lemongrass) and *C. flexuosus* (East Indian lemongrass) from India, Sri Lanka, Burma, and Thailand; palmarosa oil from *C. martinii*; citronella oil from *C. nardus* (Sri Lanka), and *C. winterianus* (Java).

This book describes the considerable ethno-botanical, phytochemical, and pharmacological knowledge that is associated with the multidimensional uses of the oils of the cymbopogons.

Lemongrass originated in Asia and is an ingredient of its herbal teas, soups, and innumerable other food recipes found in South East Asia, Cambodia, and Vietnam. In Europe the oil is used in spiced wines and herbal beers. Citronella oil gives a pleasant, refreshing aroma to personal care products including mosquito repelling lotions, etc. Palmarosa oil supplies a rose note to fine perfumes and to, for example, perfumed candles and herbal pillows. These last two oils may result in a higher income for a farmer than from the traditional food crops.

Extraction of the oils is by steam distillation and the aqueous distillates (hydrosols), after separation of the oils, are said to have antiviral, antibacterial, and antifungal properties. The spent grass, after the extraction of the oil, is used as a cattle fodder, or for paper making, or as a fuel in the next round of distillation.

These oils are a source of precursors for the production of vitamin A and potentially for other compounds; as "green" factories these plants provide alternative synthetic routes to the petrochemical ones.

For those in academia, both teachers and research students, those in agriculture and other industries, and those in business, this book provides an account of the botany, taxonomy, chemistry, and biogenesis of the oils, and their extraction and analytical methods, biotechnology, storage, legislation, and trade with all the accompanying references.

Professor Massimo Maffei is already a notable contributor to this series having edited a related Graminea volume (20), *Genus Vetiveria,* and contributed chapters to several other volumes. I thank him and all the other contributors for their kind cooperation with me and particularly for their expert data. My special thanks are due to Dr. Roland Hardman for his continuous encouragement from the very beginning of writing this book, despite his busy schedule. His help and persistent enthusiasm have been a great inspiration to me.

Anand Akhila
M.Sc., Ph.D. (University of London)

Author/Editor Page

Dr. Anand Akhila was born in 1955. He has an outstanding academic career, obtaining his Ph.D. from University College London in 1980 in the area of Natural Product Chemistry, particularly the biosynthetic pathways occurring in nature. For the last 27 years he has been working as a senior scientist at the Central Institute of Medicinal and Aromatic Plants, a reputed national laboratory of the government of India. During the course of his long career, he has published over 60 research papers, book chapters, and review articles, besides giving presentations at symposia and conferences. Anand Akhila was presented with the CSIR Young Scientist Award in 1988, a national honor in recognition of outstanding work, awarded by the Council of Scientific and Industrial Research, the premier scientific organization of the government of India. He is a Member of the Royal Society of Chemistry, United Kingdom, and many other Indian scientific societies. His current area of research has been the study of metabolic pathways of compounds of medicinal and aromatic values in plants such as *Azadirachta indica, Artemisia annua, Cymbopogon* species, *Mentha* species, and others.

Contributors

Anand Akhila
Central Institute of Medicinal and Aromatic
    Plants
Lucknow, India

Cinzia M. Bertea
Department of Plant Biology and Centre of
    Excellence CEBIOVEM
Turin, Italy

Ange Bighelli
Equipe Chimie-Biomasse
Université de Corse
Ajaccio, France

Joseph Casanova
Equipe Chimie-Biomasse
Université de Corse
Ajaccio, France

Watcharee Khunkitti
Faculty of Pharmaceutical Sciences
Khon Kaen University

Massimo E. Maffei
Department of Plant Biology and Centre of
    Excellence CEBIOVEM
Turin, Italy

Ajay K. Mathur
Division of Plant Tissue Culture
Central Institute of Medicinal and Aromatic
    Plants (CIMAP)
Lucknow, India

David A. Moyler
Fuerst Day Laws on Ltd
London, U.K.

A. K. Pandey
Tropical Forest Research Institute
(Indian Council of Forestry Research and
    Education)
NWFP Division, Tropical Forest Research
    Institute
Jabalpur, India

Hiroyuki Sumi
Kurashiki University of Science and the Arts
Department of Physiological Chemistry
Kurashiki, Japan

Rakesh Tiwari
Central Institute of Medicinal and Aromatic
    Plants
Lucknow, India

Chieko Yatagai
Kurashiki University of Science and the Arts
Department of Physiological Chemistry
Kurashiki, Japan

1 The Genus *Cymbopogon*
Botany, Including Anatomy, Physiology, Biochemistry, and Molecular Biology

Cinzia M. Bertea and Massimo E. Maffei

CONTENTS

1.1 INTRODUCTION

Among monocots forming the family of Gramineae some grasses produce essential oils that are a valuable source for the flavor industry. Two grasses are known for their industrial potential for essential oil production: *Vetiveria zizanioides* Stapf, which has been the subject of a monograph in the series *Medicinal and Aromatic Plants—Industrial Profiles* (Maffei 2002) and *Cymbopogon*, which is the subject of this book and which updates the monograph edited by Kumar et al. (2000).

Cymbopogon is a genus comprising about 180 species, subspecies, varieties, and subvarieties. It is native to warm temperate and tropical regions of the Old World and Oceania. Table 1.1 lists the several species, subspecies, varieties, and subvarieties as reported by the International Plant Names Index (2004), published on the Internet http://www.ipni.org (accessed September 10, 2007).

The name *Cymbopogon* was introduced by Sprengel in 1815 (Sprengel 1815) and at that time the genus consisted of a few species, which were then moved to the genus *Andropogon*. In fact, both *Cymbopogon* and *Andropogon* belong to the tribe Andropogoneae, a monophyletic tribe that

TABLE 1.1
List of *Cymbopogon* Species, Subspecies, Varieties, and Subvarieties as Reported in The International Plant Names Index (2004)

Cymbopogon acutispathaceus De Wild

C. afronardus Stapf

C. ambiguus A. Camus.

C. andongensis Rendle

C. angustispica Nakai

C. annamensis A. Camus.

C. arabicus Nees ex Steud.

C. arriani Aitch.

C. arundinaceus Schult.

C. bagirmicus Stapf

C. bassacensis A. Camus.

C. bequaertii De Wild

C. bhutanicus Noltie

C. bombycinus A. Camus.

C. bombycinus var. *bombycinus* (RBr) Domin.

C. bombycinus var. *townsvillensis* Domin.

C. bombycinus var. *typicus* Domin.

C. bracteatus Hitchcock

C. caesius (Hook and Arn.) Stapf

C. caesius subsp. *giganteus* (Chiov) Sales

C. calcicola C.E. Hubb.

C. calciphilus Bor

C. cambodgiensis E.G. Camus and A. Camus

C. chevalieri A. Camus

C. chrysargyreus Stapf

C. circinnatus Hochst. ex Hookf.

C. citratus Stapf

C. citriodorus Link

C. claessensii Robyns

C. clandestinus Stapf

C. coloratus Stapf

C. commutatus Stapf

C. commutatus var. *jammuensis* (Gupta) H.B. Naithani

C. condensatus Spreng.

C. confertiflorus Stapf

C. connatus Chiov.

C. cyanescens Stapf

C. cymbarius Rendle.

C. densiflorus Stapf

C. dependens B.K. Simon

C. dieterlenii Stapf ex Phillips

C. diplandrus De Wild.

C. distans (Nees ex Steud.) Will Watson.

C. divaricatus Stapf

C. eberhardtii A. Camus.

C. effusus A. Camus.

C. elegans Spreng.

C. exaltatus A. Camus.

C. exaltatus var. *ambiguus* Domin.

C. exaltatus var. *exaltatus* (RBr) Domin.

C. exaltatus var. *genuinus* Domin.

C. exaltatus var. *gracilior* Domin.

C. exaltatus var. *lanatus* (RBr) Domin.

C. exarmatus Stapf

C. excavatus Stapf

C. familiaris De Wild

C. figarianus Chiov.

C. filipendulus Rendle

C. finitimus Rendle

C. flexuosus Stapf

C. flexuosus var. *assamensis* S.C. Nath and K.K. Sarma

C. floccosus Stapf

C. foliosus Roem and Schult.

C. gazensis Rendle

C. gidarba [Buch–Ham. ex Steud.] Haines

C. giganteus Chiov.

C. glandulosus Spreng.

C. glaucus Schult.

C. globosus Henrard.

C. goeringii A. Camus

C. goeringii var. *hongkongensis* S. Soenarko

C. gratus Domin.

C. hamatulus A Camus

C. hirtus Stapf ex Burtt Davy

C. hirtus subsp. *villosum* (Pignatti) Pignatti

C. hispidus Griff.

C. hookeri (Munro ex Hackel) Stapf ex Bor

C. humboldtii Spreng.

C. iwarancusa Schult.

C. jinshaensis R. Zhang and C.H. Li

C. jwarancusa subsp. *olivieri* (Boiss.) S. Soenarko

C. kapandensis De Wild

C. khasianus (Hackel) Stapf ex Bor

C. ladakhensis B.K. Gupta

C. lanatus Roberty

C. laniger Duthie

C. lecomtei Rendle

C. lepidus (Nees) Chiov.

C. liangshanensis S.M. Phillips and S.L. Chen

C. lividus (Thwaites) Willis

C. luembensis De Wild

C. mandalaiaensis Soenarko

C. marginatus Stapf ex Burtt Davy

C. martinii Stapf

TABLE 1.1 (continued)
List of *Cymbopogon* Species, Subspecies, Varieties, and Subvarieties as Reported in The International Plant Names Index (2004)

C. martinianus Schult.

C. mekongensis A. Camus

C. melanocarpus Spreng.

C. micratherus Pilg.

C. microstachys (Hookf.) S. Soenarko

C. microthecus A. Camus

C. minor B.S. Sun and R. Zhang ex S.M. Phillips and S.L. Chen

C. minutiflorus S. Dransf.

C. modicus De Wild

C. motia B.K. Gupta

C. munroi (C.B. Clarke) Noltie

C. nardus (L.) Rendle

C. nardus subvar. *bombycinus* (R.Br.) Roberty

C. nardus var. *confertiflorus* (Steud.) Stapf ex Bor

C. nardus subvar. *exaltatus* (R.Br.) Roberty

C. nardus subvar. *grandis* Roberty.

C. nardus subvar. *lanatus* (R.Br.) Roberty

C. nardus var. *luridus* (Hookf.) Gavade and M.R. Almeida

C. nardus subvar. *procerus* (R.Br.) Roberty

C. nardus subvar. *refractus* (R.Br.) Roberty

C. nardus subvar. *schultzii* Roberty

C. nervatus A. Camus

C. nyassae Pilg.

C. obtectus S.T. Blake

C. olivieri (Boiss.) Bor

C. osmastonii R. Parker

C. pachnodes (Trin.) Will Watson

C. papillipes (Hochst. ex A. Rich) Chiov.

C. parkeri Stapf

C. pendulus Stapf

C. phoenix Rendle

C. pilosovaginatus De Wild

C. pleiarthron Stapf

C. plicatus Stapf

C. plurinodis Stapf ex Burtt Davy

C. polyneuros Stapf

C. pospischilii (K. Schum) C.E. Hubb

C. princeps Stapf

C. procerus A. Camus

C. procerus var. *genuinus* Domin.

C. procerus var. *procerus* (R.Br.) Domin.

C. procerus var. *schultzii* Domin.

C. prolixus (Stapf) Phillips

C. prostratus Sweet

C. proximus Stapf

C. proximus var. *sennarensis* (Hochst.) Tackholm

C. pruinosus Chiov.

C. pubescens (vis) Fritsch.

C. queenslandicus S.T. Blake

C. quinhonensis (A. Camus) S.M. Phillips and S.L. Chen [transferred to Andropogon (Phillips and Hua 2005)]

C. ramnagarensis B.K. Gupta

C. rectus A. Camus

C. reflexus Roem and Schult.

C. refractus A. Camus

C. rufus Rendle

C. ruprechtii Rendle

C. scabrimarginatus De Wild

C. schimperi Rendle

C. schoenanthus Spreng.

C. schoenanthus subsp. *velutinus* Cope

C. schultzii Roberty

C. sennaarensis Chiov.

C. setifer Pilg.

C. siamensis Bor

C. solutus Stapf

C. stipulatus Chiov.

C. stolzii Pilg.

C. stracheyi (Hookf.) Raizada and Jain

C. strictus Bojer.

C. stypticus Fritsch.

C. suaveolens Pilger

C. subcordatifolius De Wild

C. tamba Rendle

C. tenuis Gilli

C. thwaitesii (Hookf.) Willis

C. tibeticus Bor

C. tortilis (Presl.) A. Camus

C. tortilis subsp. *goeringii* (Steud.) TKoyama

C. traninhensis (A. Camus) SSoenarko

C. transvaalensis Stapf ex Burtt Davy

C. travancorensis Bor

C. tungmaiensis L. Liu

C. umbrosus Pilg.

C. validus Stapf ex Burtt Davy

C. vanderystii De Wild

C. versicolor (Nees ex Steud.) Will Watson

C. virgatus Stapf ex Rhind.

C. virgatus Stapf ex Bor

C. welwitschii Rendle

C. winterianus Jowitt

C. xichangensis R. Zhang and B.S. Sun

Source: Published on the Internet http://www.ipni.org [accessed September 10, 2007].

includes 85 genera. Mathews and coworkers (2002) found strong support for a core Andropogoneae that includes, among others, *Andropogon* and *Cymbopogon,* and support for its relationship with an expanded Saccharinae that includes *Microstegium.* The limited difference in the plant traits between *Andropogon* and *Cymbopogon,* has argued the possibility that species belonging to *Cymbopogon* might be a subgenus of *Andropogon.* Most of Andropogoneae have pairs of spikelets in the inflorescence, one sessile and one on a pedicel, although in some species one or the other of these spikelets appear to be suppressed. The inflorescences form is also highly variable (Mathews et al. 2002). Morphologically, the main difference in the genus *Cymbopogon* is the presence of some pair of spikelets, for each spike, with unisexual male flowers, whereas in the *Andropogon* spikelets are usually sessile and often sterile. *Cymbopogon* plants are tall (up to and above 1 m) perennial plants, with narrow and long leaves that are mostly characterized by the presence of silica thorns aligned on the leaf edges. Leaves bear glandular hairs, usually each with a basal cell that is wider than the distal cell (see Section 1.2). Representative of the Andropogoneae exhibit C_4 photosynthesis, with NADP-ME as the primary decarboxylating enzyme (Mathews et al. 2002), they usually have a chromosome number of five, with ploidy levels ranging from tetraploids to 24-ploid. Polyploidy, either as alloploidy or segmental alloploidy, is frequent. Representative specimens of various species of the genus *Cymbopogon* have been cytogenetically studied by Spies and coworkers (Spies et al. 1994). The monophyly of *Cymbopogon* has also been clearly demonstrated, and the genus is sister to *Heteropogon* (Mathews et al. 2002).

Among the several aromatic species belonging to the genus *Cymbopogon* the most important in terms of essential oil production are *C. martinii,* also known as palmarosa; *C. citratus,* better known as lemongrass; and the so-called East Indian lemongrass, *Cymbopogon flexuosus,* native to India, Sri Lanka, Burma, and Thailand; whereas, for the related West Indian *C. citratus,* a Malesian origin is generally assumed. *C. nardus* and *C. winterianus* produce the famous citronella from Sri Lanka and Java, respectively. Also known to produce essential oils are *C. schoenanthus,* or camel grass; *C. caesius,* or inchi/kachi grass; *C. afronardus, C. clandestinus; C. coloratus; C. exaltatus; C. goeringii; C. giganteus; C. jwarancusa; C. polyneuros; C. procerus; C. proximus; C. rectus; C. sennaarensis; C. stipulatus;* and *C. virgatus* (Guenther 1950b). The main constituents of *Cymbopogon* essential oils will be described in other chapters of the book.

Many laboratories in several countries are deeply involved in studying various aspects of cymbopogons, using variously derived genetic resources. The work already done covers a wide array of topics, including botanical identification, plant description, cytogenetics, and cell, tissue, and organ in vitro cultures. Physiology and biochemistry of stress tolerance and essential oil biosynthesis, genetics and biotechnology, and agrotechnology involved in crop production and disease and pest control chemistry of terpenes, biological activities of essential oil terpenoids and trade and marketing aspects (reviewed by Kumar et al. 2000).

The major objective of this monograph is to update the literature and give references on the aforementioned topics of cymbopogons, with particular attention to industrial aspects. In the following sections of this introductory chapter, we will explore the anatomy and biochemistry of the photosynthetic apparatus, also considering the most recent advances in molecular biology of *Cymbopogon.* We will conclude the chapter with some physiological and ecophysiological considerations.

1.2 ANATOMY

1.2.1 General Considerations

Higher plants can be divided into two groups, C_3 and C_4, based on the mechanism utilized for photosynthetic carbon assimilation related to anatomical and ultrastructural features. A cross section of a typical C_3 leaf reveals essentially one type of photosynthetic, chloroplast-containing cell, the mesophyll, and in these plants atmospheric CO_2 is fixed directly by the primary carbon-fixation enzyme ribulose 1,5-bisphosphate carboxylase/oxygenase (Rubisco). In contrast, a typical C_4 leaf has two distinct chloroplast-containing cell types, the mesophyll and the bundle sheath (or Kranz) cells, and they differ in photosynthetic activities (Hatch 1992; Maurino et al. 1997). The operation of the C_4 photosynthetic mechanism requires the cooperative effort of both cell types, connected by an extensive network of plasmodesmata that provides a pathway for the flow of metabolites between the cells.

The C_4 pathway is a complex adaptation of the C_3 pathway that overcomes the limitation of photo-respiration and is found in a diverse collection of species, many of which grow in hot climates. It was first discovered in tropical grasses (e.g., sugarcane and maize) and is now known to occur in 16 plant families. It occurs in both monocotyledonous and dicotyledonous plants, and is particularly prominent in species of the Gramineae, Chenopodiaceae, and Cyperaceae (Edwards and Walker 1983). About half of the species of the *Poaceae* are included among the C_4 plants (Smith and Brown 1973). The key feature of C_4 photosynthesis is the compartmentalization of activities into two specialized cell and chloroplast types. Rubisco and C_3 photosynthetic carbon reduction (PCR) cycle are found in the inner ring of bundle sheath cells. These cells are separated from the mesophyll and from the air in the intercellular spaces by a lamella that is highly resistant to the diffusion of CO_2 (Hatch 1988). Thus, by virtue of this two-stage CO_2 fixation pathway, the mesophyll-located C_4 cycle acts as a biochemical pump to increase the concentration of CO_2 in the bundle sheath an estimated 10-fold over atmospheric concentrations. The net result is that the oxygenase activity of Rubisco is effectively suppressed, and the PCR cycle operates more efficiently.

C_4 plants have two chloroplast types, each found in a specialized cell type. Leaves of C_4 plants show extensive vascularization, with a ring of bundle sheath (BS) cells surrounding each vein and an outer ring of mesophyll (M) cells surrounding the bundle sheath. CO_2 fixation in these plants is a two-step process.

There are three variants on the basic C_4 pathway, and the biochemical distinctions are correlated with the ultrastructural differences of Kranz cells (Gutierrez et al. 1974; Hatch et al. 2007; Hatch et al. 1975). The three C_4 variants can be distinguished ultrastructurally by using combinations of two characters of bundle sheath cell chloroplasts, by the degree of granal stacking, and by the chloroplasts position (Gutierrez et al. 1974).

For this reason, comparative grass leaf anatomy has become the object of intensive investigation in relation to photosynthesis along with biochemical studies.

1.2.1.1 Leaf Anatomy

First descriptions of the genus *Cymbopogon* were given by Breakwell (1914) on *C. bombycinus* and *C. refractus* under the name of *Andropogon bombycinus* R. Br. and *A. refractus*, respectively, followed by a leaf structure description done by Vickery (1935) and Prat (1937). Further studies were conducted by Metcalfe (1960). The descriptions given by these authors are very similar and still valid.

1.2.1.1.1 Generic Characters

Both adaxial and abaxial epidermises of *Cymbopogon* species contain short-cells, over the veins, solitary, paired or in short or long rows, the proportion of each type varying with the species. Silica bodies are located over the veins, mostly crossed to dumbell shaped. Microhairs are present usually each with the basal cell wider than the distal cell, the latter frequently tapering to a pointed apex, or hemispherical. Stomata with subsidiary cells range from low or tall dome-shaped to triangular, the proportions of each type varying in different species and sometimes in separate preparations from a single species. The vascular bundles are small, mostly angular, but less conspicuous in some species than in others. The mesophyll presents a distinctly radial chlorenchyma. Bundle sheaths are single (Metcalfe 1960).

Figure 1.1A shows an electron scanning micrograph of a *C. citratus* leaf blade. It is possible to observe developed prickle hairs with rather elongated bases over the veins. The abaxial epidermis also reveals several stomata, with narrow guard cells associated with subsidiary cells, typical of grasses (Figure 1.1B).

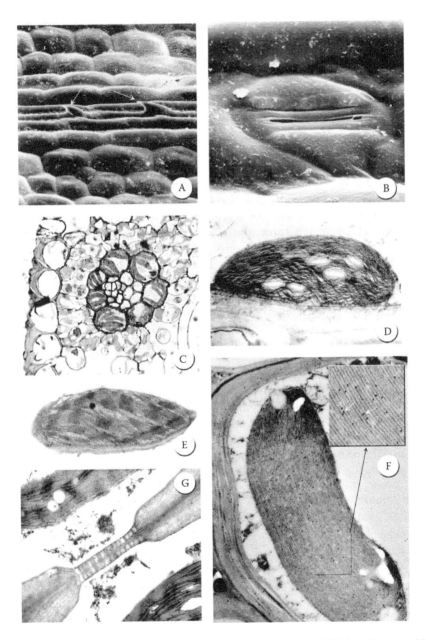

FIGURE 1.1 (A) Electron scanning micrograph of *C. citratus* leaf blade. Prickle hairs are evident on the veins (white arrows) (80×). (B) Electron scanning micrograph of a stoma seen from the surface. Guard cells are narrow in the middle and enlarged at the end, while subsidiary cells are triangular (2000×). (C) Semi-thin cross section of *C. citratus* leaf stained with toluidine blue. The vascular bundle in a minor vein is surrounded by a layer of sheath cells, with chloroplasts arranged in a centrifugal position. (D) Bundle sheath chloroplast without grana and with few starch grains. (E) Mesophyll chloroplast with most of the thylakoid stacked in grana and without starch grains. (F) Bundle sheath chloroplast showing immunolabeling against Rubisco. Colloidal gold particles (white arrows in the high magnification) strongly label the stroma. (G) Enlargement of plasmodesmata connecting mesophyll and bundle sheath cells, providing metabolite flow between the two photosynthetic tissues.

1.2.1.1.1.1   Leaf Ultrastructure of Cymbopogon citratus   The structure of parenchymatic bundle sheath (BS) cells is particularly important in distinguishing C_3 and C_4 species. A commonly mentioned anatomical feature of C_4 plants is the orderly arrangement of mesophyll cells with reference to the BS, the two together forming concentric layers around the vascular bundle. The anatomical differences between plants exhibiting a C_4 photosynthetic carbon assimilation pathway can be disclosed by electron microscopical observations of the BS (Chapman and Hatch 1983; Edwards and Walker 1983; Gutierrez et al. 1974; Hatch 1988; Hatch et al. 1975; Jenkins et al. 1989). In the NADP-ME type, chloroplasts are peripherally arranged and grana are deficient or absent in bundle sheath cells. Two other distinctive features are the presence or absence of a mestome sheath, a layer of cells intervening between metaxylem vessel elements and laterally adjacent BS Kranz cells, and the presence or absence of a cell wall suberized lamella (SL). The mestome sheath occurs in NAD-ME and PCK species, while the SL is present in bundle sheath cell walls of several NADP-ME and PCK species (Eastman et al. 1988; Hattersley and Watson 1976). The number of mitochondria in the NADP-ME subtype is lower than in the NAD-ME one because, in the latter, the enzymes involved in the transformation of aspartate to CO_2 and pyruvate are present in these organelles (Hatch et al. 1975). Anatomical studies conducted on *C. citratus* leaves indicated the presence of the C_4 Kranz anatomy in this plant along with several ultrastructural features typical of NADP-ME species.

In *C. citratus* cross sections of minor veins, the vascular bundle appears surrounded by one layer of sheath cells, in which the chloroplasts are located in a centrifugal position (Figure 1.1C). No mestome sheath between metaxylem vessel elements and laterally adjacent Kranz cells is observed. In the bundle sheath, ultrastructural analyses show a suberized cell wall lamella, and the presence of agranal chloroplasts, containing numerous starch grains (Figure 1.1D). Figure 1.1E shows mesophyll chloroplasts with most of the thylakoids stacked in grana. The two photosynthetic tissues are connected by a certain number of plasmodesmata that provide a pathway for the flow of metabolites between the mesophyll and bundle sheath cells (Figure 1.1G).

These observations are in accordance with previous anatomical descriptions related to the genus *Cymbopogon* reported by Rajendrudu and Das (1981). These authors reported on the leaf anatomy and photosynthetic carbon assimilation in five species of *Cymbopogon* (*C. flexuosus*, *C. martinii* var. *motia*, *C. nardus*, *C. pendulus*, and *C. winterianus*) a Kranz-type leaf anatomy with a centrifugal position of starch-containing chloroplasts in the bundle sheath cells. Starch was exclusively localized in the bundle sheath cells that were typically elongated parallel to the veins and nearly twice as long as wide in the species of *Cymbopogon*. A narrow leaf interveinal distance was a common feature among the five *Cymbopogon* species. A xylem-mestome sheath of cells between metaxylem vessels and laterally adjacent bundle sheath cells of primary vascular bundles was totally absent in the five *Cymbopogon* species (Rajendrudu and Das 1981).

1.2.1.1.2   Rubisco Immunolocalization

High-resolution immunolocalization of Rubisco by electron microscopy showed that labeling occurred only in the bundle sheath chloroplasts of *C. citratus*. For these experiments, purified rabbit polyclonal antibodies raised against Rubisco were employed. Bound antibodies were then visualized by linking conjugated gold-labeled goat antirabbit polyclonal antibodies. Gold particles appeared to be uniformly distributed throughout the stroma (Figure 1.1F). These studies conducted on *C. citratus* leaves provide evidence for the localization of Rubisco in the stroma of bundle sheath chloroplasts, as expected for a C_4 plant (Bertea et al. 2003).

1.3   BIOCHEMISTRY

In a preliminary physiological study conducted by (Maffei et al. 1988) on *C. citratus* grown in humid temperate climates, some of the PEP-carboxylase kinetic characteristics, and Rubisco and glycolate oxidase activities were found to be comparable to those of C_4 plants.

The C_4 mechanism was also confirmed by the $^{13}C/^{12}C$ stable isotope ratio analyses ($\delta^{13}C$ = −13.0). These results are in accordance with $\delta^{13}C$ values measured on other species of *Cymbopogon* (Rajendrudu and Das 1981) in which $\delta^{13}C$ value of −11.0 for *C. flexuosus*, −9.7 for *C. martinii*, −11.6 for *C. nardus*, −10.3 for *C. pendulus*, and −11.3 for *C. winterianus*, respectively, were recorded.

From a biochemical point of view, the three types of the basic C_4 pathway differ mainly in the C_4 acid transported into the bundle sheath cells (malate and aspartate) and in the way in which it is decarboxylated; they are named (based on the enzymes that catalyse their decarboxylation) NADP-dependent malic enzyme (NADP-ME) found in the chloroplasts, NAD-dependent malic enzyme (NAD-ME) found in mitochondria, and phosphoenolpyruvate (PEP) carboxykinase (PCK), found the cytosol of the bundle sheath cells (Edwards and Walker 1983; Ghannoum et al. 2001; Hatch et al. 1975; Jenkins et al. 1989; Huang et al. 2001). Furthermore, a characteristic leaf anatomy, biochemistry, and physiology are associated with each of the C_4 types (Dengler and Nelson 1999; Hattersley and Watson 1976). A clear indication of the C_4 photosynthetic pathway of *C. citratus* and the variant to which it belongs was obtained by estimating the activities of NADP-ME (EC 1.1.1.40), NADP-MDH (EC 1.1.1.82), PPDK (EC 2.7.9.1), NAD-ME (EC 1.1.1.39), and PCK (EC 4.1.1.49) as well as some kinetic characteristics of NADP-ME and NADP-MDH. Adaptation to a particular environment is a complex process involving a number of physiological, morphological, and ecological factors (Ghannoum et al. 2001; Huang et al. 2001).

Therefore, enzyme activities were recorded at the low and high temperatures typical of humid-temperate climates, in order to evaluate the adaptability of *C. citratus*. In order to estimate increases or decreases in the reaction rate due to changes in the protonation state, groups involved in the catalysis and/or binding of substrates as a consequence of pH fluctuations, activities were also recorded at different pH values.

1.3.1 CHARACTERIZATION OF THE PHOTOSYNTHETIC VARIANT

Further studies dealing with the characterization of the C_4 variant indicated an NADP-dependent malic enzyme photosynthetic pathway in *C. citratus*.

The biochemical subtype was established through the estimation of the highest activities of NADP-dependent malic enzyme (NADP-ME, EC 1.1.1.40), NADP-dependent malate dehydrogenase (NADP-MDH, E.C. 1.1.1.82), pyruvate, orthophosphate dikinase (PPDK, E.C. 2.7.9.1), NAD-dependent malic enzyme (NAD-ME, E.C. 1.1.1.39), and phosphoenolpyruvate carboxykinase (PCK, E.C. 4.1.1.49) and some kinetic, along with some chemical-physical parameters of NADP-ME and NADP-MDH.

Extraction and partial purification sequentially involved precipitation with crystalline ammonium sulfate, dialysis, and anion exchange (DEAE-Sephacell). Both, extraction and assays were conducted according to Ashton (1990). The low activity values of PPDK (90.28 nKat mg^{-1} prot), PCK (<1 nKat mg^{-1} prot), and NAD-ME (52.51 nKat mg^{-1} prot) in *C. citratus* leaf extracts did not allow to determine the kinetic characteristics, such as *Km* and *Vmax*, and/or other chemical-physical parameters of these enzymes. NADP-MDH and NADP-ME presented relatively high activity values (15.93 mKat mg^{-1} and 12.56 mKat mg^{-1} prot, respectively). NADP-ME activity was 239-fold greater than NAD-ME activity. The kinetics of NADP-MDH and NADP-ME were therefore measured to gain a clearer picture of the photosynthetic pathway. The low activity of PPDK found in plant extracts of *C. citratus* agrees with the literature data for C_4 plants (Ashton 1990) and with data on species of the same photosynthetic subtype (Ashton 1990; Bertea et al. 2001). The low levels of NAD-ME and PCK activities found in our extracts clearly indicated the absence of a C_4 variant utilizing these two pathways. The relatively high activities found for NADP-MDH allowed us to determine some of its kinetic characteristics (*Km* and *Vmax*), which were comparable to those of plants belonging to the NADP-ME photosynthetic variant (Ashton 1990; Bertea et al. 2003).

Km values obtained for OAA and NADPH (NADP-MDH) were 29.0 (±0.014) mM and 31,67 (±0.09) mM, respectively. With regard to NADP-ME, the apparent *Km* values for $NADP^+$ and malate were 19.40 (±0.08) and 242.0 (±0.008) mM, respectively. In the case of NADP-MDH, *Vmax* values for OAA and NADPH were 12.52 (±0.021) and 14.97 (±0.012) mKat mg^{-1} prot, respectively.

NADP-ME *Vmax* values for malate and $NADP^+$ were 8.63 (±0.507) and 18.60 (±0.007) mKat mg^{-1} prot, respectively. In general, relatively high activities of NADP-MDH and NADP-ME allowed a partial characterization of these enzymes and provided evidence for an NADP-ME subtype for *C. citratus*. The apparent kinetic properties of both enzymes were comparable to those of plants belonging to this subtype (Ashton 1990; Hatch et al. 2007; Hatch et al. 1975), and were consistent with a high photosynthetic activity, even when the plant was cultivated in a temperate climate.

1.3.2 NADP⁺ INHIBITION OF NADP-MDH

Inhibition studies were carried out by measuring the NADP-MDH-catalyzed reaction at varying concentrations of $NADP^+$ and constant concentrations of OAA (1.0 mM) and NADPH (0.2 mM). NADP-MDH activity was increasingly inhibited by increasing $NADP^+$ concentrations. The activity value recorded in the presence of 0.25 mM $NADP^+$ was only 38% of the activity measured in absence of the oxidized coenzyme.

In *Zea mays*, activation of NADP-MDH is regulated by oxidation and reduction of cysteine residues (thioredoxin-mediated system) (Lunn et al. 1995), and interconversion of the reduced and oxidized forms is influenced by the $NADPH/NADP^+$ ratio (Trevanion et al. 1997). A high NADPH/$NADP^+$ ratio leads to a more active enzyme; thus, high rates of OAA reduction only occur in reduced conditions. The percentage of inhibition caused increasing $NADP^+$ concentration in our DEAE preparations was in accordance with the observations reported earlier. The relatively high activities of NADP-ME detected enable us to characterize the enzyme in *C. citratus*. A low *Km* value was calculated for free $NADP^+$. These results confirm the high affinity of $NADP^+$ for its binding site in all isoforms of this enzyme (Rothermel and Nelson 1989). A higher *Km* value was calculated for malate in accordance with literature data.

1.3.3 pH AND TEMPERATURE DEPENDENCE OF NADPH-MDH AND NADP-ME

pH studies were carried out by using a buffer system that contained an equimolar mixture of buffers adjusted to different pH values with KOH (Bertea et al. 2001).

Assays were performed at different pH values, 6.0 to 10.5 for NADP-MDH, and 6.0 to 10 for NADP-ME, using the DEAE-preparation. Maximal activity of NADP-MDH enzyme was observed at pH 8.3, in agreement with the published data (Ashton 1990). An increase in activity was observed starting from the lowest pH value (6.0) up to pH 8.3. At pH 7.0–7.5, the activity was comparable to that at pH values ranging between 9.5 and 10.0. At pH 10.5 the activity was comparable to that recorded at pH 6.0. Temperature changes also affected the reaction rate. The influence of temperature on enzyme activities was determined for both NADP-MDH and NADP-ME by adding the substrates to standard assay mixtures equilibrated at the appropriate temperatures. The assays were performed by using the DEAE-preparation. Apparent activation energy was calculated from Arrhenius plots.

When enzyme activity was measured using the standard assay system at temperatures ranging from 20°C to 49°C, maximal activity was detected at 35°C, while at 49°C activity was lower, but still much higher than that at 20°C, in accordance with the typical behavior of C_4 photosynthetic enzymes. From a linear Arrhenius plot of the data, in a temperature range from 20°C to 38°C, the activation energy of the reaction was calculated to be 6970.8 cal mol^{-1}. The highest NADP-ME activity was recorded at pH 8.3, in accordance with the enzyme characteristics (Edwards and

Andreo 1992), whereas at pH 10.0 the activity was higher than the activity recorded at pH 6.0 and 6.5. With regard to temperature, maximal activity was measured at 45°C, the lowest at 20°C. Also in this case, activity response to temperature changes was typical of C_4 plants (Edwards and Andreo 1992). The activation energy of the reaction, calculated in a temperature range from 20°C to 45°C, was 7605.2 cal mol^{-1}.

Climatic conditions exert an evident effect on the physiological status of photosynthetic enzymes, and variations in light, temperature, moisture, etc., may influence the cytosolic and stromal pH (Ashton 1990). In C_4 plants, NADP-MDH has subunits of 42 kDa, and the native enzyme apparently occurs as either a tetramer or a dimer. The tetramer is the more active form; it is stable at alkaline pH values and at high temperatures (Ashton 1990). NADP-ME is a tetramer with a molecular weight of about 280 kDa, and it is more stable at pH values above 8.0 (Edwards and Andreo 1992).

This enzyme exists as a dimer and a monomer, both of which are active. Differences in pH can dramatically alter the activities of these photosynthetic enzymes. Temperature is another critical parameter. When it is low, photosynthetic rates of C_4 plants may fall below those of C_3 ones. The response of enzyme activities to such changes depends on the photosynthetic pathway adopted, resulting in a different optimum range of temperatures over which the highest growth rate can be maintained (Fitter and Hay 1987).

Because they originated in tropical and subtropical areas, the optimum temperature for photosynthesis in C_4 plants is 30°C–40°C, which is approximately 10°C higher than in C_3 plants (Leegood 1993). However, C_4 photosynthesis is usually sensitive to low temperature; the minimum temperature for photosynthesis in several C_4 tropical grasses is 5°C–10°C (Casati et al. 1997). Activities at different temperatures and pH values of *C. citratus* NADP-MDH and NADP-ME indicated that this species is a C_4 NADP-ME plant, which is able to retain its photosynthetic mechanism even when cultivated in temperate climates.

1.3.4 CO$_2$ ASSIMILATION AND STOMATAL CONDUCTANCE

A very low compensation point (between 8 and 15 ppm CO_2) was calculated for *C. citratus*. This result is typical for a C_4 plant. Stomatal opening increased in response to CO_2 concentration up to 157 ppm. However, at higher CO_2 values a decrease was recorded, thus indicating a clear effect of the CO_2-concentrating mechanism present in C_4 plants (data not shown). In order to evaluate changes in photosynthesis as a function of leaf age, CO_2 assimilation and stomatal conductance were also measured at different developmental stages of *C. citratus* leaves. A general increase for both parameters was observed, starting from primordial up to mature leaves, while a decrease in CO_2 assimilation and stomatal conductance was recorded in old leaves. Thus, primordial leaves presented the lowest CO_2 assimilation value (9.01 mmol CO_2 dm^{-2} s^{-1}), while the highest values were recorded in young and mature leaves, without appreciable differences (22.71 and 23.94 mmol CO_2 dm^{-2} s^{-1}, respectively). With regard to stomatal conductance, the lowest value was measured in old leaves (55.00 mM H_2O dm^{-2} s^{-1}), while the highest occurred in mature ones (165.27 mM H_2O dm^{-2} s^{-1}).

From a physiological point of view, the remarkable differences between the photosynthetic responses of C_3 and C_4 plants to CO_2 concentration become apparent when calculating the CO_2 compensation point. In plants with CO_2-concentrating mechanisms, including C_4 plants, CO_2 concentrations at the carboxylation sites are often saturating. Plants with C_4 metabolism have a CO_2 compensation point of or close to zero, reflecting their very low levels of photorespiration. The results obtained in *C. citratus* are in accordance with the values previously recorded on other species of *Cymbopogon* (Rajendrudu and Das 1981). In addition, the C_4 mechanism allows the plant to maintain high photosynthetic rates at lower partial CO_2 pressures in the intercellular spaces of the leaf, which require lower rates of stomatal conductance for a given rate of photosynthesis. For these reasons, measuring the CO_2 compensation point and stomatal conductance can be useful to distinguish between C_3 and C_4 pathways.

1.4 MOLECULAR BIOLOGY

The morphological variation and oil characteristics of various species and varieties of *Cymbopogon* have been reported, but such information is not sufficient to precisely define the relatedness among the morphotypes and chemotypes. For instance, *C. martinii* var. *sofia* and *C. martinii* var. *motia* are morphologically almost indistinguishable, but show distinct chemotypic characteristics in terms of oil constituents (Guenther 1950a). Conversely, phenotypically and taxonomically well distinguishable species produce oils of almost identical chemical compositions, such as lemongrass oils from *C. citratus* and *C. flexuosus* (Khanuja et al. 2005). Such phenotypic traits, whether morphological or chemotypic, are basically the phenotypic expression of the genotype, while DNA markers are independent of environment, age, and tissue, and expected to reveal the genetic variation more conclusively in assessing such variations. Introgression of various traits, intermittent mutations, and selection through human intervention may lead to variation in chemotypic characters across geographical distributions (Kuriakose 1995). While natural hybridization may lead to the formation of morphological or chemotypic intermediates, defining taxa purely on this basis may not be appropriate. Molecular markers provide extensive polymorphism at DNA level used for differentiating closely related genotypes (Pecchioni et al. 1996) and also to find out the extent of genetic diversity (Jain et al. 2003).

Different types of molecular markers have been developed and used in various plant species including grasses in the recent past years.

1.4.1 RANDOMLY AMPLIFIED POLYMORPHIC DNA (RAPD) MARKERS

DNA-based markers such as randomly amplified polymorphic DNA (RAPD) (Welsh and McClelland 1990) have been employed not only for cultivar identification but also for phylogenetic and pedigree studies in a number of food, forage, and fiber crops (Chalmers et al. 1992; Kresovich et al. 1994). RAPDs have provided rare biotype specific markers in medicinal and aromatic plants such as vetiver (Adams and Dafforn 1997) and *Artemisia annua* (Sangwan et al. 1999).

Randomly primed polymerase chain reaction provides a simple and fast approach to detecting DNA polymorphism, with allelic RAPD marker variations being detected as a plus or minus allele (Welsh and McClelland 1990). In particular, the approach provides multilocus profiling of DNA sequence differences of genotypes when genetic knowledge is lacking. Several studies have been carried out on *Cymbopogon* species by employing the RAPD approach.

A study using RAPD markers was carried out by Shasany and coworkers (2000) to trace the ancestors of cultivar Java II within *C. winterianus*. The species *C. winterianus* Jowitt is believed to have originated from the well-known species *C. nardus*, type Maha Pengiri, referred to as Ceylonese (Sri Lankan) commercial citronella. It was introduced into Indonesia and became commercially known as the Javanese citronella. The Javanese type *C. winterianus* material was introduced into India for the commercial cultivation of this crop during 1959. Varieties of this species have been developed later by the use of breeding procedures from the same introduced material. The authors carried out a comparative analysis of the morphological characters, chemical traits (oil percentage and constituents), and RAPD profiles to assess the diversity and relationships among the Java citronella cultivated forms, which were systematically developed for their suitability to different climatic regions, and also their differences and similarities to the believed parent species *C. nardus* (Purseglove 1975). All these accessions were analyzed at the molecular level for the similarity and genetic distances through RAPD profiling, using 20 random primers. More than 50% divergence was observed for all the *C. winterianus* accessions in relation to *C. nardus* accession CN2. The clustering based on the similarity matrices showed a major cluster of six accessions, consisting of two subclusters. The accession *C. nardus* CN2 got carved out along with two *C. winterianus* accessions, CW2 and CW6. On the other hand, the accessions CW2 and CW6 demonstrated distinct identities compared to CN2 at the DNA level (Shasany et al. 2000).

The same approach was used by Sangwan et al. (2001) on eleven elite and popular Indian cultivars of *Cymbopogon* aromatic grasses of essential oil trade types—citronella, palmarosa, and lemongrass. They were characterized by means of RAPDs to discern the extent of diversity at the DNA level between and within the oil biotypes. Primary allelic variability and the genetic bases of the cultivated germplasm were computed through parameters of gene diversity, expected heterozygosity, allele number per locus, SENA, and Shannon's information indices. The allelic diversity was found to be in this order: lemongrass > palmarosa > citronella. Lemongrasses displayed higher (1.89) allelic variability per locus than palmarosa (1.63) and citronella (1.40). Also, RAPDs of diagnostic and curatorial importance were discerned as "stand-along" molecular descriptors. Principal component analysis (PCA) resolved the cultivars into four clusters: one each of citronella and palmarosa, and two of lemongrasses (one of *C. flexuosus* and another of *C. pendulus* and its hybrid with *C. khasianus*). Proximity of the two species-groups of lemongrasses was also revealed as they shared the same dimension in the three-dimensional PCA (Sangwan et al. 2001).

The same authors analyzed the elite and popular cultivars of *C. martinii* for genomic and expressed molecular diversity using RAPD, enzyme, and SDS-PAGE protein polymorphisms. The allelic score at each locus of the enzymes, as well as presence and absence profiling in RAPDs, and overall occurrence of band types were subjected to computation of gene diversity, expected heterozygosity, allele number per locus, and similarity matrix. These, in turn, provide inputs to derive primary account of allelic variability, genetic bases of the cultivated germplasm, putative need for gene/trait introgression from the wild or geographically diverse habitat in elite selections. 'PRC1' possessed the highest number of unique bands based on RAPD polymorphism. In variety 'IW31245E,' diaphorase and glutamate oxaloacetate transaminase isozymes generated two unique bands as dia-III2 and got-II4. 'RRL(B)77' exhibited three unique bands; one produced by esterase as allele est-II1 and two by malic enzyme (me-III1,3). Only one unique band was generated by malic enzyme in variety 'Trishna.' But *sofia* had three unique bands, two contributed by diaphorase (dia-II3 and dia-II4) and one by glutamate oxaloacetate transaminase (got-II2). SDS-PAGE analysis revealed the presence of unique polypeptide fragments (97.7 to 31.6 kDa) in varieties 'IW31245E,' 'RRL(B)77,' 'Tripta,' 'Trishna,' 'PRC1,' and *sofia*, generated as a diagnostic marker. In general, molecular distinctions associated with varieties. *motia* and *sofia* were clearly noticed in *C. martinii* (Sangwan et al. 2003).

Khanuja et al. (2005) analyzed 19 *Cymbopogon* taxa belonging to 11 species, 2 varieties, 1 hybrid taxon, and 4 unidentified species for their essential oil constituents and RAPD profiles to determine the extent of genetic similarity and thereby the phylogenetic relationships among them. Remarkable variation was observed in the essential oil yield ranging from 0.3% in *Cymbopogon travancorensis* Bor to 1.2% in *Cymbopogon martinii* (Roxb.) Wats. var. *motia*. Citral, a major essential oil constituent, was employed as the base marker for chemotypic clustering. Based on genetic analysis, elevation of *Cymbopogon flexuosus* var. *microstachys* (Hook. F.) Soenarko to species status and separate species status for *C. travancorensis* Bor, which has been merged under *C. flexuosus* (Steud.) Wats., were suggested toward resolving some of the taxonomic complexes in *Cymbopogon*. The separate species status for the earlier proposed varieties of *C. martinii* (*motia* and *sofia*) is further substantiated by these analyses. The unidentified species of *Cymbopogon* have been observed as intermediate forms in the development of new taxa (Khanuja et al. 2005).

Somaclonal variants that arise through the tissue culture have been reported in a large number of species. The significance of somaclonal variation in crop improvement depends upon establishing a genetic basis for variation (Larkin and Scowcroft 1981). The use of the molecular marker is becoming widespread for the identification of somaclonal variant. In particular, RAPD markers have proved useful for this purpose owing to its ability to analyze DNA variation at many loci using small amounts of tissue (Munthali et al. 1996; Wallner et al. 1996). Screening of somaclonal variants with improved oil yield and quality have been reported in two species of *Cymbopogon, C. winterianus* and *C. martinii* (Mathur et al. 1988; Patnaik et al. 1999), but RAPDs were not used to establish the genetic basis of this somaclonal variation. The paper of Nayak and coworkers (2003) reports the quantitative and qualitative analysis of selected somaclones of jamrosa (a hybrid *Cymbopogon*) in the field, screening and selection of agronomically useful somaclonal variants with high oil yield and desirable quality, and detection of gross genetic changes through RAPD analysis.

In this study, the high oil yield somaclonal variants SC1 and SC2 were subjected to RAPD analysis and the result was compared with RAPD profile of the control. A total of 22 arbitrary primers were utilized for initial screening for their amplifying ability. Of these, 12 primers successfully amplified jamrosa DNA with reproducible banding pattern. In general, 2-11 amplified fragments were scored, depending upon primers, ranging in molecular sizes from 266 bp to 1.9 Kb. The test samples SC1, SC2, and the control could be suitably distinguished by the presence of specific markers or by their absence. Out of the two somaclones analyzed, relatively less distinctness in the amplified DNA of SC2 was detected using the primers tested. Banding pattern of this somaclone SC2 and the control plant was similar whereas in other variants, somaclone (SC1) DNA polymorphism was observed by having distinct banding pattern. As indicated by RAPDs gross genetic changes have occurred in somaclone (SC1). The results obtained by Nayak and coworkers are in agreement with detection of somaclonal variants by RAPD analysis in *Populus deltoides* (Rani et al. 1995), garlic (Al-Zahim et al. 1999) and in rice (Yang et al. 1999). Taylor et al. (1995) also reported that RAPD analysis proved suitable for detecting gross genetic changes occurring in sugarcane tissues subjected to prolonged in vitro culture. This work has demonstrated the scope of selecting improved clones of jamrosa with high oil yield and quality through somaclonal variation and suitability of RAPDs for detecting gross genetic changes in somaclonal variants at DNA level.

1.4.2 SIMPLE SEQUENCE REPEAT MARKERS (SSRs)

Kumar and coworkers (2007) developed a set of simple sequence repeat markers from a genomic library of *Cymbopogon jwarancusa* to help in the precise identification of the species (including accessions) of *Cymbopogon*. For this purpose, they isolated 16 simple sequence repeats containing genomic deoxyribonucleic acid clones of *C. jwarancusa*, which contained a total of 32 simple sequence repeats with a range of 1 to 3 simple sequence repeats per clone. The majority (68.8%) of the 32 simple sequence repeats comprised dinucleotide repeat motifs followed by simple sequence repeats with trinucleotide (21.8%) and other higher-order repeat motifs. Eighteen (81.8%) of the 22 designed primers for the above simple sequence repeats amplified products of expected sizes, when tried with genomic DNA of *C. jwarancusa*. Thirteen (72.2%) of the 18 functional primers detected polymorphism among the three species of *Cymbopogon* (*C. flexuosus*, *C. pendulus*, and *C. jwarancusa*) and amplified a total of 95 alleles (range 1–18 alleles) with a PIC value of 0.44 to 0.96 per simple sequence repeat. Thus, the higher allelic range and high level of polymorphism demonstrated by the developed simple sequence repeat markers are likely to have many applications such as in improvement of essential oil quality by authentication of *Cymbopogon* species and varieties, and mapping or tagging the genes controlling agronomically important traits of essential oils, which can further be utilized in marker assisted breeding (Kumar et al. 2007). Considering the high reproducibility and polymorphic nature of the SSRs, the SSRs developed by Kumar and coworkers (2007) may be utilized for identification/authentication of superior accessions/species of the genus *Cymbopogon* with correctness and certainty to ensure production of high-quality oil. The SSR markers developed during this kind of study might also be used to resolve the taxonomic disputes, study the genetic diversity, and for genetic mapping and QTL (quantitative trait loci) analysis. The SSRs due to their codominant nature may be specifically useful for the identification of interspecific hybrids, which have been shown to be superior in terms of both their yielding high quantity and better quality of essential oils.

1.5 PHYSIOLOGY AND ECOPHYSIOLOGY

As discussed, *Cymbopogon* has a photosynthetic machinery that allows the plant to perform high rates of carbon assimilation and, at the same time, save water. In species that produce essential oil, the biogenesis of terpenoids relies on photosynthetic carbon dioxide reduction on the one hand and availability of water and nutrients, on the other. For this reason several studies have been conducted in order to assess which nutrients and at what conditions were required for an optimal production of both biomass and essential oils. In this section, we will discuss the most important *Cymbopogon* species in terms of yield of biomass and essential oil production as related to nutrition. Furthermore, when available, references to biotechnological applications will be also reported.

1.5.1 *CYMBOPOGON MARTINII*

Water requirement, productivity, and water use efficiency of palmarosa (*C. martinii*) were studied under different levels of irrigation (0.1, 0.3, 0.5, 0.7, 0.9, 1.1, 1.3, and 1.5 IW:CPE ratio). Growth, herb, and essential oil yield increased significantly up to 0.5 IW:CPE ratio. At 0.5 IW:CPE ratio palmarosa produced 47.3 t ha^{-1} yr^{-1} of fresh herb and 227.3 kg ha^{-1} yr^{-1} of essential oil. Further increase in irrigation levels caused an adverse effect on growth and yield of palmarosa. Irrigation levels did not affect the quality of oil in terms of its geraniol and geranyl acetate contents. Water requirement of palmarosa was worked out to be 89.1 cm. The highest water use efficiency of 2.97 kg ha^{-1} cm^{-1} oil was recorded at 0.1 IW:CPE ratio, at 0.5 IW:CPE ratio (optimum) it was 2.55 kg ha^{-1} cm^{-1} oil. Irrigation scheduled at 0.5 IW:CPE ratio gave the highest net return of Rs 51 963 ha^{-1} yr^{-1} (Singh et al. 1997). In *C. martinii* the application of 160 kg N/ha per year produced the highest amount of biomass and essential oil, and increased the net profit and NPK uptake by the crop (Rao et al. 1988); furthermore, dressing of 40 kg K/ha enhanced the yield of biomass by 13.6% and 6.5% and that of oil by 12.9% and 6.1%, compared with 20 and 80 kg K/ha, respectively (Singh et al. 1992). In the same species, harvesting the crop at early seeding (112–115 days after planting) gave 25% more herbage and 51% more oil yield over harvesting vegetative stage, while the oil so produced had higher content (90.1%) geraniol (Maheshwari et al. 1992). Highest dry-matter yield, essential oil yield, and maximum net return of palmarosa were recorded by applying Azotobacter at 2 kg/ha together with 20 kg N + 20 kg P/ha under rainfed condition in a shallow black soil (Maheshwari et al. 1998). Intercropping of blackgram–blackgram or sorghum fodder–ratoon with palmarosa gave additional yields of 660 kg/ha seed and 16.6 t/ha fodder, respectively, compared with the sole crop of palmarosa (Rao et al. 1994). Moreover, sowing of pigeon pea in alternate rows parallel to palmarosa proved most efficient and economic, as it provided higher economic returns, bonus income, and monetary advantage, and the oil content and quality in terms of total geraniols of palmarosa were not adversely affected by adoption of intercropping (Maheshwari et al. 1995). However, in palmarosa–pigeon pea intercropping systems, competition exists mainly for light rather than for nutrients and moisture, possibly because the two crop components acquire their nutrients and moisture from different soil layers (Singh et al. 1998). Concerning essential oil production of palmarosa, changes in fresh weight, dry weight, chlorophyll, and essential oil content and its major constituents, such as geraniol and geranyl acetate, were examined for both racemes and spathe at various stages of spikelet development (Dubey et al. 2000). The essential oil content was maximal at the unopened spikelets stage and decreased significantly thereafter. At unopened spikelets stage, the proportion of geranyl acetate (58.6%) in the raceme oil was relatively greater compared with geraniol (37.2%), whereas the spathe oil contained more geraniol (61.9%) compared with geranyl acetate (33.4%). The relative percentage of geranyl acetate in both the oils, however, decreased significantly with development, and this is accompanied by a corresponding increase in the percentage of geraniol. Analysis of the volatile constituents from racemes and spathes (from mature spikelets) and seeds by capillary GC indicated 28 minor constituents besides the major constituent geraniol. (E)-Nerolidol was detected for the first time in an essential oil from this species. The geraniol content predominated in the seed oil, whereas the geranyl acetate content was higher in the raceme oil (Dubey et al. 2000).

Biotechnology is a powerful and consolidated technique for understanding plant growth and development as well as for improving biomass and yield of crops. Callus could be induced from nodal explant of mature tillering plant of *C. martinii* in different basal media supplemented with 2,4-dichlorophenoxy acetic acid (2,4-D) and kinetin (Kin). Shoot bud was regenerated from such calli in MS and B5 basal media modified with various combinations of phytohormones, vitamins, and amino acids. Root formation was induced either in white basal medium or half-strength MS or B5 media containing naphthalene acetic acid (NAA) or indole-3-butyric acid (IBA). High survival percentage of regenerated plants in soil was obtained after acclimatization in normal environment (Baruah and Bordoloi 1991). A detailed characterization of chromosomal status was carried out in callus, somatic embryos, and regenerants derived from in vitro cultured nodal and inflorescence explants of *C. martinii* ($2n = 20$). Both the callus lines revealed considerable ploidy variations (tetraploids to octoploids and hyperoctoploids), and the degree of polyploidization increased with the culture age. Frequencies of various polyploid cells were significantly higher in nodal callus lines (3.6% to 46.3%) than the inflorescence callus lines (1.9% to 23.6%) when analyzed over 520 days of culture. Somatic embryos derived from both the callus lines retained a predominantly diploid chromosome status throughout (99.0% to 93.1%). Root tip analysis of about 70 regenerants randomly taken from cultures of various ages (days 20 to 520) revealed only diploid chromosome numbers ($2n = 20$) implying a strong relative stability of diploidy among the regenerants (Patnaik et al. 1996). Chromosome counts of cells in suspensions, calli, and somatic embryos derived from cultures of different ages revealed the presence of diploids, tetraploids, and octoploids (Patnaik et al. 1997). Sodium chloride tolerant callus lines of *C. martinii* were obtained by exposing the callus to increasing concentrations of NaCl (0–350 mM) in the MS medium. The tolerant lines grew better than the sensitive wild-type lines in all concentrations of NaCl tested up to 300 mM. Callus survival and growth were completely inhibited, resulting in tissue browning and subsequent death at 350 mM NaCl. The selected lines retained their salt tolerance after 3–4 subcultures on salt-free medium, indicating the stability of the induced salt tolerance. The growth behavior, the Na^+, K^+, and proline contents of the selected callus lines were characterized and compared with those of the NaCl-sensitive lines. The Na^+ levels increased sharply, while the K^+ level declined continuously with the corresponding increase in external NaCl concentrations in both lines, but the NaCl-tolerant callus lines always maintained higher Na^+ and K^+ levels than that of the sensitive lines. The NaCl-selected callus line accumulated high levels of proline under salt stress. The degree of NaCl tolerance of the selected lines was in negative correlation with the K^+/Na^+ ratio and in positive correlation with proline accumulation (Patnaik and Debata 1997a). The embryogenic potential of NaCl-tolerant callus selected even at 300 mM could be improved significantly by the incorporation of gibberellic acid ($GA_{(3)}$) and abscisic acid (ABA), in the medium where, with 2 mg/L of $GA_{(3)}$ and 1 mg/L of ABA, the highest rates of embryogenesis (44.5%, 28.8%, and 18.6%) were achieved against 17.5%, 8.2%, and 1.8% on medium devoid of $GA_{(3)}$, and ABA at 50%, 150%, and 250 mM of NaCl, respectively (Patnaik and Debata 1997b). Finally, plants regenerated from cell suspension cultures of palmarosa were analyzed for somaclonal variation in five clonal generations. A wide range of variation in important quantitative traits, for example, plant yield, height, tiller number, oil content and qualitative changes in essential oil constituents geraniol, geranyl acetate, geranyl formate, and linalool, were observed among the 120 somaclones screened. Eight somaclones were selected on the basis of high herb and oil yield over the donor line and high geraniol content in the oil. Based on performance in the field trials, three superior lines were selected, and maintained for five clonal generations. The superior lines exhibited a reasonable degree of stability in the traits selected (Patnaik et al. 1999).

Palmarosa was also found to be associated with a vesicular-arbuscular mycorrhizal (VAM) fungus, *Glomus aggregatum*. Glasshouse experiments showed that inoculation of palmarosa with *G. aggregatum* caused a twofold and threefold biomass production as compared to nonmycorrhizal plants. These findings indicate the potential use of VAM-fungi for improving the production of this essential oil-bearing plant (Gupta and Janardhanan 1991). Furthermore, when the interactive effects of phosphate solubilizing bacteria, N-2 or fixing bacteria, and arbuscular mycorrhizal fungi (AMF) were studied in a low phosphate alkaline soil amended with a tricalcium insoluble source of inorganic phosphate on the growth of *C. martinii*. The rhizobacteria behaved as a "mycorrhiza helper" and enhanced root colonization by *G. aggregatum* in presence of tricalcium phosphate at the rate of 200 mg kg^{-1} soil (P1 level) (Ratti et al. 2001).

A dramatic increase in PEP carboxylase activity and oil biosynthesis was observed under drought conditions in *C. martinii* (Sangwan et al. 1993). The physiological and biochemical basis of drought tolerance in *C. martinii* has been elucidated on the basis of growth and metabolic responses (Fatima et al. 2002).

1.5.2 *Cymbopogon flexuosus*

Cymbopogon flexuosus (also known as lemongrass) is a perennial, multicut aromatic grass that yields an essential oil used in perfumery and pharmaceutical industries and vitamin A. It has a long initial lag phase. The growth and herbage and oil production of *C. flexuosus* in response to different levels of irrigation water (IW) 0.1, 0.3, 0.5, 0.7, 0.9, 1.1, 1.3, and 1.5 times cumulative pan evaporation CPE evaluated on deep sandy soils showed that an increment in the level of irrigation increased the plant height up to 0.7 IW:CPE ratio. However, the response of irrigation levels on tiller production of lemongrass differed with the season of harvest. Oil content had an inverse relationship with the levels of irrigation, whereas significantly higher herb and essential oil yields were recorded at 0.7 IW:CPE ratio, irrespective of season of harvest (Singh et al. 2000). Application of nitrogen (0, 50, 100, and 150 kg N ha^{-1} yr^{-1}) and phosphorus to *C. flexuosus* crops maintained the fertility of the soil, while potassium depletion was noticed (Singh 2001). When the effects of phosphorus (at 0, 17.75, and 35.50 kg ha^{-1} yr^{-1}), potassium (at 0, 33.2, 66.4, and 99.6 kg ha^{-1} yr^{-1}) and nitrogen (at 100 and 200 kg ha^{-1} yr^{-1}) and potassium (at 0, 33.2 and 66.4 kg ha^{-1} yr^{-1}) were studied on herbage and oil yield of *C. flexuosus*, it was found that plants produced significantly higher herbage and oil yields compared with controls (Singh et al. 2005; Singh and Shivaraj 1999). Spraying of iron-complexed additives on *C. flexuosus* increased iron translocation and the dry-matter production. Application of iron chelates and salts increased the vegetative herb yield, and oil and citral content. While maximum geraniol and less citral were obtained in the chlorotic plants, Fe recovered plants possessed more citral and less geraniol. The maximum recovery of total chlorophyll and nitrate reductase activity were recorded in the crop when Fe-EDTA chelates were sprayed at 22.4 ppm (Misra and Khan 1992). In *C. flexuosus*, a closer plant spacing of 45 × 45 cm resulted in higher herb and oil yields compared to wider spacing of 60 × 60 cm. Application of 150 kg N ha^{-1} yr^{-1} resulted in higher herb and oil yields. Higher nitrogen applications also increased the plant height and number of tillers per clump. The oil content and quality were not influenced by spacing and nitrogen levels (Singh et al. 1996b).

As for *C. martinii*, intercropping of *C. flexuosus* with the food legumes such as blackgram (*Vigna mungo* (L) Hepper), cowpea (*Vigna unguiculata* (L) Walp), or soybean (*Glycine max* (L) Merr.) prompted extra yields over and above that of pure cultures, without affecting the oil yield (Singh and Shivaraj 1998).

The influence of different foliar applications of the triacontanol (Tria.)-based plant growth regulator Miraculan on growth, CO_2 exchange, and essential oil accumulation in *C. flexuosus* showed increased rates in plant height, tillers per plant, biomass yield, accumulation of essential oil, net CO_2, and exchange and transpiration compared to the untreated control, but the number of leaves per tiller remained unaffected. Application of Miraculan also increased micronutrient uptake and total chlorophyll and citral content but decreased chlorophyll a/b ratio and stomatal resistance. Increase in shoot biomass, photosynthesis, and chlorophyll were significantly correlated with essential oil content (Misra and Srivastava 1991). Only young and rapidly expanding *C. flexuosus* leaves were found to have the capacity to synthesize and accumulate essential oil and citral. The pattern of the ratio of the label incorporated in citral to that in geraniol, during leaf ontogeny, evinced parallelism with the geraniol dehydrogenase activity. The elevated levels of glucose-6-phosphate dehydrogenase, 6-phosphogluconate dehydrogenase, NADP$^+$-malic enzyme, and NADP$^+$-isocitrate dehydrogenase coincided with the period of active essential oil biogenesis accompanying early leaf growth (Singh et al. 1990). Thus, there is an active involvement of oxidative pathways in essential oil biosynthesis. The time-course (12 h light followed by 12 h dark) monitoring of the C-14 radioactivity in starch and essential oil, after exposure of the immature (15 days after emergence) leaf to (CO)-C-14, revealed a progressive loss of label from starch and a parallel increase in radioactivity in essential oil. Thus, there was indication of a possible degradation of transitory starch serving as the source of carbon precursor for essential oil (monoterpene) biogenesis in the tissue (Singh et al. 1991).

Biotechnological applications revealed that *C. flexuosus* plants derived from somatic embryoids were more uniform in all the characteristics examined when compared with the field performance of plants raised through slips by standard propagation procedures (Nayak et al. 1996).

A fungal endophyte, *Balansia sclerotica* (Pat.) Hohn., has been found to establish a perennial association with the commercially grown East Indian *C. flexuosus* cv. Kerala local (syn. = OD-19). Endophyte-infected plants produced 195% more shoot biomass and 185% more essential oil than the endophyte-free control plants when grown experimentally under glasshouse conditions. The essential oil extracted from the endophyte-infected plants is qualitatively identical with that of endophyte-free plants and is free of toxic ergot alkaloids. Thus, *B. sclerotica*-infected East Indian *C. flexuosus* has potential for agricultural exploitation (Ahmad et al. 2001).

1.5.3 *Cymbopogon winterianus*

Java citronella (*C. winterianus*) is a perennial, multiharvest aromatic grass, the shoot biomass of which, on steam distillation, yields an essential oil extensively used in fragrance and flavor industries. Fresh *C. winterianus* (Java citronella) herbage and essential oil yields were significantly influenced by application of N up to 200 kg ha^{-1} yr^{-1}, while tissue N concentration and N uptake increased only to 150 kg N ha^{-1}. The oil yields with neem cake-coated urea (urea granules coated with neem cake) and urea super granules were 22 and 9% higher over that with prilled urea, and urea supergranules were significantly increased up to 200 kg N ha^{-1} while with neem cake-coated urea, response was observed only to 150 kg N ha^{-1}! Estimated recovery of N during two years from neem cake-coated urea, urea supergranules, and prilled urea were 38%, 31%, and 21%, respectively (Singh and Singh 1992). The interaction between N doses and nitrification inhibitors was also significant. Nitrification inhibitors performed better at the highest N dose (450 kg N ha^{-1} yr^{-1}), and the increase in the essential oil yields was to an extent of 27.3% to 34.6% when compared with "N alone" treatment. The nitrification inhibitors also increased the apparent N recoveries by citronella considerably. The oil content in the herb and its quality were not affected by the treatments. The nitrification inhibitors increased citronella yields and improved N economy (Puttanna et al. 2001). In Java citronella significant positive correlations were observed between fresh matter, citronellol content, dry and fresh matter yields, and total essential oil content (Omisra and Srivastava 1994). When the effect of depth (25, 37.5, and 50 mm) and methods (ridge and furrow, and broad bed and furrow method) of irrigation acid nitrogen levels (0, 200, and 400 kg N ha^{-1} yr^{-1}) were studied on herb and oil yields of Java citronella, highest herb and oil yields were achieved with the application of 400 kg N, maintaining 25 mm depth of irrigation, while the content and quality of oil were not affected either by irrigation or nitrogen (Singh et al. 1996a).

Among food legumes, greengram (*Vigna radiata* (L.) Wilez.), and among vegetables, clusterbean (*Cyamopsis psoraloides* D. C., syn. *Cyamopsis tetragonoloba* (L.) Taub.), tomato (*Lycopersicon esculentum* Mill.) and lady's finger (*Abelmoschus esculentus* Moench.) as intercrops of *C. winterianus* did not decrease its biomass and essential oil yield and produced bonus yields of these crops over and above that of Java citronella. Maximum monetary returns were recorded by Java citronella intercropped with tomato or greengram. However, Java citronella intercropped with red-gram (*Cajanus cajan* (L.) Millsp.), horsegram (*Macrotyloma uniflorum* (Lam.) Verd, syn. *Dolichos biflorus* Roxb.), and brinjal (*Solanum melongena* L.) suffered significant biomass and essential oil yield reductions. Horsegram proved to be the most competitive intercrop, producing least yields and minimum monetary returns (Rao 2000).

Changes in the utilization pattern of primary substrate, viz. [U-C-14] acetate, (CO_2)-C-14 and [U-C-14] saccharose, and the contents of C-14 fixation products in photosynthetic metabolites (sugars, amino acids, and organic acids) were determined in Fe-deficient Java citronella in relation to the essential oil accumulation. An overall decrease in photosynthetic efficiency of the Fe-deficient plants as evidenced by lower levels of incorporation into the sugar fraction and essential oil after (CO_2)-C-14 had been supplied was observed. When acetate and saccharose were fed to the Fe-deficient plants, despite a higher incorporation of label into sugars, amino acids, and organic acids, there was a lower incorporation of these metabolites into essential oils than in control plants. Thus, the availability of precursors and the translocation to a site of synthesis/accumulation, severely affected by Fe deficiency, is equally important for the essential oil biosynthesis in citronella (Srivastava et al. 1998). Lal and coworkers (2001) observed that improvement of oil quality with high citronellal content and low elemol content in Java citronella is believed to be achievable, although some compromise will have to be made in oil yield.

Nutrient acquisition and growth of Java citronella was also studied in a P-deficient sandy soil to determine the effects of mycorrhizal symbiosis and soil compaction. When a pasteurized sandy loam soil was inoculated either with rhizosphere microorganisms excluding VAM fungi (nonmycorrhizal) or with the VAM fungus, *Glomus intraradices* Schenck and Smith (mycorrhizal) and supplied with 0, 50, or 100 mg P kg^{-1} soil, *G. intraradices* was found to substantially increase root and shoot biomass, root length, nutrient (P, Zn, and Cu) uptake per unit root length, and nutrient concentrations in the plant, compared to inoculation with rhizosphere microorganisms when the soil was at the low bulk density and not amended with P. Little or no plant response to the VAM fungus was observed when the soil was supplied with 50 or 100 mg P kg^{-1} soil and/or compacted to the highest bulk density. At higher soil compaction and P supply, the VAM fungus significantly reduced root length. Nonmycorrhizal plants at higher soil compaction produced relatively thinner roots and had higher concentrations and uptake of P, Zn, and Cu than at lower soil compaction, particularly under conditions of P deficiency (Kothari and Singh 1996).

Pythium aphanidermatum was the predominant fungus recovered from the roots of Java citronella showing lethal yellowing in the northern part of India. Roots of infected plants showed marked discoloration, and the cortical region was completely disintegrated and sloughed from the vascular tissue. Diseased plants were chlorotic and stunted. Rotting was often found to spread from roots to stem, leading to severe chlorosis and death of the infected plants. The pathogenicity of the fungus was established. The disease is a potential constraint to citronella cultivation in nonarid climates where the crop is irrigated extensively (Alam et al. 1992). Another disease affecting commercial plantations of Java citronella is a collar rot and wilt disease. The causal organism was identified as *Fusarium moniliforme*, anamorph of *Gibberella fujikuroi*. Isolates of the pathogen differed in their pathogenicity on the host plant under glasshouse conditions. Differences were also observed in growth rates, pigment production, and sporulation between isolates (Alam et al. 1994).

1.5.4 OTHER *CYMBOPOGON* SPECIES

Application of graded levels of lime up to 10 t/ha on acid soil (pH 4.2) raised the pH up to 6.7. It increased the dry herbage of *C. khasianus* linearly. Increase of soil pH decreased N, P, K, Fe, and Zn contents in dry herbage significantly but increased the Ca and Mg contents. Liming showed a positive effect on the uptake of N, P, K, and Ca. However, Fe and Mg declined beyond lime levels of 23 and 5.0 t, respectively. Uptake of Zn was found fluctuating. Oil content (2.00%–2.07%; DWB) and geraniol (80.2%–81.0%) in the oil were unaffected by the lime treatments (Choudhury and

Bordoloi 1992). Experiments were also conducted to measure the rate of *C. caesius* litter decomposition and to identify fungal flora associated with the litter during different stages of decomposition in a tropical grassland. Rate of litter decomposition was several times higher than in temperate grasslands. Buried litter decayed more rapidly, and this rate was not influenced by climatic conditions. In contrast, surface litter recorded a lower decomposition rate, which was dependent on temporal (seasonal) fluctuations. Total nitrogen, available phosphorus, and potassium contents of the stem litter decreased during the initial stages of incubation.

Thirty-five species of fungi were isolated from the litter during the different stages of litter degradation. Most belonged to Hyphomycetes, which are active decomposers (Senthilkumar et al. 1992). *C. nardus* var. *confertiflorus* and *C. pendulus* were grown under mild and moderate water stress for 45 and 90 d to investigate the impact of in situ drought stress on plants in terms of relative water content, psi, concentration of proline, activities of PEP carboxylase and geraniol dehydrogenase, and geraniol and citral biogenesis. The results revealed that the species exhibited differential responses under mild and moderate stress treatments. In general, plant growth was reduced considerably, while the level of essential oils was maintained or enhanced.

Significant induction in catalytic activity of PEP carboxylase under water stress was one of the consistent metabolic responses of the aromatic grasses. The major oil constituents, geraniol and citral, increased substantially in both the species. Activity of geraniol dehydrogenase was also modulated under moisture stress. The responses varied depending upon the level and duration of moisture stress. The observations have been analyzed in terms of possible relevance of some of these responses to their drought stress adaptability/tolerance (Singhsangwan et al. 1994). In vitro plants of *C. citratus* were established, starting from shoot apices derived from plants cultivated under field conditions. The effect of the immersion frequency (two, four, and six immersions per day) on the production of biomass in temporary immersion systems (TIS) of 1 L capacity was studied. The highest multiplication coefficient (12.3) was obtained when six immersions per day were used. The maximum values of fresh weight (FW; 62.2 and 66.2 g) were obtained with a frequency of four and six immersions per day, respectively. However, the values for dry weight (DW; 6.4 g) and height (8.97 cm) were greater in the treatment with four immersions per day. The TIS used in this work for the production of lemongrass biomass may offer the possibility of manipulating the culture parameters, which can influence the production of biomass and the accumulation of secondary metabolites. We describe for the first time the in vitro production of *Cymbopogon citratus* biomass in TIS. In vitro regeneration of *C. polyneuros* was obtained through callus culture using leaf base, node, and root as explants. Callus was induced from different explants with 2–5 mg/L alpha-naphthalene acetic acid (NAA) and 1–2 mg/L kinetin in Murashige and Skoog's (MS) basal medium. High frequency shoots were noticed from leaf-base callus supplemented with 3.5 mg/L 6-benzylaminopurine (BA), l-arginine, adenine, and a low level of NAA (0.2 mg/L). About 80–85 shoot buds were obtained from ca. 200 mg of callus per culture. The individual shoots produced root in the presence of 0.5–3 mg/L indole 3-butyric acid or its potassium salt.

Regenerated plants were cytologically and phenotypically stable. Regenerants were transplanted into soil and subsequently transferred to the field (Das 1999). *C. nardus* could be propagated via tissue culture using axillary buds as explants. The aseptic bud explants obtained using double sterilization methods produced stunted abnormal multiple shoots when they were cultured on Murashige and Skoog (MS) medium supplemented with 1.0 mg L^{-1} or 2.0 mg L^{-1} benzyladenine (BA). Stunted shoots that cultured on MS + 1.0 mg L^{-1} RA + 1.0 mg L^{-1} N-6-isopentenyl-adenine (2iP) could induce elongation of shoots from about 60% of the stunted shoots. Normal multiple shoots could be induced at the highest (19.7 shoots per bud) from the bud explants within 6 weeks when cultured on proliferation medium consisted of MS supplemented with 0.3 mg L^{-1} BA and 0.1 mg L^{-1} indole-3-butyric acid (IBA). The separated individual shoot produced roots when transferred to basic MS solid medium. The essential oils that were contained in the mature plants namely citronellal, geraniol, and citronellol, were also found in the in vitro *C. nardus* plantlets. Citronellal was the main essential oil component in the matured plants, while geraniol was the main component in the in vitro plantlets (Chan et al. 2005). The occurrence, mode of infection, and the extent of damage caused by *Psilocybe kashmeriensis* sp. nov. Abraham on oil grass *C. jawarancusa* in Kashmir valley is discussed herein. A brief description of the new agaric species is also offered (Abraham 1995). Dormant vegetative slips of jamrosa (*C. nardus* var. *confertiflorus* × *C. jwarancusa*) were subjected to various doses of gamma rays. Plants raised from them were screened with a view to isolate improved clones of the crop. Five mutant clones isolated exhibited variation in quality/quantity of essential oil. These changes in oil characters were attributed to microlevel mutations induced by gamma rays (Kak et al. 2000).

REFERENCES

Abraham SP. 1995. Notes on the occurrence of an unusual agaric on Cymbopogon in Kashmir Valley. *Nova Hedwigia* **60**: 227–232.

Adams RP, Dafforn MR. 1997. DNA sampling of the pan-tropical vetiver grass uncovers genetic uniformity in erosion control germplasm. *Diversity* **13**: 27, 28.

Ahmad A, Alam M, Janardhanan KK. 2001. Fungal endophyte enhances biomass production and essential oil yield of East Indian lemongrass. *Symbiosis* **30**: 275–285.

Al-Zahim MA, Ford-Lloyd BV, Newbury HJ. 1999. Detection of somaclonal variation in garlic (*Allium sativum* L.) using RAPD and cytological analysis. *Plant Cell Reports* **18**: 473–477.

Alam M, Chourasia HK, Sattar A, Janardhanan KK. 1994. Collar rot and wilt—a new disease of Java-citronella (*Cymbopogon winterianus*) caused by *Fusarium-moniliforme* sheldon. *Plant Pathology* **43**: 1057–1061.

Alam M, Sattar A, Janardhanan KK, Husain A. 1992. Lethal yellowing of Java citronella (*Cymbopogon winterianus*) caused by *Pythium aphanidermatum*. *Plant Disease* **76**: 1074–1076.

Ashton AR. 1990. Enzymes of C_4 photosynthesis. In Dey PMHJB, Ed. *Methods in Plant Biochemistry*. London: Academic Press, pp. 39–72.

Baruah A, Bordoloi DN. 1991. Growth and regeneration of palmarosa (*Cymbopogon martinii* [Roxb.] Wats.) callus tissues under varied nutritional status. *Indian Journal of Experimental Biology* **29**: 582, 583.

Bertea CM, Scannerini S, D'Agostino G, Mucciarelli M, Camusso W, Bossi S, Buffa G, Maffei M. 2001. Evidence for a C4 NADP-ME photosynthetic pathway in *Vetiveria zizanioides* Stapf. *Plant Biosystems* **135**: 249–262.

Bertea CM, Tesio M, D'Agostino G, Buffa G, Camusso W, Bossi S, Mucciarelli M, Scannerini S, Maffei M. 2003. The C–4 biochemical pathway, and the anatomy of lemongrass (*Cymbopogon citratus* (DC) Stapf.) cultivated in temperate climates. *Plant Biosystems* **137**: 175–184.

Breakwell E. 1914. A study of the leaf anatomy of some native species of the genus *Andropogon*, N.O. Gramineae. *Proceedings of the Linnean Society, N.S.W.* **39**: 385–394.

Casati P, Spampinato CP, Andreo CS. 1997. Characteristics and physiological function of NADP-malic enzyme from wheat. *Plant Cell Physiology* **38**: 928–934.

Chalmers KJ, Waugh R, Simons AJ, Powell W. 1992. Detection of genetic variations between and within populations of *Gliricidia sepium* and *G. maculata* using RAPD markers. *Heredity* **69**: 465–472.

Chan LK, Dewi PR, Boey PL. 2005. Effect of plant growth regulators on regeneration of plantlets from bud cultures of *Cymbopogon nardus* L. and the detection of essential oils from the in vitro plantlets. *Journal of Plant Biology* **48**: 142–146.

Chapman KSR, Hatch MD. 1983. Intracellular location of phosphoenolpyruvate carboxykinase and other C_4 photosynthetic enzymes in mesophyll and bundle sheath protoplasts of *Panicum maxymum*. *Plant Science Letters* **29**: 145–154.

Choudhury SN, Bordoloi DN. 1992. Effect of liming on the uptake of nutrients and yield performance of *Cymbopogon khasianus* in acid soils of Northeast India. *Indian Journal of Agronomy* **37**: 518–522.

Das AB. 1999. Effects of different physiochemical factors on regeneration of *Cymbopogon polyneuros* Stapf. via callus culture and subsequent chromosomal stability. *Israel Journal of Plant Sciences* **47**: 195–198.

Dengler NG, Nelson T. 1999. Leaf structure and development in C4 plants. In Sage RF, Monson RK, Eds. *The Biology of C4 Plants*. San Diego: Academic Press, pp. 133–172.

Dubey VS, Mallavarapu GR, Luthra R. 2000. Changes in the essential oil content and its composition during palmarosa (*Cymbopogon martinii* (Roxb.) Wats. var. *motia*) inflorescence development. *Flavour and Fragrance Journal* **15**: 309–314.

Eastman PAK, Dengler NG, Peterson CA. 1988. Suberized bundle sheaths in grasses (*Poaceae*) of different photosynthetic types I. Anatomy, ultrastructure, and histochemistry. *Protoplasma* **142**: 92–111.

Edwards GE, Andreo CS. 1992. NADP-Malic enzyme from plants. *Phytochemistry* **31**: 1845–1857.

Edwards GE, Walker OA. 1983. *C3, C4: Mechanism, and Cellular and Environmental Regulation of Photosynthesis*. London: Blackwell Scientific Publications.

Fatima S, Farooqi AHA, Sharma S. 2002. Physiological and metabolic responses of different genotypes of *Cymbopogon martinii* and *C. winterianus* to water stress. *Plant Growth Regulation* 37: 143–149.

Fitter AH, Hay RKM. 1987. *Environmental Physiology of Plants*. London: Academic Press.

Ghannoum O, van Caemmerer S, Conroy JP. 2001. Carbon and water economy of Australian NAD-ME and NADP-ME C4 grasses. *Australian Journal of Plant Physiology* 28: 213–223.

Guenther E. 1950a. *The Essential Oils* (IV). New York, Van Nostrand. Ref Type: Serial (Book, Monograph).

Guenther E. 1950b. *The Essential Oils*. Princeton: Van Nostrand.

Gupta ML, Janardhanan KK. 1991. Mycorrhizal association of *Glomus aggregatum* with palmarosa enhances growth and biomass. *Plant and Soil* 131: 261–263.

Gutierrez M, Gracen VE, Edwards GE. 1974. Biochemical and cytological relationships in C_4 plants. *Planta* 119: 279–300.

Hatch MD. 1988. C_4 photosynthesis: A unique blend of modified biochemistry, anatomy and ultrastructure. *Biochimica et Biophysica Acta* 895: 81–106.

Hatch MD. 1992. C_4 photosynthesis: An unlikely process full of surprises. *Plant and Cell Physiology* 33: 333–342.

Hatch MD, Kagawa T, Craig S. 1975. Subdivision of C_4-pathway species based on differing C_4 acid decarboxylating systems and ultrastructural features. *Australian Journal of Plant Physiology* 2: 111–128.

Hatch MD, Kagawa T, Craig S. 2007. Subdivision of C_4-pathway species based on differing C_4 acid decarboxylating systems and ultrastructural features. *Australian Journal of Plant Physiology* 2: 111–128.

Hattersley PW, Watson L. 1976. C_4 grasses: An anatomical criterion for distinguishing between NADP-malic enzyme species and PCK or NAD-malic enzyme species. *Australian Journal of Botany* 24: 297–308.

Huang Y, Street-Perrot FA, Metcalfe SE, Brenner M, Moreland M, Freeman KH. 2001. Climate change as the dominant control on glacial–interglacial variations in C3 and C4 plant abundance. *Science* 293: 1647–1651.

Jain N, Shasany AK, Sundaresan V, Rajkumar S, Darokar MP, Bagchi GD, Gupta AK, Kumar S, Khanuja SPS. 2003. Molecular diversity in *Phyllanthus amarus* assessed through RAPD analysis. *Current Science* 85: 1454–1458.

Jenkins CLD, Furbank RT, Hatch MD. 1989. Mechanism of C_4 photosynthesis. *Plant Physiology* 91: 1372–1381.

Kak SN, Bhan MK, Rekha K. 2000. Development of improved clones of jamrosa (*Cymbopogon nardus* [L.] Rendle var. *confertiflorus* [Steud.] Bor. x *C. jwarancusa* Jones Schult.) through induced mutations. *Journal of Essential Oil Research* 12: 108–110.

Khanuja SPS, Shasany AK, Pawar A, Lal RK, Darokar MP, Naqvi AA, Rajkumar S, Sundaresan V, Lal N, Kumar S. 2005. Essential oil constituents and RAPD markers to establish species relationship in Cymbopogon Spreng. (Poaceae). *Biochemical Systematics and Ecology* 33: 171–186.

Kothari SK, Singh UB. 1996. Response of Java citronella (*Cymbopogon winterianus* Jowitt) to VA mycorrhizal fungi and soil compaction in relation to P supply. *Plant and Soil* 178: 231–237.

Kresovich S, Lamboy RL, Li R, Szewc-McFadden AK, Blick SM. 1994. Application of molecular methods and statistical analysis for discrimination of accessions and clones of vetiver grass. *Crop Science* 34: 805–809.

Kumar J, Verma V, Shahi AK, Qazi GN, Balyan HS. 2007. Development of simple sequence repeat markers in *Cymbopogon* species. *Planta Medica* 73: 262–266.

Kumar S, Dwivedi S, Kukreja AK, Sharma JR, Bagchi GD. 2000. *Cymbopogon: The Aromatic Grass*. Lucknow, India: Central Institute of Medicinal and Aromatic Plants.

Kuriakose KP. 1995. Genetic variability in East Indian lemongrass (*Cymbopogon flexuosus* Stapf). *Indian Perfumer* 39: 76–83.

Lal RK, Sharma JR, Misra HO, Sharma S, Naqvi AA. 2001. Genetic variability and relationship in quantitative and qualitative traits of Java citronella (*Cymbopogon winterianus* Jowitt). *Journal of Essential Oil Research* 13: 158–162.

Larkin PJ, Scowcroft WR. 1981. Somaclonal variation a novel source of variability from cell cultures for plant improvement. *Theoretical and Applied Genetics* 60: 197–214.

Leegood RC. 1993. Carbon dioxide-concentrating mechanisms. In Lea PJ, Leegood RC, Eds. *Plant Biochemistry and Molecular Biology*. Chichester, U.K: Wiley & Sons, pp. 47–72.

Lunn JE, Agostino A, Hatch MD. 1995. Regulation of NADP-malate dehydrogenase in C_4 plants—activity and properties of maize thioredoxin M and the significance of non-active site thiol groups. *Australian Journal of Plant Physiology* 55: 577–584.

Maffei M. 2002. *Vetiveria—The Genus Vetiveria*. London: Taylor & Francis.

Maffei M, Codignola A, Fieschi M. 1988. Photosynthetic enzyme activities in lemongrass cultivated in temperate climates. *Biochemical Systematics and Ecology* **16**: 263, 264.

Maheshwari SK, Chouhan GS, Trivedi KC, Gangrade SK. 1992. Effect of irrigation and stage of crop harvest on oil yield and quality of palmarosa (*Cymbopogon martinii*) oil grass. *Indian Journal of Agronomy* **37**: 514–517.

Maheshwari SK, Sharma RK, Gangrade SK. 1995. Effect of spatial arrangement on performance of palmarosa (*Cymbopogon martinii* var. *motia*)—pigeonpea (*Cajanus cajan*) intercropping on a black cotton soil (vertisol). *Indian Journal of Agronomy* **40**: 181–185.

Maheshwari SK, Sharma RK, Gangrade SK. 1998. Response of palmarosa (*Cymbopogon martinii* var. *motia*) to biofertilizers, nitrogen, and phosphorus in a shallow black soil under rainfed condition. *Indian Journal of Agronomy* **43**: 175–178.

Mathews S, Spangler RE, Mason-Gamer RJ, Kellogg EA. 2002. Phylogeny of Andropogoneae inferred from phytochrome B, GBSSI, and NDHF. *International Journal of Plant Sciences* **163**: 441–450.

Mathur AK, Ahuja PS, Pandey B, Kukreja AK, Mandal S. 1988. Screening and evaluation of somaclonal variations for quantitative and qualitative traits in aromatic grass, *Cymbopogon winterianus* Jowitt. *Plant Breeding* **101**: 321–334.

Maurino VG, Drincovich MF, Casati P, Andreo CS, Edwards GE, Ku MSB, Gupta SK, Franceschi VR. 1997. NADP-malic enzyme: Immunolocalization in different tissues of the C_4 plant maize and the C_3 plant wheat. *Journal of Experimental Botany* **48**: 799–811.

Metcalfe CR. 1960. *Anatomy of the Monocotyledons. I. Gramineae*. London: Oxford University Press.

Misra A, Khan A. 1992. Correction of iron-deficiency chlorosis in lemongrass (*Cymbopogon flexuosus* Steud.) Watts. *Agrochimica* **36**: 349–360.

Misra A, Srivastava NK. 1991. Effect of the triacontanol formulation miraculan on photosynthesis, growth, nutrient uptake, and essential oil yield of lemongrass (*Cymbopogon flexuosus*) Steud. Watts. *Plant Growth Regulation* **10**: 57–63.

Munthali MT, Newbury HJ, Ford-Loyd BV. 1996. The detection of somaclonal variants of beet using RAPD. *Plant Cell Reports* **15**: 474–478.

Nayak S, Debata BK, Sahoo S. 1996. Rapid propagation of lemongrass (*Cymbopogon flexuosus* [Nees] Wats.) through somatic embryogenesis in vitro. *Plant Cell Reports* **15**: 367–370.

Nayak S, Debata BK, Srivastava VK, Sangwan NS. 2003. Evaluation of agronomically useful somaclonal variants in jamrosa (a hybrid *Cymbopogon*) and detection of genetic changes through RAPD. *Plant Science* **164**: 1029–1035.

Omisra A, Srivastava NK. 1994. Influence of iron nutrition on chlorophyll contents, photosynthesis, and essential monoterpene oil(s) in Java citronella (*Cymbopogon winterianus* Jowitt). *Photosynthetica* **30**: 425–434.

Patnaik J, Debata BK. 1997a. In vitro selection of NaCl tolerant callus lines of *Cymbopogon martinii* (Roxb.) Wats. *Plant Science* **124**: 203–210.

Patnaik J, Debata BK. 1997b. Regeneration of plantlets from NaCl tolerant callus lines of *Cymbopogon martinii* (Roxb.) Wats. *Plant Science* **128**: 67–74.

Patnaik J, Sahoo S, Debata BK. 1996. Cytology of callus, somatic embryos and regenerated plants of palmarosa grass, *Cymbopogon martinii* (Roxb.) Wats. *Cytobios* **87**: 79–88.

Patnaik J, Sahoo S, Debata BK. 1997. Somatic embryogenesis and plantlet regeneration from cell suspension cultures of palmarosa grass (*Cymbopogon martinii*). *Plant Cell Reports* **16**: 430–434.

Patnaik J, Sahoo S, Debata BK. 1999. Somaclonal variation in cell suspension culture-derived regenerants of *Cymbopogon martinii* (Roxb.) Wats. var. *motia*. *Plant Breeding* **118**: 351–354.

Pecchioni N, Faccioli P, Monetti A, Stanca AM, Terzi V. 1996. Molecular markers for genotype identification in small grain cereals. *Journal of Genetic Breeding* **50**: 203–219.

Phillips SM, Hua P. 2005. Notes on grasses (Poaceae) for the flora of China, V. New species in Cymbopogon. *Novon* **15**: 471–473.

Prat H. 1937. Caractères anatomique et histologiques de quelques Andropogonées de l'Afrique occidentale. *Ann. Mus. Colon. Marseille* **5**: 25–28.

Purseglove JW. 1975. *Tropical Crops: Monocotyledons*. London: Longman Group Ltd.

Puttanna K, Gowda NMN, Rao EVSP. 2001. Effects of applications of N fertilizers and nitrification inhibitors on dry matter and essential oil yields of Java citronella (*Cymbopogon winterianus* Jowitt.). *Journal of Agricultural Science* **136**: 427–431.

Quiala E, Barbon R, Jimenez E, De Feria M, Chavez M, Capote A, Perez N. 2006. Biomass production of *Cymbopogon citratus* (DC) Stapf., a medicinal plant, in temporary immersion systems. *In Vitro Cellular and Developmental Biology–Plant* **42**: 298–300.

Rajendrudu G, Das VSR. 1981. C4 photosynthetic carbon metabolism in the leaves of aromatic tropical grasses—I. Leaf anatomy, CO_2 compensation point and CO_2 assimilation. *Photosynthesis Research* **2:** 225–233.

Rani V, Parida A, Raina SN. 1995. Random amplified polymorphic DNA (RAPD) micropropagated plants of *Populus deltoides* Marsh. *Plant Cell Reports* **14:** 459–462.

Rao BRR. 2000. Biomass yield and essential oil yield variations in Java citronella (*Cymbopogon winterianus* Jowitt.), intercropped with food legumes and vegetables. *Journal of Agronomy and Crop Science—Zeitschrift fur Acker und Pflanzenbau* **185:** 99–103.

Rao EVSP, Singh M, Chandrasekhara G. 1988. Effect of nitrogen application on herb yield, nitrogen uptake and nitrogen recovery in Java citronella (*Cymbopogon Winterianus* Jowitt). *Indian Journal of Agronomy* **33:** 412–415.

Rao EVSP, Singh M, Rao RSG. 1994. Performance of intercropping systems based on palmarosa (*Cymbopogon martinii* var. *motia*). *Indian Journal of Agricultural Sciences* **64:** 442–445.

Ratti N, Kumar S, Verma HN, Gautam SP. 2001. Improvement in bioavailability of tricalcium phosphate to *Cymbopogon martinii* var. *motia* by rhizobacteria, AMF and Azospirillum inoculation. *Microbiological Research* **156:** 145–149.

Rothermel BA, Nelson T. 1989. Primary structure of the maize NADP-dependent malic enzyme. *Journal of Biological Chemistry* **264:** 19587–19592.

Sangwan NS, Yadav U, Sangwan RS. 2001. Molecular analysis of genetic diversity in elite Indian cultivars of essential oil trade types of aromatic grasses (*Cymbopogon* species). *Plant Cell Reports* **20:** 437–444.

Sangwan NS, Yadav U, Sangwan RS. 2003. Genetic diversity among elite varieties of the aromatic grasses, *Cymbopogon martinii*. *Euphytica* **130:** 117–130.

Sangwan RS, Farooqi AHA, Bansal RP, Singhsangwan N. 1993. Interspecific variation in physiological and metabolic responses of 5 species of Cymbopogon to water stress. *Journal of Plant Physiology* **142:** 618–622.

Sangwan RS, Sangwan NS, Jain DC, Kumar S, Ranade SA. 1999. RAPD profile-based genetic characterization of chemotypic variants of *Artemisia annua* L. *Biochemistry and Molecular Biology International* **47:** 935–944.

Senthilkumar K, Udaiyan K, Manian S. 1992. Rate of litter decomposition in a tropical grassland dominated by *Cymbopogon caesius* in Southern India. *Tropical Grasslands* **26:** 235–242.

Shasany AK, Lal RK, Patra NK, Darokar MP, Garg A, Kumar S, Khanuja SPS. 2000. Phenotypic and RAPD diversity among *Cymbopogon winterianus* Jowitt accessions in relation to *Cymbopogon nardus* Rendle. *Genetic Resources and Crop Evolution* **47:** 553–559.

Singh A, Singh M, Singh K. 1998. Productivity and economic viability of a palmarosa-pigeonpea intercropping system in the subtropical climate of North India. *Journal of Agricultural Science* **130:** 149–154.

Singh K, Singh DV. 1992. Effect of rates and sources of nitrogen application on yield and nutrient-uptake of Java citronella (*Cymbopogon winterianus* Jowitt). *Fertilizer Research* **33:** 187–191.

Singh M. 2001. Long-term studies on yield, quality, and soil fertility of lemongrass (*Cymbopogon flexuosus*) in relation to nitrogen application. *Journal of Horticultural Science and Biotechnology* **76:** 180–182.

Singh M, Rao RSG, Ramesh S. 2005. Effects of nitrogen, phosphorus, and potassium on herbage, oil yield, oil quality, and soil fertility status of lemongrass in a semi-arid tropical region of India. *Journal of Horticultural Science and Biotechnology* **80:** 493–497.

Singh M, Rao RSG, Rao EVSP. 1996a. Effect of depth and method of irrigation and nitrogen application on herb and oil yields of Java citronella (*Cymbopogon winterianus* Jowitt) under semi-arid tropical conditions. *Journal of Agronomy and Crop Science—Zeitschrift fur Acker und Pflanzenbau* **177:** 61–64.

Singh M, Shivaraj B. 1998. Intercropping studies in lemongrass (*Cymbopogon flexuosus*) (Steud, Wats.). *Journal of Agronomy and Crop Science—Zeitschrift fur Acker und Pflanzenbau* **180:** 23–26.

Singh M, Shivaraj B. 1999. Effect of irrigation regimes on growth, herbage and oil yields of lemongrass (*Cymbopogon flexuosus*) under semi-arid tropical conditions. *Indian Journal of Agricultural Sciences* **69:** 700–702.

Singh M, Shivaraj B, Sridhara S. 1996b. Effect of plant spacing and nitrogen levels on growth, herb and oil yields of lemongrass (*Cymbopogon flexuosus* (Steud) Wats. var. *cauvery*). *Journal of Agronomy and Crop Science—Zeitschrift fur Acker und Pflanzenbau* **177:** 101–105.

Singh N, Luthra R, Sangwan RS. 1990. Oxidative pathways and essential oil biosynthesis in the developing *Cymbopogon flexuosus* leaf. *Plant Physiology and Biochemistry* **28:** 703–710.

Singh N, Luthra R, Sangwan RS. 1991. Mobilization of starch and essential oil biogenesis during leaf ontogeny of lemongrass (*Cymbopogon flexuosus* Stapf). *Plant and Cell Physiology* **32:** 803–811.

Singh RS, Bhattacharyya TK, Kakti MC, Bordoloi DN. 1992. Effect of nitrogen, phosphorus and potash on essential oil production of palmarosa (*Cymbopogon martinii* var. *motia*) under rainfed condition. *Indian Journal of Agronomy* **37:** 305–308.

Singh S, Ram M, Ram D, Sharma S, Singh DV. 1997. Water requirement and productivity of palmarosa on sandy loam soil under a sub-tropical climate. *Agricultural Water Management* **35:** 1–10.

Singh S, Ram M, Ram D, Singh VP, Sharma S, Tajuddin. 2000. Response of lemongrass (*Cymbopogon flexuosus*) under different levels of irrigation on deep sandy soils. *Irrigation Science* **20:** 15–21.

Singhsangwan N, Farooqi AHA, Sangwan RS. 1994. Effect of drought stress on growth and essential oil metabolism in lemongrasses. *New Phytologist* **128:** 173–179.

Smith BN, Brown WV. 1973. The Kranz syndrome in the Gramineae as indicated by carbon isotopic ratios. *American Journal of Botany* **60:** 505–513.

Spies JJ, Troskie TH, Vandervyver E, Vanwyk SMC. 1994. Chromosome studies on African plants—11. The tribe Andropogoneae (Poaceae, Panicoideae). *Bothalia* **24:** 241–246.

Sprengel CPJ. 1815. *Cymbopogon*. *Plantarum Minus Cognitarum Pugillus* **2:** 14.

Srivastava NK, Misra A, Sharma S. 1998. The substrate utilization and concentration of C-14 photosynthates in citronella under Fe deficiency. *Photosynthetica* **35:** 391–398.

Taylor PWJ, Geijskes JR, Ko HL, Fraser TA, Henry RJ, Birch RG. 1995. Sensitivity of random amplified polymorphic DNA analysis to detect genetic change in sugarcane during tissue culture. *Theoretical and Applied Genetics* **90:** 1165–1173.

Trevanion SJ, Furbank RT, Ashton AR. 1997. NADP-Malate dehydrogenase in the C_4 plant *Flaveria bidentis*. *Plant Physiology* **113:** 1153–1165.

Vickery JW. 1935. The leaf anatomy and vegetative characters of the indigenous grasses of N.S. Wales. *Proceedings of the Linnean Society, N.S.W.* **60:** 340–373.

Wallner E, Weising R, Rompf G, Kahl G, Kopp B. 1996. Oligonucleotide finger printing and RAPD analysis of Achillea species: characterisation and long term monitoring of micropropagated clones. *Plant Cell Reports* **15:** 647–652.

Welsh JM, McClelland M. 1990. Fingerprinting genomes using PCR with arbitrary primers. *Nucleic Acids Research* **18:** 7213–7218.

Yang H, Tabei Y, Kamad H, Kayano T, Takaiwa F. 1999. Detection of somaclonal variation in cultured rice cell using digoxigenin-based random amplified polymorphic DNA. *Plant Cell Reports* **18:** 520–526.

2 Chemistry and Biogenesis of Essential Oil from the Genus *Cymbopogon*

Anand Akhila

CONTENTS

2.1   INTRODUCTION

Aromatic grasses are one of the chief sources of essential oils. The genus *Cymbopogon* comprises a large number of species, out of which lemongrass, citronella, palmarosa, and few others produce oil of commercial importance (Gupta 1969; Gupta and Deniel 1982; Gupta and Jain 1978; Gupta et al. 1975). The chemical compounds present in the essential oils of *Cymbopogon* do not reflect the actual olfactory or other properties of the species (Gildemeister and Hoffmann 1956). There are instances when distinct species such as lemongrass oils from *C. pendulus, C. citratus,* and *C. flexuosus* (Anonymous 1958) produce oil of almost identical chemical composition. "Chemical characters are like other characters; they work when they work and they don't work when they don't work. Like all taxonomic characters they attain their value through correlation with other characters," has been rightly quoted by Cronquits (1980). However, most of the *Cymbopogon* species found all around the world produce essential oils that differ widely in their physical properties and chemical constituents. The varieties *motia* and *sofia* of *C. martinii* are good examples of such variable characters.

    The *Cymbopogon* species has great prospects for producing quality essential oils (Arctander 1960; Han et al. 1971), and it has direct relevance to the perfumery industry with economic benefit to humankind. However, the actual potential of cymbopogons has not been exploited to the fullest. Though tremendous work has been done regarding *Cymbopogon* chemistry, a lot more needs to be done to make use of the major and minor constituents present in its essential oils, particularly the mono- and sesquiterpenes (Chopra et al. 1956). An effort has been made in this chapter to present a comprehensive overview of most of the cultivated and wild species of cymbopogons.

2.2   CHEMISTRY AND BIOGENESIS OF ESSENTIAL OIL FROM CYMBOPOGONS

About 25 to 30 species are reported in genus *Cymbopogon*, and many of them are very good sources of essential oils of commercial importance. The compounds present in these oils are characteristic, but cannot necessarily be used for identification, of the species. Several botanical races of these species produce essential oils that are entirely different in their constituents. The essential oils of the *Cymbopogon* species mainly comprises of mono- and sesquiterpenoids and, despite their importance, very few high-tech identification techniques (such as GC-MS high resolution) have been utilized to identify the minor and trace constituents present in them; further, only a few reports are available in the literature. An attempt has been made in this chapter to present a complete analysis of most of the essential oils obtained from these *Cymbopogon* species. Biosynthetic pathways to most of the mono- and sesquiterpenes have been discussed. Efforts have also been made to provide complete data on the physicochemical properties of the oil and spectroscopic data ($^1H_2$ $^{13}CNMR$) of its individual constituents (Table 2.2 and Table 2.3). During the years 1950 to 1980, when GC and GC-MS techniques were not frequently available for the analysis of essential oils, the most significant data that could have been used was the density, specific gravity, refractive index, specific rotation, and solubility of the oil in aqueous alcohol.

2.3 PHYSICOCHEMICAL CHARACTERISTICS OF THE ESSENTIAL OILS FROM CYMBOPOGONS

Physical and chemical properties of any essential oil are of prime importance, and chemists are now working in an era when highly sophisticated instruments are available for quality and quantity analysis. Still, the specific gravity, optical rotation, solubility in dilute alcohol, and the refractive index must be determined for all oils and liquid isolates. Before the availability of modern analytical techniques, the essential oil chemists were working using their ingenuity, a highly developed sense of smell and taste, and analytical ability. Besides the determination of physical and chemical properties, other tests have also been carried out, such as ester content, total alcohol determination, congealing point, and melting points in the case of solids, which is of great importance. The reported values of these constants for the essential oils of *Cymbopogon* species are shown in Table 2.1, which is self-explanatory. This chapter will cater to the needs of students besides researchers and, therefore, brief definitions of the physicochemical characteristics, which are highly relevant in testing the quality of the essential oils, have been provided.

Specific gravity—Specific gravity is defined as the ratio of the density of a given solid or liquid substance to the density of water at a specific temperature and pressure, typically at 4°C (39°F) and 1 atm (14.7 psia). Substances with a specific gravity greater than 1 are denser than water, and so (ignoring surface tension effects) will sink in it, and those with a specific gravity less than 1 are less dense than water, and hence will float in it. Specific gravity is a special case of relative density, with the latter term often preferred in modern scientific writing. Specific gravity (SG) is expressed mathematically as

$$SG = \frac{\text{density of the substance } \rho}{\text{density of water } \rho l}$$

where ρ is the density of the substance, and ρ*l* is the density of water. (By convention ρ, the Greek letter rho, denotes density.)

Refractive index—The refractive index (or index of refraction) of a medium is a measure of how much the speed of light (or other waves such as sound waves) is reduced inside the medium. For example, typical glass has a refractive index of 1.5, which means that, in glass, light travels at $1/1.5 = 0.67$ times the speed of light in a vacuum. Two common properties of glass and other transparent materials are directly related to their refractive index. First, light rays change direction when they cross the interface from air to the material, an effect that is used in lenses and glasses. Second, light reflects partially from surfaces that have a refractive index different from that of their surroundings.

Definition: The refractive index *n* of a medium is defined as the ratio of the phase velocity *c* of a wave phenomenon such as light or sound in a reference medium to the phase velocity *vp* in the medium itself: $n = c/vp$.

Specific rotation—The specific rotation of a chemical compound [α] is defined as the observed angle of optical rotation α when plane-polarized light is passed through a sample with a path length of 1 dm and a sample concentration of 1 g/dL. The specific rotation of a pure material is an intrinsic property of that material at a given wavelength and temperature. Values should always be accompanied by the temperature at which the measurement was performed and the solvent in which the

Table 2.1 Major and Minor Constituents and Physiochemical Properties of Essential Oils Obtained from Cymbopogon Species

Cymbopogon flexuosus (Steud.) Wats.

Major terpenes—citral-a (geranial; 40%–50%), citral-b (neral; 30%–35%), borneol (0%–2%), citronellal (0.37%–8.04%), citronellol (0.44%–4.58%), citronellyl acetate (1.2%–3.6%), geraniol (1.73%–40.0%), geranyl acetate (1.95%–5.1%), limonene (2.4%–3.7%), and methyl eugenol (20%)	Atal and Bradu 1976a; De Martinez 1977
Myrcene (0.1%–14.2%)	Formacek and Kubeczka 1982; Borovik and Kuravskaya 1977; Bhattacharya et al. 1997; Gonzalo and Villarrubia 1973; Guenther 1950; Gonzalo 1973; Taskinen et al. 1983; Thapa and Agarwal 1989; Thapa et al. 1976; Thapa et al. 1981
Traces—α-bergamotene, β-bisabolene, τ-cadinene, α-cadinol, camphene, δ-3-carene, β-caryophyllene, β-caryophyllene oxide, 1,8-cineole, α-curcumene, *p*-cymene, *n*-decyldehyde, dipentene, β-elemene, τ-elemene, elemicin, elemol, farnesol, geranyl formate, α-humulene, isopulegol, linalool, linalyl acetate, p-menthane, methyl heptenol, methyl heptenone, τ-murolene, nerol, nerolidol, neryl acetate, 2-nonanone, *cis*-β-ocimene, τ-β-ocimene, perillene, phellandrene, α-pinene, β-pinene, piperitone, terpinen-4-ol, α-terpineol, and terpinolene	Atal and Bradu 1976a; Chiang et al. 1981; Jyrkit 1983; Le and Chu 1976; Foda et al. 1975; Mohammad et al. 1981a, 1981b; Nair et al. 1980a, 1980b; Sobti et al. 1978c, 1982; Srikulvandhana et al. 1976; Zaki et al. 1975; Jyrkit 1983

Origin	Specific Gravity	η_D	$[\alpha]_D$	Solubility	Reference
Cochin	0.899–0.905 at 15°	1.4883–1.488 at 20°	+1°25′ to −5°0′	1.5–3 vol. of 70% alcohol	Guenther 1950; Gildemeister and Hoffmann 1956
	Aldehyde content (a) Bisulfite method 70%–80%; (b) Neutral sulfite method 65%–80%				
Ceylon	0.895–0.908 at 15°	1.483–1.489 at 20°	1° to 5°	NA	De Sylva 1959
	Aldehyde content 65% to 85%				
Jammu Regional Research Laboratory (RRL)-57	0.9212 at 18°	1.4895 at 20°	−15.2°	NA	Thapa et al. 1976
	Acid value 2.69; ester value 73.01				
Jammu Regional Research Laboratory (RRL)-59	0.9137 at 18°	1.4887 at 20°	−15.5°	NA	Thapa et al. 1976
	Acid value 2.52; ester value 62.79				
Kerala	0.899–0.905	1.480–1.486 at 35°	+1°25′ to −5°0′	75% alcohol at 35°	Chakrabarti and Ghosh 1974
Kumaon	0.9651	1.489	NA	NA	Baslas and Baslas 1968
	Acid value 22.28; ester value 87.22				
Lucknow	0.8911 at 25°	1.4816 at 30°	−1°	1.2 vol. of 70% alcohol	Virmani and Datta 1973
	Aldehyde content 89%				
Lucknow	0.892 at 30°	1.4825 at 30°	−1.5°	NA	Sharma et al. 1972
	Aldehyde content 71.3%				
Odakali	0.8886 at 20°	1.4862 at 20°	−0.40°	NA	Thapa et al. 1981
	Acid value 3.65; ester value 46.77				
Pantnagar	0.8975–0.899	1.483–1.488	1°25′ to 5°	1.2 vol. of 70% alcohol	Gulati et al. 1976
	Aldehyde content 81.39%				
West Bengal	0.90	1.4803 at 35°	+1°30′	75% alcohol at 35°	Chakrabarti and Ghosh 1974

Table 2.1 (continued) Major and Minor Constituents and Physiochemical Properties of Essential Oils Obtained from Cymbopogon Species

Origin	Specific Gravity	η_D	$[\alpha]_D$	Solubility	Reference
Travancore	0.895–0.908	1.483–1.489	+1°30′ to –5°	3 vol. of 70% alcohol	Anonymous 1950
Indian Standards Institution (ISI)	0.892–0.902 at 25°	1.4802 at 20°	–3° to +1°		Anonymous 1952

Cymbopogon jwarancusa (Jones) Schult.

Major terpenes—piperitone (20%–70%) and Δ^4-carene (20%–24%), citronellal (30%–40%), *p*-cymene (0.6%–3.5%), geraniol (0.04%–22.5%), β-pinene (3.5%), and γ-terpinene (7.5%)

Traces—alloaromadendrene, *cis*- and γ-allo-ocimene, α-bisabolene, β-bisabolene, borneol, *d*-cadinene, calamene, camphene, camphor, β-caryophyllene, β-caryophyllene oxide, α-chamigrene, 1,8-cineole, citronellol, α-cubebene, cuprene, *o*-cymene, 5,6-dimethyl-5-norbornen-2-ol, dipentene, β-elemene, *d*-elemene, elemol, eucarvone, eudesmol, α-farnesene, β-farnesene, fenchone, geranyl acetate, geranyl formate, geranyl propionate, germacrene, α-humulene, iso-borneol, kasuralcohol, lavendulol, linalool, longifolene, *p*-mentha-2,8-dien-1-ol, *cis*- and γ-*p*-mentha-2-en-1-ol methyl heptenone, methyl thymyl ether, α-muurolene, myrcene, myrtenal, phellandrene, α-pinene, γ- and *cis*-peperitol, terpinen-4-ol, α-terpineol, terpinolene, γ-thuj-2-en-4-ol, verbenone, and β-ylangene

Ansari and Quadry 1987; Balyan et al. 1979; Dev et al. 1988; Dhar and Lattoo 1985; Dhar et al. 1981; Dhar and Dhar 1997; Guenther 1950; Liu et al. 1981; Maheshwari and Mohan 1985; Mathela and Pant 1988; Mathela et al. 1986; Nair et al. 1982; Saeed et al.1978; Shahi 1992; Shahi and Sen 1989; Sobti et al. 1982; Thapa et al. 1971; Shahi and Tava 1993

Origin	Specific Gravity	η_D	$[\alpha]_D$	Solubility	Reference
Not specified	0.9203–0.9228 at 30° Acid value 0.7; ester value 12.0	1.481–1.4858 at 30°	+51°41′ to 42°48′ at 30°	NA	Guenther 1948, 1950
Hazara	0.9203 at 30°	1.481 at 30°	+51.65° at 30°	NA	Anonymous 1950
Sind	0.923 at 30°	1.4858 at 30°	+42.8° at 30°	NA	Anonymous 1950
UP	0.909 at 30°	1.4856 at 30°	+25.7° at 30°	NA	Anonymous 1950

Cymbopogon martinii (Roxb.) Wats.

Major terpenes—geraniol (65%–85%), citral (4%–12%), citronellol (6.4%), linalool (2.4%), and geranyl acetate (6%–12%)

Traces— α-amorphine, β-betulenol, α-betulenol, bicyclogermacrene, β-bisabolene, γ-bisabolene, α-cadinene, γ-cadinene, *cis*-calamene, γ-calamene, calacorene, β-curcumene, *o*-cymene, *p*-cymene, *m*-cymene, dipentene, β-elemene, γ-elemene, β-farnesene, farnesol, farnesyl acetate, formaldehyde, geranyl-*n*-butyrate, germacrene-B, germacrene-D, β-helmiscapene, α-humulene, β-humulene, isovaleraldehyde, limonene, methyl heptenone, myrcene, γ-muurolene, nerolidol, 2-nonanol, α-phellandrene, α-pinene, β-pinene, selina-4,7-diene, α-selinene, β-selinene, *d*selinene, α-terpinene, γ-terpinene, α-terpineol, β-terpineol, and terpinolene

Gaydou and Raudriamiharisoa 1987; Guenther 1950; Mallavarapu et al. 1998; Peyron 1972, 1973; Anon. 1973 Anonymous 1980; Chiang et al. 1981; De Martinez 1977; Maheshwari and Mohan 1985; Mohammad et al. 1981b; Nair et al. 1980a, 1980b; Nigam et al. 1987; Oliveros-Belardo 1989; Sobti et al. 1981, 1982; Bottani et al. 1987; Naves 1970, 1971; Opdyke 1974

(continued on next page)

Table 2.1 (continued)   Major and Minor Constituents and Physiochemical Properties of Essential Oils Obtained from Cymbopogon Species

Origin	Specific Gravity	η_D	$[\alpha]_D$	Solubility	Reference
India	0.8903–0.8911 at 15°	1.4718–1.4738 at 20°	−0°5′ to 1°20′	1.5 vol. of 70% alcohol	Anonymous 1950
Java	0.891–0.892	—	+0°30′ to 0°42′	3.5 vol. of 60% alcohol	
West Bengal	0.89	1.4702 at 35°	−4°	75% alcohol	Chakrabarti and Ghosh 1974
ISI	0.8778–0.8898 at 25°C Acid value 3.0; alcohol content 88%–94%	1.4710–1.4755 at 25°	−2° to +3°	NA	Anonymous 1952
Hyderabad	0.889–0.900	1.468–1.4729	−3° to +6°	80% alcohol	Chakrabarti and Ghosh 1974
Jammu	0.889 at 15° Acid value 0.5–3.0; ester value 12.48	1.475 at 20°	NA	NA	Sobti et al. 1981
Lucknow (Pre-winter harvest)	0.8810 at 25° Acid value 0.5; ester value 6.8; geraniol content 93.7%	1.4722–1.4702 at 25°	+1°	2 vol. of 70% alcohol	Virmani and Dutta 1973

Cymbopogon martinii Stapf.

Major terpenes—geraniol (36%–65%), perillyl alcohol (15%–27%), *p*-menthenols (40%–62%), γ- and *cis*-carveol (3%–25%), 1,8-cineole (9.8%), and isopiperitenol (13.4%)

Traces—Δ³-carene, carveyl acetate, *d,l*-carvone, caryophyllene oxide, *p*-cymene, dihydrocarveol, γ- and *cis*-dihydrocarvone, dipentene, *d*-limonene oxide, *d*-α-phellandrene, α-pinene, piperitone oxide, and tricylene

Boelens 1994; Guenther 1950; Kalia et al. 1980; Nigam et al. 1965; Sobti et al. 1978; Thapa et al. 1971, 1981

Origin	Specific Gravity	η_D	$[\alpha]_D$	Solubility	Reference
ISI	0.8997–0.9287 at 25°C Acid value 6.0; alcohol content 36%–60%	1.4760–1.4910 at 25°	−14° to +54°	NA	Anonymous 1952
Jammu (wild collection)	0.9646 at 20° Acid value 22.4; ester value 92	1.4974 at 20°	+22.56°	NA	Kalia et al. 1980
Madras	0.900–0.953 at 15° Acid value 6.2; ester value 8.0	1.4780–1.4930 at 20°	−30° to +54°	Soluble in 2.3 vol. of 70% alcohol	Gildemeister and Hoffmann 1956; Guenther 1950

Cymbopogon citratus (D.C.) Stapf.

Major terpenes—citral-a or geranial (10%–48%) and citral-b or neral (3%–43%), borneol (5%), geraniol (2.6%–40%), geranyl acetate (0.1%–3.0%), linalool (1.2%–3.4%), and nerol (0.8%–4.5%)

Traces—camphene, camphor, α-camphorene, Δ-3-carene, caryophyllene, caryophyllene oxide, 1,8-cineole, citronellal, citronellol, *n*-decyldehyde, α,β-dihydropseudoionone, dipentene, β-elemene, elemol, farnesal, farnesol, fenchone, furfural, iso-pulegol, iso-valeraldehyde, limonene, linalyl acetate, menthol, menthone, methyl heptenol, ocimene, α-oxobisabolene, β-phellandrene, α-pinene, β-pinene, terpineol, terpinolene, 2-undecanone, neral, nerolic acid, and geranic acid

Abdullah et al. 1975; Abegaz et al. 1983; Brazil et al. 1971; Baruah et al. 1995; Guenther 1950; Idrissi et al. 1993; Thapa et al. 1981; Torres 1993; Zheng et al. 1993

Beech 1977; Crawford et al. 1975; El Tawil and El Beih 1982; Hanson et al. 1976; Kusumov and Babaev 1983; Liu et al. 1981; Manjoor-i-Khuda et al. 1984; Mathela 1991; Neyberg 1953; Nigam et al. 1987; Olaniyi et al. 1975; Opdyke 1973; Oliveros-Belardo and Aureus 1978, 1979; Rabha et al. 1979; Rouesti and Voriate 1960; Sarer et al. 1983; Sargenti and Lancas 1997; Zamureenka et al. 1981

Table 2.1 (continued)   Major and Minor Constituents and Physiochemical Properties of Essential Oils Obtained from Cymbopogon Species

Origin	Specific Gravity	η_D	$[\alpha]_D$	Solubility	Reference
India	0.865–0.914 at 15°		−0°10′ to 2°40′		Anonymous 1950
Belgium	0.8847 at 20°	1.4849 at 20°	−0°18′ at 25°		Neyberg 1953
	Citral content 71.3%				
Odakali	0.8986 at 20°	1.4910 at 20°	−0.62° at 20°	75% alcohol	Thapa et al. 1981
	Acid value 5.34; ester value 44.2				

Cymbopogon pendulus (Nees ex Steud.) Wats.

Major terpenes—citral-a or geranial (30%–50%), citral-b or neral (20%–35%), geranyl acetate (3%–5%), β-caryophyllene (2.1%), elemol (2.2%), geraniol (2%–6%), and linalool (3.0%)

Atal and Bradu 1976a; Balyan et al. 1979; Gulati and Garg 1976; Manjoor-i-Khuda et al. 1984, 1986; Nigam et al. 1975; Pino et al. 1996; Rajendrudu and Rama Das 1983; Sobti et al. 1982; Thapa et al. 1981; Thapa and Agarwal 1989

Traces—camphene, Δ³-carene, caryophyllene oxide, citronellal, citronellyl acetate, *p*-cymene, dipentene, β-elemene, methyl heptenone, myrcene, β-phellandrene, α-pinene, and β-pinene

Origin	Specific Gravity	η_D	$[\alpha]_D$	Solubility	Reference
Haldwani India	0.9002–0.9152	1.4905–1.4890	−0.36°	Soluble in 1.9 vol. of alcohol	Gulati et al. 1976
	Aldehyde content 88.73%				

Cymbopogon winterianus Jowitt

Major terpenes—geraniol (20%–25%), citronellol (4%–10%), citronellal (30%–45%), caryophyllene (2.1%), citronellyl acetate (3.0%), elemol (6.0%), geranyl acetate (4.2%), linalyl acetate (2.0%), methyl-iso-eugenol (2.3%), and nerol (7.7%)

Baslas 1970; Iruthayathas et al. 1977; Pino et al. 1996; Razdan and Koul 1973; Siddiqui et al. 1975; Wijesekera et al. 1973a, 1973b

Traces—borneol, cadinene, *l*-cadinol, *l*-camphene, 1-carvone, citral, citronellyl butyrate, cymbopol, dipentene, eugenol, farnesol, geranyl formate, *l*-limonene, linalool, methyl heptenone, methyl eugenol, α-pinene, sesquicitronellene, terpinene, terpinen-4-1, and thujyl alcohol

Anonymous 1973; Chiang et al. 1981; Ganguly et al. 1979; Kaul et al. 1977; Liu et al. 1981; Singh et al. 1970; Sobti et al. 1982.

Origin	Specific Gravity	η_D	$[\alpha]_D$	Solubility	Reference
ISI	0.8710–0.8870 at 30°	1.4610–1.4700 at 30°	−0°30′–6°		Anonymous 1952
	Total geraniol 85%–97%				
Nilgiri Hills	0.900–0.929		+2°11′ to +12°12′		Anonymous 1950
Wild	0.885–0.901 at 15°	1.463–1.475 at 20°	−4° to +1°47′	Soluble in 1–2 vol. of alcohol	Gildemeister and Hoffmann 1956; Guenther 1950
	Total geraniol content 85%–96%; citronellol 25%–54%				
Java	0.897 at 27°	1.4654 at 27.5°	−2°34′ at 28°	Soluble in 2.5–7.5 vol. of 70% alcohol	Guenther 1948, 1950
	Total geraniol 88.8%; citronellol content 42.7%				
Java	0.887–0.895 at15°	1.4685–1.4728 at 20°	−0°35′ to 5°6′	Soluble in 1–2 vol. of 80% alcohol	Guenther 1950
	Total geraniol 82.3%–89.4%; citronellol content 28.8%–43.9%				
Java	0.900–0.920	1.479–1.494 at 20°	−7° to +22°	Soluble in 80% alcohol	Chakrabarti and Ghosh 1974

(continued on next page)

Table 2.1 (continued)   Major and Minor Constituents and Physiochemical Properties of Essential Oils Obtained from Cymbopogon Species

Origin	Specific Gravity	η_D	$[\alpha]_D$	Solubility	Reference
Pantnagar	0.9009	1.471			Baslas 1970
	Total geraniol content 85.4%; citronellol 32.47%				
Kumaon	0.9095	1.471			Baslas 1968
	Total geraniol content 83.2%; citronellol 32%				
West Bengal	0.912	1.470 at 35°	−14°5′	Soluble in 80% alcohol	Chakrabarti and Ghosh 1974
Bangalore	0.8870 at 24°	1.4660 at 24°	−0°30′	Soluble in 1 vol. of 80% alcohol	Virmani and Dutta 1971
	Total geraniol content 65%				
Jorhat	0.890–0.892 at 25°	1.4650–1.4714 at 25°	−3°3′	Soluble in 1.5 vol. of 80% alcohol	Virmani and Dutta 1971
	Total geraniol content 85 to 90%				
Lucknow	0.875 at 25°	1.462 at 20°	−1°2′	Soluble in 1–2 vol. of 80% alcohol	Virmani and Dutta 1971
	Total geraniol content 96%				

Cymbopogon nardus (L.) Rendle

Major terpenes—limonene (9%–28%), geraniol (20%–30%), citronellol (8%–20%), citronellal (5%–16%), caryophyllene (1.4%), camphene (5%–6%), citral (18%), citronellol (20%), geranyl acetate (7%–8%), linalool (8.0%), methyl eugenol (4.1%), α-phellandrene (16.2%), β-pinene (15%), sesquicitronellene (5%–6%), and thujyl alcohol (8%–30%)
Bruns et al. 1981; Guenther 1950; Gupta and Chauhan 1970; Krishnarajah et al. 1985; Manjoor-i-Khuda et al. 1984; Opdyke 1976; Razdan 1984

Traces—α-bergamotene, *l*-borneol, citronellyl acetate, citronellyl butyrate, dipentene, elemol, ethyl *iso*-eugenol, eugenol, farnesol, *n*-heptyl alcohol, *iso*-pulegol, *iso*-valeraldehyde, methyl heptenone, nerol, *cis*-ocimene, pelargonaldehyde, and tricyclene
Bruns et al. 1981; Gulati and Sadgopal 1972; Herath et al. 1979; Lucius and Adler 1971; Thieme et al. 1980; Wijesekera et al. 1973a, 1973b

Origin	Specific Gravity	η_D	$[\alpha]_D$	Solubility	Reference
ISI	0.8870–0.9080 at 30°	1.4745–1.4805 at 30°	−9° to −18°		Anonymous 1952
	Total alcohol 55%–65%				
Haldwani	0.9233	1.4820 at 25°	+4.06°		Gulati and Sadgopal 1972
	Acid value 6.5; ester value 28.3				
Nainital	0.8632 at 20°	1.479 at 20°			Gupta and Chauhan 1970
	Acid value 2.83; ester value 26.38				
Lucknow	0.895 at 30°	1.478 at 30°	−12°		Sharma et al. 1972
	0.900–0.920	1.479–1.494 at 20°	−7° to −22°	Soluble in 1–2 vol. of 80% alcohol	Gildemeister and Hoffmann 1956; Guenther 1950
Ceylon	0.899–0.908 at 15°	1.4792–1.4842 at 20°	−9°40′ to −14°40′ at 20°	Soluble in 1 vol. of 80% alcohol	Guenther 1950
Ceylon	0.898–0.908 at 15°	1.4785–1.4900 at 20°	−7° to −14°		Anonymous 1950
Java	0.885–0.900 at 15.5°	1.465–1.473 at 20°	−5° to +1° at 20°	Soluble in 3 vol. of 80% alcohol	Anonymous 1950

Table 2.1 (continued)   Major and Minor Constituents and Physiochemical Properties of Essential Oils Obtained from Cymbopogon Species

Cymbopogon schoenanthus (L.) Spreng subsp. *proximus* Hochst.

Major terpenes—piperitone (80% amongst monoterpenes), elemol (39%), eudesmol (20%) amongst sesquiterpene alcohol, *cis*-carveol (4.8%), citral-a (2.4%), citral-b (3.3%), dihydrocarveol (35%), limonene (3.12%), linalool (21.6%)

Dawidar et al. 1990; Elagamal and Wolff 1987; El Tawil and El Beih 1982; Modawi et al. 1984

Traces—α-pinene, β-elemene, β-selinene, calamenene, cadalene phydroxycinnamic acid, and several sesquiterpene alcohols

Ahmed et al. 1970; Evans et al. 1982; Shahi et al. 1990; Siddiqui et al. 1980

Origin	Specific Gravity	η_D	$[\alpha]_D$	Solubility	Reference
	0.9169 at 15°C	1.4831 at 20°	−59°22′	Soluble in 0.8 vol.	Gildemeister and
	Acid value 6.0; alcohol content 36%–60%			of 70% alcohol	Hoffmann 1956

Cymbopogon caesius (Nees) Stapf

Major terpenes—Perillyl alcohol (25.6%), geraniol (19.8%), limonene (7.2%), citronellol (6.8%), and citronellal (6.7%)

Trace—Carvone (30%)

Liu et al. 1981

α-Thujene, α-pinene, terpinolene, linalool, isopulegol, borneol, terpineol, geraniol, bornyl acetate, eugenol, citronellyl acetate, geranyl acetate, β-caryophyllene, perillaldelyde, caryophyllene oxide, elemol, and guaiol

Kanjilal et al. 1995

Origin	Specific Gravity	η_D	$[\alpha]_D$	Solubility	Reference
Bangalore	0.9267–0.9339 at 15°C	1.484 to 1.4856 at 25°	−18.3° to −5.6° at 25°	Soluble in 70% alcohol	Gildemeister and Hoffmann 1956
	Acid value 0.9–2.5; saponification value 13.2–24.0; sap. value after acetylation 15.0–164.0				

Cymbopogon coloratus (Nees) Stapf.

Monoterpenes—myrcene, limonene, *trans*-β-ocimene, linalool, neral, geranial, geraniol (69.11%), geranyl acetate, and elemol

Mallavarapu et al. 1992

Origin	Specific Gravity	η_D	$[\alpha]_D$	Solubility	Reference
Malabar District	0.911–0.920 at 15°C	NA	−7°43′ to −10° 20′ at 25°	Soluble in 1 vol. of 80% alcohol	Gildemeister and Hoffmann 1956

Cymbopogon densiflorus (Steud.) Stapf.

Monoterpenes—Flower-limonene (52.1%), *trans*-p-menth-2,8-dien-1-ol (10%), verbenol (9.7%), and perillyl alcohol (7.2%)

Boelens 1994; Chisowa 1997

Monoterpenes—Leaf—*trans*-p-mentha-2,8-dien-1-ol (22.4%), verbenol (18%), perillyl alcohol (17.2%), and *cis*-p-mentha-1-(7)-dien-2-ol (11.1%)

Origin	Specific Gravity	η_D	$[\alpha]_D$	Solubility	Reference
Congo	0.9304 at 15°C	1.4683 at 20°	+59°30′	Soluble in 0.5 vol. of 80% alcohol	Gildemeister and Hoffmann 1956
	Sap. value 2.1; ester value 19.6; ester value 8.42 after acetylation				

Cymbopogon distans (Nees) Wats.

Major terpenes—terpineol (20%)

Piperitone (30%–40%) and geraniol (10%)

Sobti et al. 1978
Liu et al. 1981;
Thapa et al. 1971

Limonene (29%) and methyl eugenol (13%), β-bisabolene (5.4%), α-bisabolol (3.0%), bornyl acetate (4.8%), γ-cadinene (3.6%), caryophyllene (4.7%), *d*-citronellal (4.0%), *p*-cymene (5.1%), farnesol (5.1%), *d*-menthone (10.4%), geranyl acetate (10%–12%), α-humulene (3.5%), limonene (5.8%–29.0%), methyl eugenol (13.4%), α-phellandrene (2.3%–6.0%), and α-pinene (3.5%)

Singh and Sinha 1976

(continued on next page)

Table 2.1 (continued)   Major and Minor Constituents and Physiochemical Properties of Essential Oils Obtained from Cymbopogon Species

Traces—amorphine, γ-α-bergamotene, borneol, *d*-cadinene, camphene, 1-carbomenthone, citronellol, β-farnesene (a keto compound), β-muurolene, myrcene, neryl propionate, octanol β-phellandrene, β-pinene, sabinene, β-selinene, α-terpinene, γ-terpinene, and terpinolene	Balyan et al. 1979; Gupta and Daniel 1982; Liu et al. 1981; Mathela et al. 1988; Mathela and Joshi 1981; Mathela et al. 1990a; Melkani et al. 1985; Singh and Sinha 1976; Sobti et al. 1978c

Origin	Specific Gravity	η_D	$[\alpha]_D$	Solubility	Reference
Nainital	0.801	—	—	NA	Mathela and Joshi 1981
	Acid value 1.15; ester value after acetylation 80.95				

Cymbopogon nervatus (Hohst.) Chiov.

Terpenes—β-selinene, β-elemene, β-bergamotene, and germacrene-D	Modawi et al. 1984

Origin	Specific Gravity	η_D	$[\alpha]_D$	Solubility	Reference
Kordofan Sudan	0.9405 at 15°	1.4946 at 20°	+26°22′	Soluble in 0.5 vol. of 80% alcohol	Gildemeister and Hoffmann 1956
	Ester value 9.3, ester value after acetylation 99.1				

material was dissolved. Often the temperature is not specified; in these cases it is assumed to be room temperature. The formal unit for specific rotation values is deg cm² g⁻¹, but scientific literature uses just degrees. A negative value means levorotatory rotation, and a positive value means dextrorotatory rotation.

Optical rotation is measured with an instrument called a *polarimeter*. There is a linear relationship between the observed rotation and the concentration of optically active compound in the sample. There is a nonlinear relationship between the observed rotation and the wavelength of light used. Specific rotation is calculated using either of two equations, depending on the sample you are measuring. For pure liquids,

$$[\alpha]_\lambda^T = \alpha/l \times d$$

In this equation, l is the path length in decimeters, and d is the density of the liquid in g/mL, for a sample at a temperature T (given in degrees Celsius) and wavelength λ (in nanometers). If the wavelength of the light used is 589 nanometer (the sodium D line), the symbol "D" is used. The sign of the rotation (+ or −) is always given: $[\alpha]_D^{20} = +6.2°$. For solutions, a different equation is used:

$$[\alpha]_\lambda^T = 100\alpha/l \times d$$

When using this equation, the concentration and the solvent are always provided in parentheses after the rotation. The rotation is reported using degrees, and no units of concentration are given (it is assumed to be g/100 mL).

Solubility—Solubility is a characteristic physical property referring to the ability of a given substance, the solute, to dissolve in a solvent. It is measured in terms of the maximum amount of solute dissolved in a solvent at equilibrium. The resulting solution is called a saturated solution. Certain liquids are soluble in all proportions with a given solvent, such as ethanol in water. This property is known as miscibility. Under certain conditions the equilibrium solubility can be exceeded to give a so-called supersaturated solution, which is metastable.

Table 2.2 ^{1}H and ^{13}C-NMR Data of Monoterpenes Found in Cymbogon Essential Oils

Acyclic Monoterpene Hydrocarbons

^{1}H-NMR   **^{13}C-NMR**

cis-Ocimene—(4Z,6E)-2,6-dimethylocta-2,4,6-triene; chemical formula: $C_{10}H_{16}$; exact mass: 136.13; molecular weight: 136.13 (100.0%), 137.13 (11.0%); elemental analysis: C, 88.16; H, 11.84

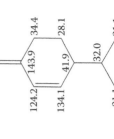

^{1}H-NMR   **^{13}C-NMR**

trans-Ocimene—(4Z,6E)-2,6-dimethylocta-2,4,6-triene; chemical formula: $C_{10}H_{16}$; exact mass: 136.13; molecular weight: 136.23, m/z: 136.13 (100.0%), 137.13 (11.0%); elemental analysis: C, 88.16; H, 11.84

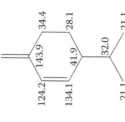

α-Phellandrene—5-isopropyl-2-methylcyclohexa-1,3-diene; chemical chemical formula: $C_{10}H_{16}$; exact mass: 136.13; molecular weight: 136.23, m/z: 136.13 (100.0%), 137.13 (11.0%); elemental analysis: C, 88.16; H, 11.84

β-Phellandrene—3-isopropyl-6-methylenecyclohex-1-ene; chemical formula: $C_{10}H_{16}$; exact mass: 136.13; molecular weight: 136.13; 136.13 (100.0%), 137.13 (11.0%); elemental analysis: C, 88.16; H, 11.84

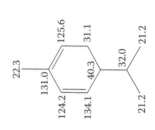

(continued on next page)

Table 2.2 (continued) ¹H and ¹³C-NMR Data of Monoterpenes Found in Cymbogon Essential Oils

Acyclic Oxygenated Monoterpenes

Citral-a-(Z)-3,7-dimethylocta-2,6-dienal; chemical formula: $C_{10}H_{16}O$; exact mass: 152.12; molecular weight: 152.23, m/z: 152.12 (100.0%), 153.12 (10.9%); elemental analysis: C, 78.90; H, 10.59; O, 10.51

Citral-b-(E)-3,7-dimethylocta-2,6-dienal; chemical formula: $C_{10}H_{16}O$; exact mass: 152.12; molecular weight: 152.23, m/z: 152.12 (100.0%), 153.12 (10.9%); elemental analysis: C, 78.90; H, 10.59; O, 10.51

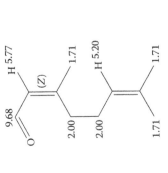

Citronellol—3,7-dimethyloct-6-en-1-ol; chemical formula: $C_{10}H_{20}O$; exact mass: 156.15; molecular weight: 156.27, m/z: 156.15 (100.0%), 157.15 (10.8%); elemental analysis: C, 76.86; H, 12.90; O, 10.24

Citronellal—3,7-dimethyloct-6-enal; chemical formula: $C_{10}H_{18}O$; exact mass: 154.14; molecular weight: 154.25, m/z: 154.14 (100.0%), 155.14 (11.1%); elemental analysis: C, 77.87; H, 11.76; O, 10.37

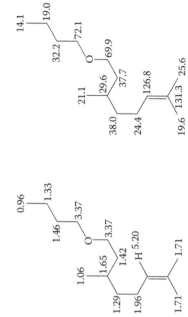

Citronellyl butyrate—8-butoxy-2,6-dimethyloct-2-ene; chemical formula: $C_{14}H_{28}O$; exact mass: 212.21; molecular weight: 212.37, m/z: 212.21 (100.0%), 213.22 (15.5%), 214.22 (1.3%); elemental analysis: C, 79.18; H, 13.29; O, 7.53

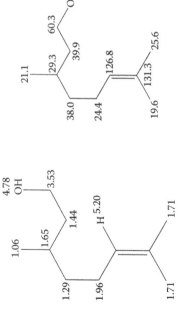

Citronellyl acetate—3,7-dimethyloct-6-enyl acetate; chemical formula: $C_{12}H_{22}O_2$; exact mass: 198.16; molecular weight: 198.3, m/z: 198.16 (100.0%), 199.17 (13.3%), 200.17 (1.2%); elemental analysis: C, 72.68; H, 11.18; O, 16.14

(continued on next page)

Table 2.2 (continued) ¹H and ¹³C-NMR Data of Monoterpenes Found in Cymbogon Essential Oils

Geraniol—(*E*)-3,7-dimethylocta-2,6-dien-1-ol; chemical formula: $C_{10}H_{18}O$; exact mass: 154.25; molecular weight: 154.14 (100.0%), 155.14 (11.1%); elemental analysis: C, 77.87; H, 11.76; O, 10.37

Lavandulol—5-methyl-2-(prop-1-en-2-yl)hex-4-en-1-ol; chemical formula: $C_{10}H_{18}O$; exact mass: 154.14; molecular weight: 154.25, m/z: 154.14 (100.0%), 155.14 (11.1%); elemental analysis: C, 77.87; H, 11.76; O, 10.37

Geranyl acetate—(*Z*)-3,7-dimethylocta-2,6-dienyl acetate; chemical formula: $C_{12}H_{20}O_2$; exact mass: 196.15; molecular weight: 196.15 (100.0%), 197.15 (13.3%), 198.15 (1.2%); elemental analysis: C, 73.43; H, 10.27; O, 16.30

Geranyl formate—(*Z*)-3,7-dimethylocta-2,6-dienyl formate; chemical formula: $C_{11}H_{18}O_2$; exact mass: 182.13; molecular weight: 182.26, m/z: 182.13 (100.0%), 183.13 (12.0%); elemental analysis: C, 72.49; H, 9.95; O, 17.56

Linalool—3,7-dimethylocta-1,6-dien-3-ol, chemical formula: $C_{10}H_{18}O$; exact mass: 154.14; molecular weight: 154.14 (100.0%), 155.14 (11.1%); elemental analysis: C, 77.87; H, 11.76; O, 10.37

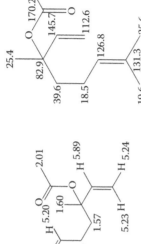

Linalyl acetate—3,7-dimethylocta-1,6-dien-3-yl acetate, chemical formula: $C_{12}H_{20}O_2$; exact mass: 196.15; molecular weight: 196.29, m/z: 196.15 (100.0%), 197.15 (13.3%), 198.15 (1.2%); elemental analysis: C, 73.43; H, 10.27; O, 16.30

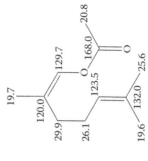

Nerol—(Z)-3,7-dimethylocta-2,6-dien-1-ol; chemical formula: $C_{10}H_{18}O$; exact mass: 154.14; molecular weight: 154.25, m/z: 154.14 (100.0%), 155.14 (11.1%); elemental analysis: C, 77.87; H, 11.76; O, 10.3

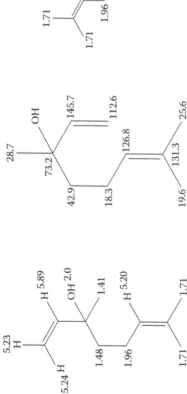

Neryl acetate—(E)-3,7-dimethylocta-2,6-dien-1-ol; chemical formula: $C_{10}H_{18}O$; exact mass: 154.14; molecular weight: 154.25, m/z: 154.14 (100.0%), 155.14 (11.1%); elemental analysis: C, 77.87; H, 11.76; O, 10.37

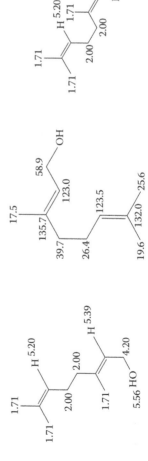

(continued on next page)

Table 2.2 (continued)   ¹H and ¹³C-NMR Data of Monoterpenes Found in Cymbogon Essential Oils

6-methylhept-5-en-2-one—chemical formula: $C_8H_{14}O$; exact mass: 126.1; molecular weight: 126.2, m/z: 126.10 (100.0%), 127.11 (8.9%); elemental analysis: C, 76.14; H, 11.18; O, 12.6

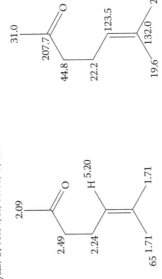

Cyclic Monoterpene Hydrocarbons

p-Cymene—chemical formula: $C_{10}H_{14}$; exact mass: 134.11; molecular weight: 134.22, m/z: 134.11 (100.0%), 135.11 (10.8%); elemental analysis: C, 89.49; H, 10.51

o-Cymene—chemical formula: $C_{10}H_{14}$; exact mass: 134.11; molecular weight: 134.22, m/z: 134.11 (100.0%), 135.11 (10.8%); elemental analysis: C, 89.49; H, 10.51

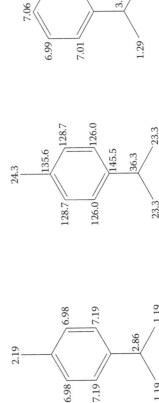

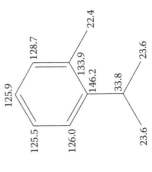

(+)-Limonene—(*R*)-1-methyl-4-(prop-1-en-2 yl)cyclohex-1-ene; chemical formula: $C_{10}H_{16}$; exact mass: 136.13; molecular weight: 136.23, m/z: 136.13 (100.0%), 137.13 (11.0%); elemental analysis: C, 88.16; H, 11.84

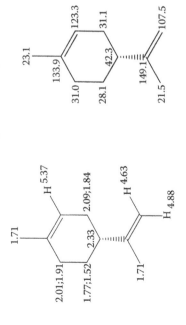

Myrcene—7-methyl-3-methyleneocta-1,6-diene; chemical formula: $C_{10}H_{16}$; exact mass: 136.13; molecular weight: 136.23, m/z: 136.13 (100.0%), 137.13 (11.0%); elemental analysis: C, 88.16; H, 11.84

(−)-Limonene—(*S*)-1-methyl-4-(prop-1-en-2 yl)cyclohex-1-ene; chemical formula: $C_{10}H_{16}$; exact mass: 136.13; molecular weight: 136.23, m/z: 136.13 (100.0%), 137.13 (11.0%); elemental analysis: C, 88.16; H, 11.84

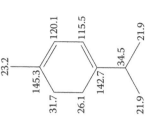

α-Terpinene—1-isopropyl-4-methylcyclohexa-1,3-diene; chemical formula: $C_{10}H_{16}$; exact mass: 136.13; molecular weight: 136.23, m/z: 136.13 (100.0%), 137.13 (11.0%); elemental analysis: C, 88.16; H, 11.84

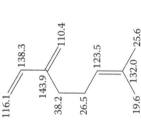

(continued on next page)

Table 2.2 (continued)   ¹H and ¹³C-NMR Data of Monoterpenes Found in Cymbogon Essential Oils

γ-Terpinene—1-isopropyl-4-methylcyclohexa-1,4-diene; chemical formula: $C_{10}H_{16}$; exact mass: 136.13; molecular weight: 136.23; m/z: 136.13 (100.0%), 137.13 (11.0%); elemental analysis: C, 88.16; H, 11.84

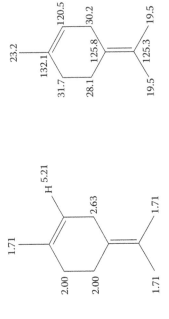

Terpinolene—1-methyl-4-(propan-2-ylidene) cyclohex-1-ene; chemical formula: $C_{10}H_{16}$; exact mass: 136.13; molecular weight: 136.23, m/z: 136.13 (100.0%), 137.13 (11.0%); elemental analysis: C, 88.16; H, 11.84

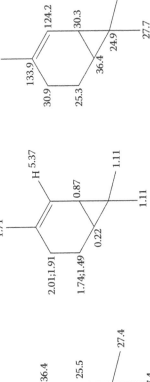

Bicyclic Monoterpene Hydrocarbons

Δ³-Carene—3,7,7-trimethylbicyclo[4.1.0]hept-3-ene; chemical formula: $C_{10}H_{16}$; exact mass: 136.13; molecular weight: 136.23; m/z: 136.13 (100.0%), 137.13 (11.0%); elemental analysis: C, 88.16; H, 11.84

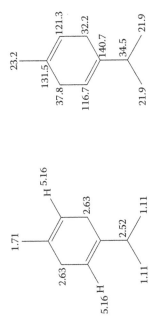

Δ⁴-Carene —3,7,7-trimethylbicyclo[4.1.0]hept-2-ene; chemical formula: $C_{10}H_{16}$; exact mass: 136.13; molecular weight: 136.23; m/z: 136.13 (100.0%), 137.13 (11.0%); elemental analysis: C, 88.16; H, 11.84

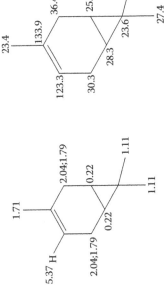

Camphene—2,2-dimethyl-3 methylenebicyclo[2.2.1]heptane; chemical formula: $C_{10}H_{16}$; exact mass: 136.13; molecular weight: 136.23, m/z: 136.13 (100.0%), 137.13 (11.0%); elemental analysis: C, 88.16; H, 11.84

α-Pinene—2,6,6-trimethylbicyclo[3.1.1]hept-2-ene; chemical formula: $C_{10}H_{16}$; exact mass: 136.13; molecular weight: 136.23, m/z: 136.13 (100.0%), 137.13 (11.0%); elemental analysis: C, 88.16; H, 11.84

β-Pinene—6,6-dimethyl-2 methylenebicyclo[3.1.1]heptane; chemical formula: $C_{10}H_{16}$; exact mass: 136.13; molecular weight: 136.23, m/z: 136.13 (100.0%), 137.13 (11.0%); elemental analysis: C, 88.16; H, 11.84

Sabienene—1-isopropyl-4 methylenebicyclo[3.1.0]hexane; chemical formula: $C_{10}H_{16}$; exact mass: 136.13; molecular weight: 136.23, m/z: 136.13 (100.0%), 137.13 (11.0%); elemental analysis: C, 88.16; H, 11.84

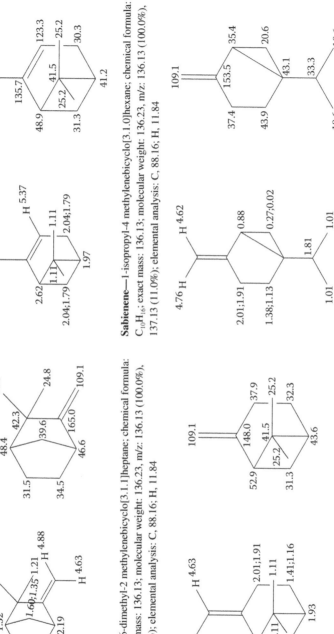

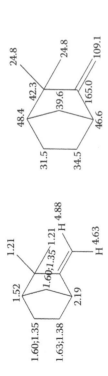

(continued on next page)

Table 2.2 (continued) ¹H and ¹³C-NMR Data of Monoterpenes Found in Cymbogon Essential Oils

Oxygenated Monoterpenes

Carvacrol—(R)-2-methyl-5-(prop-1-en-2-yl)cyclohex-1-enol; chemical formula: $C_{10}H_{16}O$; exact mass: 152.12; molecular weight: 152.23, m/z: 152.12, (100.0%), 153.12 (10.9%); elemental analysis: C, 78.90; H, 10.59; O, 10.51

cis-Carveol—2-methyl-5-(prop-1-en-2-yl)cyclohex-2-enol; chemical formula: $C_{10}H_{16}O$; exact mass: 152.12; molecular weight: 152.23, m/z: 152.12, (100.0%), 153.12 (10.9%); elemental analysis: C, 78.90; H, 10.59; O, 10.51

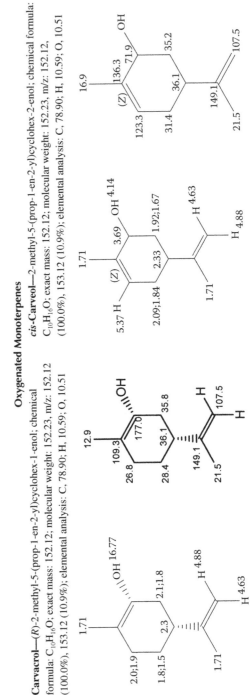

Carvyl acetate—2-methyl-5-(prop-1-en-2-yl) cyclohex-2-enyl acetate; chemical formula: $C_{12}H_{18}O_2$; exact mass: 194.13; molecular weight: 194.27; m/z: 194.13 (100.0%), 195.13 (13.1%); elemental analysis: C, 74.19; H, 9.34; O, 16.47

Carvotanacetone—5-isopropyl-2-methylcyclohex-2-enone; chemical formula: $C_{10}H_{16}O$; exact mass: 152.12; molecular weight: 152.23, m/z: 152.12 (100.0%), 153.12 (10.9%); elemental analysis: C, 78.90; H, 10.59; O, 10.51

Carvone—2-methyl-5-(prop-1-en-2-yl)cyclohex-2-enone; chemical formula: $C_{10}H_{14}O$; exact mass: 150.1; molecular weight: 150.22; m/z: 150.10 (100.0%), 151.11 (11.0%); elemental analysis: C, 79.96; H, 9.39; O, 10.65

Eucarvone—(2Z,4Z)-2,6,6-trimethylcyclohepta-2,4-dienone; chemical formula: $C_{10}H_{14}O$; exact mass: 150.1; molecular weight: 150.22; m/z: 150.10 (100.0%), 151.11 (11.0%); elemental analysis: C, 79.96; H, 9.39; O, 10.65

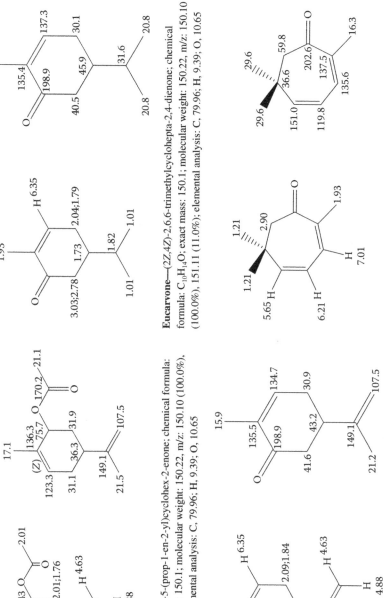

(continued on next page)

Table 2.2 (continued) ^{1}H and ^{13}C-NMR Data of Monoterpenes Found in Cymbogon Essential Oils

1,8-Cineole—1,3,3-trimethyl-2 oxabicyclo [2.2.2]octane; chemical formula: $C_{10}H_{18}O$; exact mass: 154.14; molecular weight: 154.25, m/z: 154.14 (100.0%), 155.14 (11.1%); elemental analysis: C, 77.87; H, 155.14 (11.1%); elemental analysis: C, 77.87; H, 11.76; O, 10.37

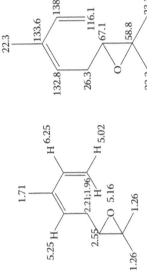

(−)-Dihdrocarveol—(1R,5R)-2-methyl-5-(prop-1-en-2-yl)cyclohexanol; chemical formula: $C_{10}H_{18}O$; exact mass: 154.14; molecular weight: 154.25, m/z: 154.14 (100.0%), 155.14 (11.1%); elemental analysis: C, 77.87; H, 11.76; O, 10.37

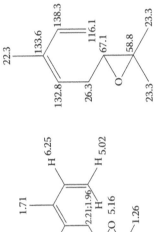

3,4-Epoxy-3,7-dimethyl-1,6-octadiene—2-methyl-3-(3-methylbut-2-enyl)-2-vinyloxirane; chemical formula: $C_{10}H_{16}O$; exact mass: 152.12; molecular weight: 152.23, m/z: 152.12 (100.0%), 153.12 (10.9%); elemental analysis: C, 78.90; H, 10.59; O, 10.5

6,7-Epoxy-3,7-dimethyl-1,3-octadiene—(Z)-2,2-dimethyl-3-(3-methylpenta-2,4-dienyl)oxirane; chemical formula: $C_{10}H_{16}O$; exact mass: 152.12; molecular weight: 152.23, m/z: 152.12 (100.0%), 153.12 (10.9%); elemental analysis: C, 78.90; H, 10.59; O, 10.51

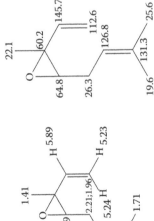

Isopiperitenone—(*R*)-3-methyl-6-(prop-1-en-2-yl)cyclohex-2-enone; chemical formula: $C_{10}H_{14}O$; exact mass: 150.1; molecular weight: 150.22, m/z: 150.10 (100.0%), 151.11 (11.0%), elemental analysis: C, 79.96; H, 9.39; O, 10.65

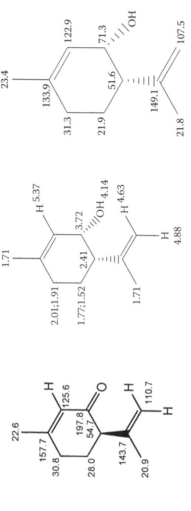

cis-Isopiperitenol—(1*S*,6*S*)-3-methyl-6-(prop-1-en-2-yl)cyclohex-2-enol, chemical formula: $C_{10}H_{16}O$; exact mass: 152.12; molecular weight: 152.23, m/z: 152.12 (100.0%), 153.12 (10.9%); elemental analysis: C, 78.90; H, 10.59; O, 10.51

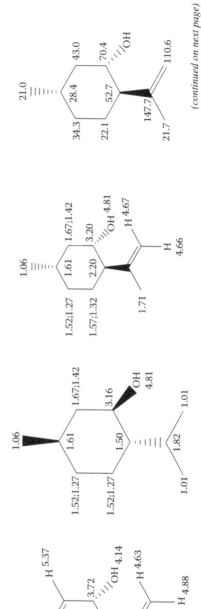

Isopulegol—(1*S*,2*R*,5*S*)-5-methyl-2-(prop-1-en-2-yl)cyclohexanol; chemical formula: $C_{10}H_{18}O$; exact mass: 154.14; molecular weight: 154.25, m/z: 154.14 (100.0%), 155.14 (11.1%); elemental analysis: C, 77.87; H, 11.76; O, 10.37

trans-Isopiperitenol—(1*S*,6*R*)-3-methyl-6-(prop-1-en-2-yl)cyclohex-2-enol; chemical formula: $C_{10}H_{16}O$; exact mass: 152.12; molecular weight: 152.23, m/z: 152.12 (100.0%), 153.12 (10.9%); elemental analysis: C, 78.90; H, 10.59; O, 10.51

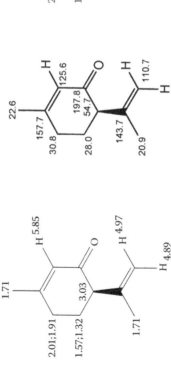

(continued on next page)

Table 2.2 (continued)   ¹H and ¹³C-NMR Data of Monoterpenes Found in Cymbogon Essential Oils

Limonene oxide—(1R,4R,6S)-1-methyl-4-(prop-1-en-2-yl)-7-oxabicyclo[4.1.0] heptane; chemical formula: C₁₀H₁₆O; exact mass: 152.12; molecular weight: 152.23, m/z: 152.12 (100.0%), 153.12 (10.9%); elemental analysis: C, 78.90; H, 10.59; O, 10.51

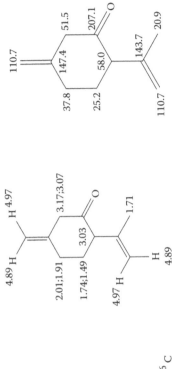

Δ³-Menthene-3-one—5-methylene-2-(prop-1-en-2-yl)cyclohexanone; chemical formula: C₁₀H₁₄O; exact mass: 150.1; molecular weight: 150.22, m/z: 150.10 (100.0%), 151.11 (11.0%); elemental analysis: C, 79.96; H, 9.39; O, 10.65

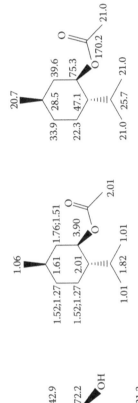

(−)-Menthol—(1R,2S,5R)-2-isopropyl-5 methylcyclohexanol; chemical formula: C₁₀H₂₀O: exact mass: 156.15; molecular weight: 156.27, m/z: 156.15 (100.0%), 157.15 (10.8%); elemental analysis: C, 76.86; H, 12.90; O, 10.24

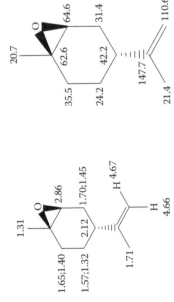

Menthyl acetate—(1R,2S,5R)-2-isopropyl-5-methylcyclohexyl acetate; chemical formula: C₁₂H₂₂O₂; exact mass: 198.16; molecular weight: 198.3, m/z: 198.16 (100.0%), 199.17 (13.3%), 200.17 (1.2%); elemental analysis: C, 72.68; H, 11.18; O, 16.14

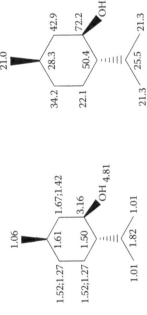

(−)-**Menthone**—(2*S*,5*R*)-2-isopropyl-5-methylcyclohexanone; chemical formula: $C_{10}H_{18}O$; exact mass: 154.25, m/z: 154.14 Molecular Weight: 154.14, (100.0%), 155.14 (11.1%); elemental analysis: C, 77.87; H, 11.76; O, 10.37

p-**Menth-2-en-1-ol**—(1*R*,4*R*)-4-isopropyl-1 methyl cyclohex-2-enol; chemical formula: $C_{10}H_{18}O$; exact mass: 154.14; molecular weight: 154.25, m/z: 154.14 (100.0%), 155.14 (11.1%); elemental analysis: C, 77.87; H, 11.76; O, 10.37

p-**Menth-8-en-1-ol**—(1*s*,4*s*)-1-methyl-4-(prop-1-en-2-yl)cyclohexanol; chemical formula: $C_{10}H_{18}O$; exact mass: 154.25; molecular weight: 154.14; m/z: 154.14 (100.0%), 155.14 (11.1%); elemental analysis: C, 77.87; H, 11.76; O, 10.37

cis-p-Menth-1(7),8-dien-2-ol—(1*S*,5*S*)-2-methylene-5-(prop-1-en-2-yl) cyclohexanol; chemical formula: $C_{10}H_{16}O$; exact mass: 152.12; molecular weight: 152.23, m/z: 152.12 (100.0%), 153.12 (10.9%); elemental analysis: C, 78.90; H, 10.59; O, 10.51

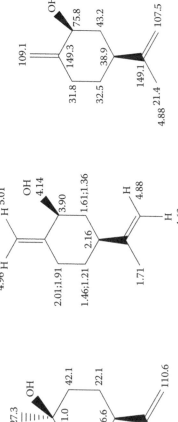

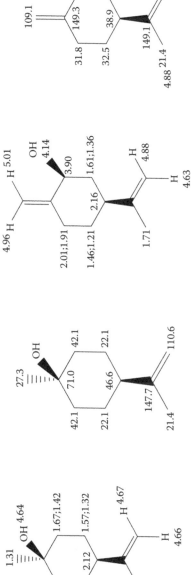

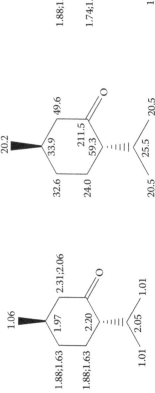

(continued on next page)

Table 2.2 (continued)   ¹H and ¹³C-NMR Data of Monoterpenes Found in Cymbogon Essential Oils

p-**Menth-2,8-dien-1-ol**—(1*R*,4*S*)-1-methyl-4-(prop-1-en-2-yl)cyclohex-2-enol; chemical formula: $C_{10}H_{16}O$; exact mass: 152.12; molecular weight: 152.23, m/z: 152.12 (100.0%), 153.12 (10.9%); elemental analysis: C, 78.90; H, 10.59; O, 10.51

Perillaldehyde—(*R*)-4-(prop-1-en-2-yl)cyclohex-1-enecarbaldehyde; chemical formula: $C_{10}H_{14}O$; exact mass: 150.1; molecular weight: 150.22, m/z: 150.10 (100.0%), 151.11 (11.0%); elemental analysis: C, 79.96; H, 9.39; O, 10.65

Methyl thymyl ether—1-isopropyl-2-methoxy-4-methylbenzene; chemical formula: $C_{11}H_{16}O$; exact mass: 164.12; molecular weight: 164.24, m/z: 164.12 (100.0%), 165.12 (11.9%); elemental analysis: C, 80.44; H, 9.82; O, 9.74

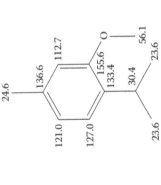

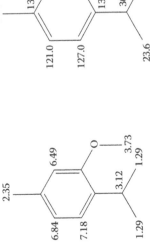

Perillene—3-(4-methylpent-3-enyl)furan; chemical formula: $C_{10}H_{14}O$; exact mass: 150.1; molecular weight: 150.22, m/z: 150.10 (100.0%), 151.11, (11.0%); elemental analysis: C, 79.96; H, 9.39; O, 10.65

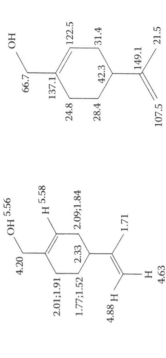

Perillyl alcohol—(4-(prop-1-en-2-yl)cyclohex-1-enyl)methanol; chemical formula: $C_{10}H_{16}O$; exact mass: 152.12; molecular weight: 152.23, m/z: 152.12 (100.0%), 153.12 (10.9%); elemental analysis: C, 78.90; H, 10.59; O, 10.51

cis-Piperitenol—(1S,6R)-3-methyl-6-(prop-1-en-2-yl)cyclohex-2-enol; chemical formula: $C_{10}H_{16}O$; exact mass: 152.12; molecular weight: 152.23, m/z: 152.12 (100.0%), 153.12 (10.9%); elemental analysis: C, 78.90; H, 10.59; O, 10.51

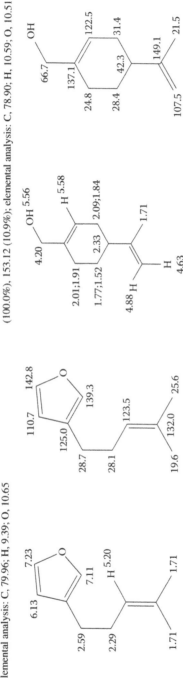

Piperitenone—3-methyl-6-(propan-2-ylidene)cyclohex-2-enone; chemical formula: $C_{10}H_{14}O$; exact mass: 150.1; molecular weight: 150.22, m/z: 150.10 (100.0%), 151.11 (11.0%); elemental analysis: C, 79.96; H, 9.39; O, 10.65

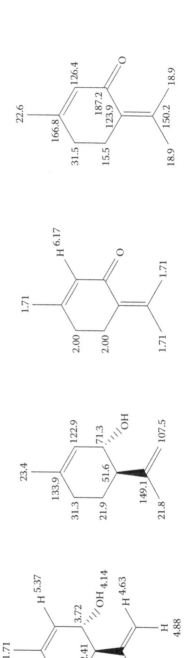

(continued on next page)

Table 2.2 (continued) ¹H and ¹³C-NMR Data of Monoterpenes Found in Cymbogon Essential Oils

cis-Piperitol—(1S,6R)-6-isopropyl-3-methylcyclohex-2-enol; chemical formula: C₁₀H₁₈O; exact mass: 154.25, m/z: 154.14; molecular weight: 154.14 (100.0%), 155.14 (11.1%); elemental analysis: C, 77.87; H, 11.76; O, 10.3

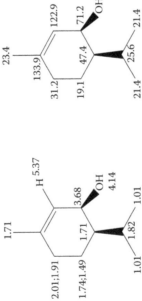

trans-Piperitol—(1R,6R)-6-isopropyl-3-methylcyclohex-2-enol; chemical formula: C₁₀H₁₈O; exact mass: 154.25, m/z: 154.14; molecular weight: 154.14 (100.0%), 155.14 (11.1%); elemental analysis: C, 77.87; H, 11.76; O, 10.37

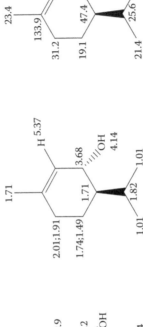

Piperitone—6-isopropyl-3-methylcyclohex-2-enone; chemical formula: C₁₀H₁₆O; exact mass: 152.12; molecular weight: 152.23, m/z: 152.12 (100.0%), 153.12 (10.9%); elemental analysis: C, 78.90; H, 10.59; O, 10.51

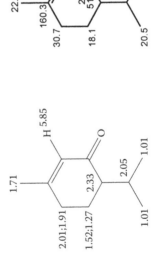

Piperitone oxide—3-isopropyl-6-methyl-7-oxabicyclo[4.1.0]heptan-2-one; chemical formula: C₁₀H₁₆O₂; exact mass: 168.23; molecular weight: 168.23, m/z: 168.12 (100.0%), 169.12 (11.1%); elemental analysis: C, 71.39; H, 9.59; O, 19.02

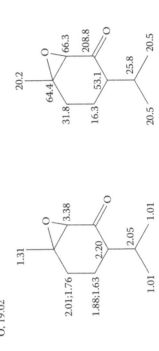

Pulgeone—5-methyl-2-(propan-2-ylidene)cyclohexanone; chemical formula: $C_{10}H_{16}O$; exact mass: 152.12; molecular weight: 152.12 (100.0%), 153.12 (10.9%); elemental analysis: C, 78.90; H, 10.59; O, 10.51

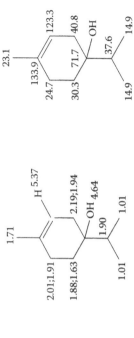

Terpin-4-ol—1-isopropyl-4-methylcyclohex-3-enol; chemical formula: $C_{10}H_{18}O$; exact mass: 154.14, molecular weight: 154.25, m/z: 154.14 (100.0%), 155.14 (11.1%); elemental analysis: C, 77.87; H, 11.76; O, 10.37

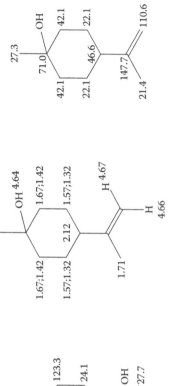

α-Terpineol—2-(4-methylcyclohex-3-enyl)propan-2-ol; chemical formula: $C_{10}H_{18}O$; exact mass: 154.14; molecular weight: 154.14 (100.0%), 155.14 (11.1%); elemental analysis: C, 77.87; H, 11.76; O, 10.37

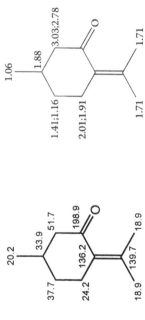

β-Terpineol—1-methyl-4-(prop-1-en-2-yl)cyclohexanol; chemical formula: $C_{10}H_{18}O$; exact mass: 154.14; molecular weight: 154.25, m/z: 154.14 (100.0%), 155.14 (11.1%); elemental analysis: C, 77.87; H, 11.76; O, 10.37

(continued on next page)

Table 2.2 (continued)   ¹H and ¹³C-NMR Data of Monoterpenes Found in Cymbogon Essential Oils

Bicyclic Oxygenated Monoterpenes

Borneol—1,7,7-trimethylbicyclo[2.2.1]heptan-2-ol; chemical formula: $C_{10}H_{18}O$; exact mass: 154.14; molecular weight: 154.25, m/z: 154.14 (100.0%), 155.14 (11.1%); elemental analysis: C, 77.87; H, 11.76; O, 10.37

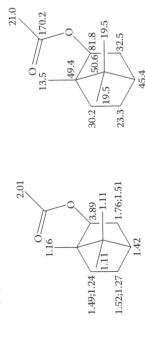

Bornyl acetate—1,7,7-trimethylbicyclo[2.2.1]heptan-2-yl-acetate; chemical formula: $C_{12}H_{20}O_2$; exact mass: 196.15; molecular weight: 196.29, m/z: 196.15 (100.0%), 197.15 (13.3%), 198.15 (1.2%); elemental analysis: C, 73.43; H, 10.27; O, 16.30

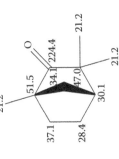

Camphor—(1S,4S)-1,7,7 trimethylbicyclo[2.2.1]heptan-2-one; chemical formula: $C_{10}H_{16}O$; exact mass: 152.12; molecular weight: 152.23, m/z: 152.12 (100.0%), 153.12 (10.9%); elemental analysis: C, 78.90; H, 10.59; O, 10.51

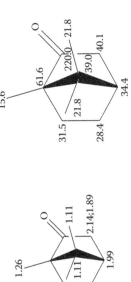

Fenchone—(1S,4R)-1,3,3-trimethylbicyclo[2.2.1]heptan-2-one; chemical formula: $C_{10}H_{16}O$; exact mass: 152.12; molecular weight: 152.23, m/z: 152.12 (100.0%), 153.12 (10.9%); elemental analysis: C, 78.90; H, 10.59; O, 10.51

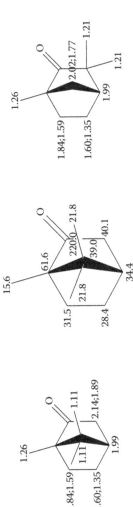
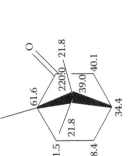

Isoborneol—(1S,2S,4S)-1,7,7-trimethylbicyclo[2.2.1]heptan-2-ol; chemical formula: $C_{10}H_{18}O$; molecular weight: 154.14 (100.0%), 155.14, (11.1%); elemental analysis: C, 77.87; H, 11.76; O, 10.37

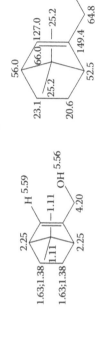

Thujyl alcohol—(1S,3S,4S,5R)-1-isopropyl-4-methylbicyclo[3.1.0]hexan-3-ol; chemical formula: $C_{10}H_{18}O$; exact mass: 154.14; molecular weight: 154.25 m/z: 154.14 (100.0%), 155.14 (11.1%); elemental analysis: C, 77.87; H, 11.76; O, 10.37

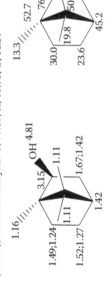

Myrtenol—7,7-dimethylbicyclo[2.2.1]hept-2-en-2-yl)methanol; chemical formula: $C_{10}H_{16}O$; exact mass: 152.12; molecular weight: 152.23, m/z: 152.12 (100.0%), 153.12 (10.9%); elemental analysis: C, 78.90; H, 10.59; O, 10.51

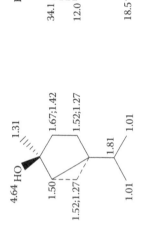

β-Thujene alcohol—(1S,2S,5R)-5-isopropyl-2-methylbicyclo[3.1.0]hexan-2-ol; chemical formula: $C_{10}H_{18}O$; exact mass: 154.14; molecular weight: 154.25, m/z: 154.14 (100.0%), 155.14 (11.1%); elemental analysis: C, 77.87; H, 11.76; O, 10.37

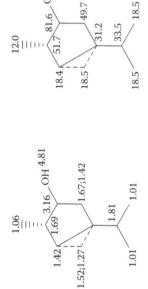

(continued on next page)

Table 2.2 (continued)　¹H and ¹³C-NMR Data of Monoterpenes Found in Cymbogon Essential Oils

Verbenone—(1R,5R)-4,6,6-trimethylbicyclo[3.1.1]hept-3-en-2-one; chemical formula: $C_{10}H_{14}O$; exact mass: 150.1; molecular weight: 150.22; m/z: 150.10 (100.0%), 151.11 (11.0%); elemental analysis: C, 79.96; H, 9.39; O, 10.65

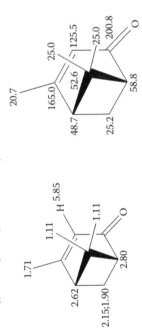

TABLE 2.3
Sesquiterpenes Found in Major, Minor, and Trace Amounts in *Cymbopogon* Oils (in alphabetical order)

Acoradiene—(4*S*)-1,8-dimethyl-4-(prop-1-en-2-yl)spiro[4.5]dec-7-ene; chemical formula: $C_{15}H_{24}$; exact mass: 204.19; molecular weight: 204.35, m/z: 204.19 (100.0%), 205.19 (16.5%), 206.19 (1.2%); elemental analysis: C, 88.16; H, 11.84

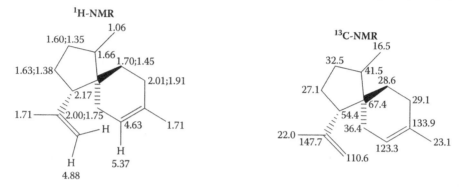

Alloaromadendrene—(1*aR*,4*aS*,7*R*,7*bS*)-1,1,7-trimethyl-4-methylenedecahydro-1H-cyclopropa[e]azulene; chemical formula: $C_{15}H_{24}$; exact mass: 204.19; molecular weight: 204.35, m/z: 204.19 (100.0%), 205.19 (16.5%), 206.19 (1.2%); elemental analysis: C, 88.16; H, 11.84

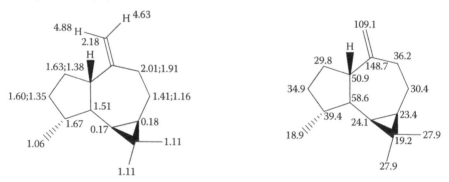

Aromadendrene—(1*aR*,4*aR*,7*R*,7*bS*)-1,1,7-trimethyl-4-methylenedecahydro-1H-cyclopropa[e]azulene; chemical formula: $C_{15}H_{24}$; exact mass: 204.19; molecular weight: 204.35, m/z: 204.19 (100.0%), 205.19 (16.5%), 206.19 (1.2%); elemental analysis: C, 88.16; H, 11.84

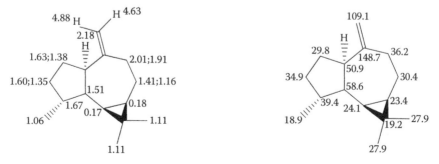

(continued on next page)

TABLE 2.3 (continued)
Sesquiterpenes Found in Major, Minor, and Trace Amounts in *Cymbopogon* Oils (in alphabetical order)

α-Bergamotene—2,6-dimethyl-6-(4-methylpent-3-enyl)bicyclo[3.1.1]hept-2-ene; chemical formula: $C_{15}H_{24}$; exact mass: 204.19; molecular weight: 204.35, m/z: 204.19 (100.0%), 205.19 (16.5%), 206.19 (1.2%); elemental analysis: C, 88.16; H, 11.84

α-Bisabolene—(Z)-1-methyl-4-(6-methylhept-5-en-2-ylidene)cyclohex-1-ene; chemical formula: $C_{15}H_{24}$; exact mass: 204.19; molecular weight: 204.35, m/z: 204.19 (100.0%), 205.19 (16.5%), 206.19 (1.2%); elemental analysis: C, 88.16; H, 11.84

β-Bisabolene—1-methyl-4-(6-methylhepta-1,5-dien-2-yl)cyclohex-1-ene; chemical formula: $C_{15}H_{24}$; exact mass: 204.19; molecular weight: 204.35, m/z: 204.19 (100.0%), 205.19 (16.5%), 206.19 (1.2%); elemental analysis: C, 88.16; H, 11.84

TABLE 2.3 (continued)
Sesquiterpenes Found in Major, Minor, and Trace Amounts in *Cymbopogon* Oils (in alphabetical order)

Bisabolol—(2S)-6-methyl-2-(4-methylcyclohex-3-enyl)hept-5-en-2-ol; chemical formula: $C_{15}H_{26}O$; exact mass: 222.2; molecular weight: 222.37, m/z: 222.20 (100.0%), 223.20 (16.6%), 224.21 (1.3%); elemental analysis: C, 81.02; H, 11.79; O, 7.20

α-Butenol—(5R,Z)-2,10,10-trimethyl-6-methylenebicyclo[7.2.0]undec-2-en-5-ol; chemical formula: $C_{15}H_{24}O$; exact mass: 220.18; molecular weight: 220.35, m/z: 220.18 (100.0%), 221.19 (16.5%), 222.19 (1.5%); elemental analysis: C, 81.76; H, 10.98; O, 7.26

β-Butenol—(Z)-6,10,10-trimethyl-2-methylenebicyclo[7.2.0]undec-5-en-3-ol; chemical formula: $C_{15}H_{24}O$; exact mass: 220.18; molecular weight: 220.35, m/z: 220.18 (100.0%), 221.19 (16.5%), 222.19 (1.5%); elemental analysis: C, 81.76; H, 10.98; O, 7.26

(continued on next page)

TABLE 2.3 (continued)
Sesquiterpenes Found in Major, Minor, and Trace Amounts in *Cymbopogon* Oils (in alphabetical order)

Calamenene—8-isopropyl-2,5-dimethyl-1,2,3,4-tetrahydronaphthalene; chemical formula: $C_{15}H_{22}$; exact mass: 202.17; molecular weight: 202.34, m/z: 202.17 (100.0%), 203.18 (16.5%), 204.18 (1.3%); elemental analysis: C, 89.04; H, 10.96

α-Cadinene—1-isopropyl-4,7-dimethyl-1,2,4a,5,6,8a-hexahydronaphthalene; chemical formula: $C_{15}H_{24}$; exact mass: 204.19; molecular weight: 204.35, m/z: 204.19 (100.0%), 205.19 (16.5%), 206.19 (1.2%); elemental analysis: C, 88.16; H, 11.84

β-Cadinene—(1S)-1-isopropyl-7-methyl-4-methylene-1,2,3,4,4a,5,6,8a-octahydronaphthalene; chemical formula: $C_{15}H_{24}$; exact mass: 204.19; molecular weight: 204.35, m/z: 204.19 (100.0%), 205.19 (16.5%), 206.19 (1.2%); elemental analysis: C, 88.16; H, 11.84

TABLE 2.3 (continued)
Sesquiterpenes Found in Major, Minor, and Trace Amounts in *Cymbopogon* Oils (in alphabetical order)

γ-Cadinene—(1*S*)-1-isopropyl-4,7-dimethyl-1,2,3,5,6,8a-hexahydronaphthalene; chemical formula: $C_{15}H_{24}$; exact mass: 204.19; molecular weight: 204.35, m/z: 204.19 (100.0%), 205.19 (16.5%), 206.19 (1.2%); elemental analysis: C, 88.16; H, 11.84

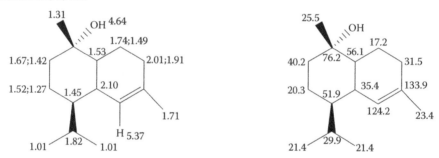

γ-Cadinol—(1*S*,4*R*)-4-isopropyl-1,6-dimethyl-1,2,3,4,4a,7,8,8a-octahydronaphthalen-1-ol; chemical formula: $C_{15}H_{26}O$; exact mass: 222.2; molecular weight: 222.37, m/z: 222.20 (100.0%), 223.20 (16.6%), 224.21 (1.3%); elemental analysis: C, 81.02; H, 11.79; O, 7.20

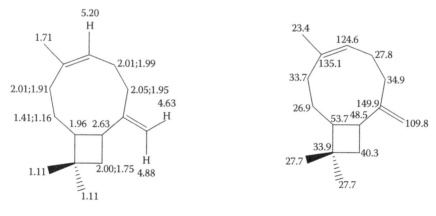

β-Caryophyllene—(*Z*)-4,11,11-trimethyl-8-methylenebicyclo[7.2.0]undec-4-ene; chemical formula: $C_{15}H_{24}$; exact mass: 204.19; molecular weight: 204.35, m/z: 204.19 (100.0%), 205.19 (16.5%), 206.19 (1.2%); elemental analysis: C, 88.16; H, 11.84

(continued on next page)

TABLE 2.3 (continued)
Sesquiterpenes Found in Major, Minor, and Trace Amounts in *Cymbopogon* Oils (in alphabetical order)

β-Caryophyllene alcohol—(*E*)-4,11,11-trimethyl-8-methylenebicyclo[7.2.0]undec-4-en-5-ol; chemical formula: $C_{15}H_{24}O$; exact mass: 220.18; molecular weight: 220.35, m/z: 220.18 (100.0%), 221.19 (16.5%), 222.19 (1.5%); elemental analysis: C, 81.76; H, 10.98; O, 7.26

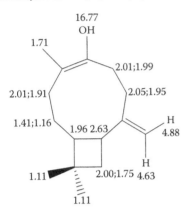 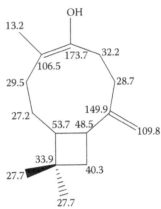

β-Caryophyllene oxide—chemical formula: $C_{15}H_{24}O$; exact mass: 220.18; molecular weight: 220.35, m/z: 220.18 (100.0%), 221.19 (16.5%), 222.19 (1.5%); elemental analysis: C, 81.76; H, 10.98; O, 7.26

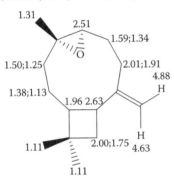 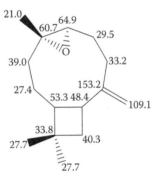

α-Chamigrene—1,5,5,9-tetramethylspiro[5.5]undeca-1,8-diene; chemical formula: $C_{15}H_{24}$; exact mass: 204.19; molecular weight: 204.35, m/z: 204.19 (100.0%), 205.19 (16.5%), 206.19 (1.2%); elemental analysis: C, 88.16; H, 11.84

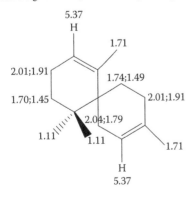 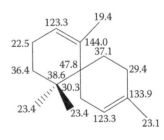

TABLE 2.3 (continued)
Sesquiterpenes Found in Major, Minor, and Trace Amounts in *Cymbopogon* Oils (in alphabetical order)

α-**Cubebene**—chemical formula: $C_{15}H_{24}$; exact mass: 204.19; molecular weight: 204.35, m/z: 204.19 (100.0%), 205.19 (16.5%), 206.19 (1.2%); elemental analysis: C, 88.16; H, 11.84

α-**Cuparene**—(*R*)-1-methyl-4-(1,2,2-trimethylcyclopentyl)cyclohexa-1,3-diene; chemical formula: $C_{15}H_{24}$; exact mass: 204.19; molecular weight: 204.35, m/z: 204.19 (100.0%), 205.19 (16.5%), 206.19 (1.2%); elemental analysis: C, 88.16; H, 11.84

Dihydro-alpha-copaene-8-ol—chemical formula: $C_{15}H_{26}O$; exact mass: 222.2; molecular weight: 222.37, m/z: 222.20 (100.0%), 223.20 (16.6%), 224.21 (1.3%); elemental analysis: C, 81.02; H, 11.79; O, 7.20

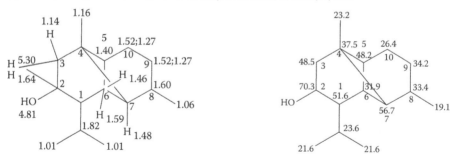

(continued on next page)

TABLE 2.3 (continued)
Sesquiterpenes Found in Major, Minor, and Trace Amounts in *Cymbopogon* Oils (in alphabetical order)

β-Elemene—(1*R*,2*S*,4*R*)-1-methyl-2,4-di(prop-1-en-2-yl)-1-vinylcyclohexane; chemical formula: $C_{15}H_{24}$; exact mass: 204.19; molecular weight: 204.35, m/z: 204.19 (100.0%), 205.19 (16.5%), 206.19 (1.2%); elemental analysis: C, 88.16; H, 11.84

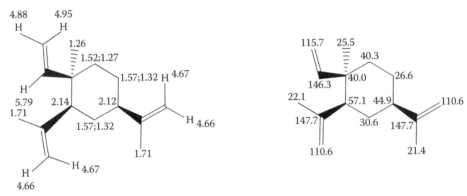

γ-Elemene—(*S*)-1-methyl-2,4-di(propan-2-ylidene)-1-vinylcyclohexane; chemical formula: $C_{15}H_{24}$; exact mass: 204.19; molecular weight: 204.35, m/z: 204.19 (100.0%), 205.19 (16.5%), 206.19 (1.2%); elemental analysis: C, 88.16; H, 11.84

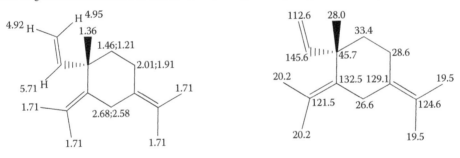

β-Eudesmol—2-((4a*R*)-4a-methyl-8-methylenedecahydronaphthalene-2-yl)propan-2-ol; chemical formula: $C_{15}H_{26}O$; exact mass: 222.2; molecular weight: 222.37, m/z: 222.20 (100.0%), 223.20 (16.6%), 224.21 (1.3%); elemental analysis: C, 81.02; H, 11.79; O, 7.20

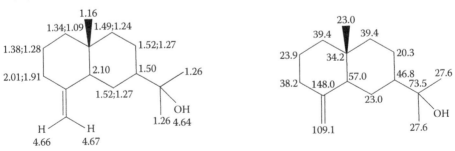

TABLE 2.3 (continued)
Sesquiterpenes Found in Major, Minor, and Trace Amounts in *Cymbopogon* Oils (in alphabetical order)

α-Farnesene—(3Z,6E)-3,7,11-trimethyldodeca-1,3,6,10-tetraene; chemical formula: $C_{15}H_{24}$; exact mass: 204.19; molecular weight: 204.35, m/z: 204.19 (100.0%), 205.19 (16.5%), 206.19 (1.2%); elemental analysis: C, 88.16; H, 11.84

β-Farnesene—(E)-7,11-dimethyl-3-methylenedodeca-1,6,10-triene; chemical formula: $C_{15}H_{24}$; exact mass: 204.19; molecular weight: 204.35, m/z: 204.19 (100.0%), 205.19 (16.5%), 206.19 (1.2%); elemental analysis: C, 88.16; H, 11.84

(continued on next page)

TABLE 2.3 (continued)
Sesquiterpenes Found in Major, Minor, and Trace Amounts in *Cymbopogon* Oils (in alphabetical order)

Germacrene D—(1*E*,6*E*)-8-isopropyl-1-methyl-5-methylenecyclodeca-1,6-diene; chemical formula: $C_{15}H$; exact mass: 204.19; molecular weight: 204.35, m/z: 204.19 (100.0%), 205.19 (16.5%), 206.19 (1.2%); elemental analysis: C, 88.16; H, 11.84

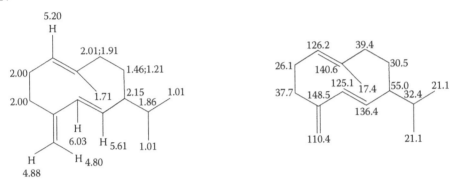

α-Himachalene—(4*aS*,9*aR*)-3,5,5-trimethyl-9-methylene-2,4a,5,6,7,8,9,9a-octahydro-1H-benzo[7]annulene; chemical formula: $C_{15}H_{24}$; exact mass: 204.19; molecular weight: 204.35, m/z: 204.19 (100.0%), 205.19 (16.5%), 206.19 (1.2%); elemental analysis: C, 88.16; H, 11.84

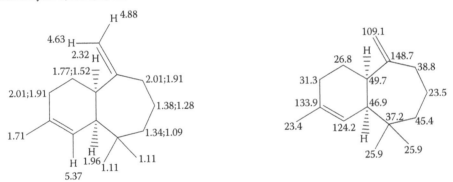

α-Humulene—(1*Z*,4*Z*,8*Z*)-2,6,6,9-tetramethylcycloundeca-1,4,8-triene; chemical formula: $C_{15}H_{24}$; exact mass: 204.19; molecular weight: 204.35, m/z: 204.19 (100.0%), 205.19 (16.5%), 206.19 (1.2%); elemental analysis: C, 88.16; H, 11.84

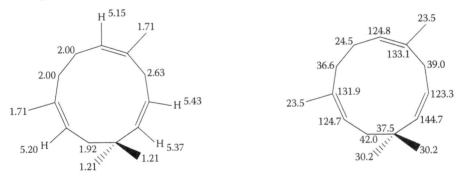

TABLE 2.3 (continued)
Sesquiterpenes Found in Major, Minor, and Trace Amounts in *Cymbopogon* Oils (in alphabetical order)

β-**Humulene**—(1Z,5Z)-1,4,4-trimethyl-8-methylenecycloundeca-1,5-diene; chemical formula: $C_{15}H_{24}$; exact mass: 204.19; molecular weight: 204.35, m/z: 204.19 (100.0%), 205.19 (16.5%), 206.19 (1.2%); elemental analysis: C, 88.16; H, 11.84

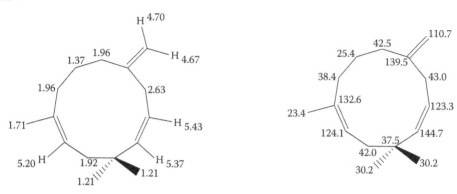

Longifolene—chemical formula: $C_{15}H_{24}$; exact mass: 204.19; molecular weight: 204.35, m/z: 204.19 (100.0%), 205.19 (16.5%), 206.19 (1.2%); elemental analysis: C, 88.16; H, 11.84

α-**Murrolene**—1-isopropyl-4,7-dimethyl-1,2,4a,5,6,8a-hexahydronaphthalene; chemical formula: $C_{15}H_{24}$; exact mass: 204.19; molecular weight: 204.35, m/z: 204.19 (100.0%), 205.19 (16.5%), 206.19 (1.2%); elemental analysis: C, 88.16; H, 11.84

2.4 CHEMISTRY AND USES OF *CYMBOPOGON* ESSENTIAL OILS

The trading of essential oils has always been dependent on the knowledge imparted regarding its quality. The components present in it and the odor value provided to it because of several physical and chemical parameters are of immense value. These have been important criteria since ancient and medieval times. Chemists had to develop methods of analysis of oils and determine their chemical composition because of possibilities of adulterations of cheaper essential oils with highly priced ones. Prior to the advent of gas liquid chromatography (GLC) (Ille 1986), chemists had to rely entirely on chemical analysis, but GLC analysis might be the most important factor in this regard, and of possible help. During the last two decades, the methodology has improved manyfold. Consequently, the data regarding major and minor constituents found in essential oils has also multiplied and helped in evaluating the quality of the oils besides providing information on the constituents that could be isolated and used in pure form.

2.4.1 LEMONGRASS OILS

Lemongrass oil is distilled from two morphologically different species of lemongrass, *C. flexuosus* (common name: East Indian lemongrass) and *C. citratus* (common name: West Indian lemongrass). The chemical composition of these oils is very similar, though the percentage of citral and other major monoterpenes vary to some extent. A high yield of citral has been reported in *C. pendulus* (common name: North Indian lemongrass), which is another wild species and under limited cultivation. It has also been a major commercial source of lemongrass oil.

2.4.1.1 *Cymbopogon flexuosus*

The East Indian lemongrass (*C. flexuosus* (Steud.) Wats.) oil is indigenous to South India, found in the Malabar and Cochin regions, and in the Malay Peninsula. A product of *C. flexuosus* comes from Ceylon, Myanmar, and adjacent countries, as well as from Mexico and the West Indies. It is commonly called Malabar lemongrass oil. The major and trace constituents reported by various workers have been tabulated in Table 2.1. In one of the reports, Nath et al. (1994) have identified 25 components after examining the essential oil. Geraniol, citronellol, neral, and geranial have been reported as the major components. Neral is also referred in the books and literature as citral-*cis*, citral-a, α-citral, (Z)-3,7-dimethylocta-2,6-dienal or citral-(Z). Similarly, geranial has also been termed as citral-*trans*, Citral-b, β-citral, (E)-3,7-dimethylocta-2,6-dienal or citral-(E). During several studies conducted in the last few years (Bhattacharya et al. 1997; Boelens 1994; Cherian et al. 1993; Choudhary and Kaul 1979; Kulkarni et al. 1997; Kuriakose 1995; Mathela et al. 1996; Nair et al. 1984; Patra and Dutta 1986; Rao et al. 1995), many chemotypes/cultivars/variants have been reported. Rao et al. (1992) have identified α-bisabolol and methyl isoeugenol as major components in a chemotype. It has also been found that the quality of the herb deteriorates on storage of the herb and also affects oil quality (Singh et al. 1994). In a GC-MS analysis of the essential oil, 32 constituents were identified (Taskinen et al. 1983). Kulkarni et al. (1997) reported a variant resembling citronella that contained citronellol (9.5%), citronellal (6%), citronellyl acetate (11.2%), geraniol (11.1%), and geranyl acetate (25.9%), along with 20 other constituents. The profiles of essential oils from five *C. flexuosus* cultivars (OD-19, *Pragati, Cauvery,* SHK-7, and GRL-1); one *C. pendulus* cultivar (*Praman*); and one hybrid *C. khasianus* × *C. pendulus* cultivar (CKP-25) have been examined on the GLC capillary column. Besides cultivated species (Atal and Bradu 1976a), a few wild-growing strains of the species have also been investigated and designated as RRL-14 and RRL-59, which contained geranial (40%) and methyl isoeugenol (20%) as the major constituents of the essential oil pool. The results revealed that cultivar GRL-1 (Patra et al. 1990) is different from other cultivars because of the presence of a high amount of geraniol (80.2%) and relatively higher concentration of myrcene (3.97%) and geranyl acetate (4.6%) (Patra et al. 1997). *C. sikkimensis,* which was a new variety of *C. flexuosus,* revealed the presence of methyl isoeugenol (20.5%), methyl eugenol (23%), and *d*-limonene (16.5%) as major constituents. The hybrid lemongrass CKP-25 differed from other citral-rich varieties with respect to a number of minor compounds (Bhattacharya et al. 1997).

The oil of lemongrass is widely used in soaps and detergents (Bhattacharya 1970; Guenther 1950; Opdyke 1976). Citral (a mixture of geranial and neral) is the major component of the oil and is isolated in bulk to be used in flavors, cosmetics, and perfumes. Ionones is another group of very important synthetic aromatics that possesses a strong and lasting odor. They are synthesized from citral and are further used in the manufacture of synthetic vitamin A.

The antifungal, antibacterial, and antioxidant properties of lemongrass oil has been widely utilized (Alam et al. 1994; Gyane 1976; Mehmood et al. 1997; Ramdan et al. 1972a, 1972b; Rao et al. 1971; Shadab-Qamar et al. 1992; Singh et al. 1978; Wannissorn et al. 1996). Allergic contact dermatitis has been reported with the oil (Selvag et al. 1995). Some other uses have been reported as preservative (Arora and Pandey 1977) and in inhibition of sensitization reactions (Opdyke 1976). The leftover residue from lemongrass leaves has been successfully utilized as a source of raw material for cellulose pulp and paper production (Ciaramello et al. 1972; Ramirez et al. 1977; Siddique-Ullah et al. 1979). The other important use of the oil has been in the preparation of an insect-repellent complex (Anonymous 1973), which has been tested for insect repellent/attractant and nematicidal activities (Ansari and Razdan 1995).

The cultivation, essential oil characteristics, chemical constituents, and uses (culinary, antibacterial, medicinal, aromatic, and others) of lemongrass, and the lemongrass industry in India have been discussed at length in an article by Gupta and Jain (1978) Ansari et al. (1996) in earlier publications. Data have been tabulated on the differences between *C. flexuosus* and *C. citratus* oils in terms of their physicochemical properties, the effects of drying the herbage in sunlight for up to 5 days before distillation on oil yields, and export of lemongrass oil from India during 2002–2007 have also been discussed in this book. Effect on quality of citral content on preservation of lemongrass oil has also been worked out (Kurian et al. 1984).

2.4.1.2 *Cymbopogon citratus*

The West Indian lemongrass oil (*C. citratus* (D.C.) Stapf) is rated inferior in quality to that of the East Indian type because of its low citral content. However, it attained importance after World War II when the latter was difficult to obtain. Its major constituents are listed in Table 2.1, but it differs from the East Indian type by the occurrence of substantial quantities of myrcene (12%–15%). The myrcene may undergo diene-condensation and polymerization on aging, and hence loses its solubility in 70% alcohol. The *C. citratus* is an important crop in Ethiopia, and its analysis has shown geraniol (40%), geranial and neral (13%–15%), and α-oxobisabolene (12%) as major constituents, which are different from usual West Indian lemongrass oil (Abegaz et al. 1983). The hydrodistilled essential oil from the leaves of *C. citratus* Stapf grown in Zambia was analyzed by GC and GC-MS. Sixteen compounds representing 93.4% of the oil were identified, of which citral-*trans* (39.0%), citral-*cis* (29.4%), and myrcene (18.0%) were the major components. Small amounts of geraniol (1.7%) and linalol (1.3%) were also detected (Esmort et al. 1998). The composition of the essential oil of the leaves growing on the campus of Lagos State University was determined by the use of GC and GC-MS. The oil gave 27 peaks, amounting to 98% of the total oil. Twenty-three constituents amounting to 97.3% of the total oil were identified. The main constituents were geranial (33.7%), neral (26.5%), and myrcene (25.3%). Small amounts of neomenthol (3.3%), linalyl acetate (2.3%), Z-β-ocimene (1.0%), and E-β-ocimene were also detected (Adeleke et al. 2001). In one of the reports, citral (69.39%) has been cited as a major component along with other minor components, such as caryophyllene, citronellol, geraniol, α- and β-pinene, ethyl laurate, 1-8-cineole, limonene, phellandrene methyl heptenone, linalool, menthol, myrcene, terpineol, and citronellol (Torres and Ragadio 1992, 1996).

In one of the studies, a method was developed to validate a HPLC procedure for the quantitative determination of citral in *C. citratus* volatile oil. The HPLC assay was performed using a Spherisorb® CN column (250 mm × 4.6 mm, 5 μm), an *n*-hexane:ethanol (85:15) mobile phase, and a UV detector (set at 233 nm). The following parameters were evaluated: linearity, precision, accuracy, specificity, quantification, and detection limits. The method showed linearity in the range of 10.0–30.0 µg mL^{-1}. Precision and accuracy were determined at the concentration of 20 µg mL^{-1}. The concentration of citral in *C. citratus* volatile oil obtained with this assay was 75%. The HPLC method developed in this study showed an excellent performance (linearity, precision, accuracy, and specificity), and can be applied to assay citral in volatile oil (Rauber et al. 2005). A total of 34 compounds were identified in Moroccan *C. citratus* oil, constituting about 89% of the total oil, with geraniol and neral (39.8% and 32%, respectively) as major constituents. The oil yield was drastically reduced under rust disease indices reduction (Baruah et al. 1995). A reduction in geraniol content in contrast to an increase in neral and myrcene was observed. The plant extract yielded tanins in herbal tea (Blake et al. 1993), with cymbopogene, cymbopogonol, triacontanol, alkaloid, and saponin (Ansari et al. 1996) being identified. The presence of alkaloids was reported; however, it needs further confirmation. The crop growing in Nagcorlan and Laguna was reported to contain 93.74% citral (citral-a, citral-b, and others in the ratio of 61:33:6) (Torres 1993).

This oil has also been reported to be antimicrobial (Chalchat et al. 1997; Handique and Singh 1990; Kokate and Verma 1971; Moris et al. 1979; Orafidiya 1993; Syeed et al. 1990; Yadav and Dubey 1994) and insecticidal (Sukari et al. 1992), insect repellant (Ansari and Razdan 1995), and found to have cytotoxic properties apart from its usual uses in perfumery, food flavor, and pharmaceutical industries (Guenther 1950; Opdyke 1976). *C. citratus* oil has also been tested for anticarcinogenic activities (Dubey et al. 1997; Zheng et al. 1993). Citral-a and citral-b have been shown to possess antibacterial activity in the oil (Onawunmi et al. 1984).

In another study, the antibacterial properties of the essential oil have been recorded. These activities are shown in two of the three main components of the oil identified through CG and MS methods. Whereas the α-citral (geranial) and β-citral (neral) components individually elicit antibacterial action on Gram-negative and Gram-positive organisms, the third component, myrcene, did not show observable antibacterial activity on its own. However, myrcene provided enhanced activities when mixed with either of the other two main components identified (Grace et al. 1984). *C. citratus* is one of the most commonly used plants in Brazilian folk medicine for the treatment of nervous and gastrointestinal disturbances. It is also used in many other places to treat feverish conditions. The usual way to use it is by ingesting an infusion made by pouring boiling water on fresh or dried leaves (which is called "abafado" in Portuguese). Abafados obtained from lemongrass harvested in three different areas of Brazil (Ceará, Minas Gerais, and São Paulo states) were tested on rats and mice in an attempt to add experimental confirmation of its popular medicinal use. Citral, the main constituent of the essential oil in Brazilian lemongrass, was also studied for comparison. Oral doses of abafados up to 40 times (C_{40}) larger than the corresponding dosage taken by humans, or 200 mg/kg of citral, were unable to reduce the body temperature of normal rats and/or rats made hyperthermic by previous administration of pyrogen. However, both compounds acted when injected via the intraperitoneal route. Oral administration of doses C_{20}–C_{100} of abafados and 200 mg/kg of citral did not change the intestinal transit of a charcoal meal in mice; neither did it decrease the defecation scores of rats in an open-field arena. Again, via the intraperitoneal route, both compounds were active. The possible central nervous system depressant effect of abafados was investigated by using batteries of 12 tests designed to detect general depressant, hypnotic, neuroleptic, anticonvulsant, and anxiolytic effects. In all the tests employed, oral doses of abafados up to C_{208} or citral up to 200 mg/kg were without effect. Only in a few instances did intraperitoneal doses demonstrate effects. These data do no lend support to the popular oral therapeutic use of lemongrass to treat nervous and intestinal ailments and feverish conditions (Carlini et al. 1986). Tea obtained from leaves of *C. citratus* (D.C.) Stapf is used for its anxiolytic, hypnotic, and anticonvulsant properties in Brazilian folk medicine. Essential oil (EO) from fresh leaves was obtained by hydrodistillation and orally administered to Swiss male mice 30 min before experimental procedures. EO at 0.5 or 1.0 g/kg was evaluated for sedative/hypnotic activity through pentobarbital sleeping time, for anxiolytic activity by elevated plus maze and light/dark box procedures, and for anticonvulsant activity through seizures induced by pentylenetetrazole and maximal electroshock. EO was effective in increasing the sleeping time, percentage of entries, and time spent in the open arms of the elevated plus maze as well as the time spent in the light compartment of the light/dark box. In addition, EO delayed clonic seizures induced by pentylenetetrazole and blocked tonic extensions induced by maximal electroshock, indicating the elevation of the seizure threshold and/or blockage of seizure spread. These effects were observed in the absence of motor impairment evaluated on the rota-rod and open-field tests. The results were in accord with the ethnopharmacological use of *C. citratus*, and, after complementary toxicological studies, it can support investigations assessing their use as an anxiolytic, sedative, or anticonvulsive agent (Blanco et al. 2007).

Studies were conducted to investigate the hypoglycemic and hypolipidemic effects of the single, daily oral dosing of 125–500 mg/kg of fresh leaf aqueous extract of *C. citratus* Stapf in normal, male Wistar rats for 42 days. The average weights of rats per group were taken at 2-week intervals for 42 days. On day 43, blood samples from the rats were collected for fasting plasma glucose (FPG), total cholesterol, triglycerides, low-density lipoproteins (LDL-c), very low-density lipoprotein (VLDL-c), and high-density lipoprotein (HDL-c) assays through cardiac puncture under halothane anesthesia. Acute oral dose toxicity study of *C. citratus* was also conducted using the limit dose test of the Up and Down Procedure statistical program (AOT425StatPgm, Version 1.0) at a dose of 5000 mg/kg body weight/oral route. These results showed *C. citrates* to lower the FPG and lipid parameters dose dependently ($p < 0.05$) while raising the plasma HDL-c level ($p < 0.05$) in the same dose-related fashion but with no effect on the plasma triglycerides level ($p > 0.05$). The results of acute oral toxicity showed CCi to be of low toxicity and, as such, could be considered relatively safe on acute exposure, thus confirming its folkloric use and safety in suspected Type 2 diabetic patients (Adeneye and Agbaje 2007).

2.4.1.3 *Cymbopogon pendulus*

The North Indian lemongrass oil (*C. pendulus* (Nees ex Steud.) Wats.) occurs in wild areas of northern India such as Saharanpur (in the state of Uttar Pradesh) (Atal and Bradu 1976a; Muthuswami and Sayed 1980). The major and minor constituents are listed in Table 2.1. This is also a major source of lemongrass oil that is used in perfumery as much as the East Indian variety. In one of the reports, elemicin (53.7%) content was found to be very high in the essential oil obtained from this plant. It is noteworthy that it is the starting material for the synthesis of the antimalarial drug trimethoxyprim. (Z)-asarone (5.3%) is another valuable component of this oil, which is used as an antiallergic compound (Shahi et al. 1997).

2.4.2 CITRONELLA OILS

C. winterianus and *C. nardus* are cultivated on a large scale, and these are closely related to each other in various aspects. Both species are distinguished morphologically by the shape and length of their leaves. The chemical composition of the essential oil obtained from them also differs considerably (Lawrence 1991; Wijesekara et al. 1973).

2.4.2.1 *Cymbopogon winterianus*

The Java citronella *(C. winterianus* Jowitt) is grown mainly in Java, Haiti, Honduras, Taiwan, Guatemala, and China, and is highly priced in comparison to the Ceylon type (see the following text) because its oil contains higher percentages of monoterpene alcohols and their esters. These are listed in Table 2.1. This oil is also known as Mahapengiri oil. The trace constituents identified from the essential oil include geranyl formate, borneol, farnesol, cadinene, *l*-cadinol, *l*-camphene, *l*-carvone, citral, citronellyl butyrate, cymbopol, dipentene, eugenol, *l*-limonene, linalool, methyl heptenone, methyl eugenol, α-pinene, sesquicitronellene, terpinene, terpinen-4-1, and thujyl alcohol. Conventional and nonconventional techniques have been applied while carrying research on these crops (Chauhan et al. 1976) because of their diversity and immense usefulness. The climatic conditions and time of harvesting have been important factors in determining the chemical profile of the citronella oil (Singh et al. 1996), and this has been discussed in another chapter in this book. The pattern of accumulation of monoterpenes and effect of storage of herb prior to distillation have been reported (Singh et al. 1994, 1996). Soulari and Fanghaenel (1971) made a detailed study of essential oil of *C. winterianus* produced in Cuba. An improved variety of Java citronella was released by Ganguly et al. (1979).

The hydrodistilled essential oil from the aerial parts of *C. winterianus* Jowitt, cultivated in Southern Brazil, was analyzed by GC-MS. Thirty-one components, representing 96% of the oil, were characterized. Enantiomeric ratios of limonene, linalool, citronellal, and β-citronellol were obtained by multidimensional gas chromatography, using a developmental model set up with two GC ovens. The enantiomeric distributions are discussed as indicators of the origin authenticity and quality of this oil (Lorenzo et al. 2000).

Java citronella oil is one of the most important essential oils because of the high content of citronellal and is mainly used for the isolation of citronellal, which is converted into citronellol. Citronellol is further converted into citronellol esters, hydroxy citronellal, and synthetic menthol (Dev Kumar et al. 1977). Java citronella oil is usually employed in the scenting of soaps and all kinds of technical preparations as well as for the extraction of aromatic isolates.

The essential oil steam-distilled from *C. winterianus* (Java citronella) of Cuban origin was ana-lyzed by GC and GC-MS. Thirty-six compounds were identified, of which citronellal (25.04%), citronellol (15.69%), and geraniol (16.85%) were the major constituents (Pino et al. 1996). Using enantioselective multidimensional chromatography (enantio-MDGC) and the column combination polyethylene glycol/heptakis (2,3-di-*O*-acetyl-6-*O*-*tert*-butyldimethylsilyl)-beta-cyclodextrin in OV 1701-vi, the chiral monoterpenoids *cis/trans*-rose oxides, linalol [linalool], citronellol, and terpinen-4-ol were stereoanalyzed simultaneously. The method was applied to chirality evaluation of these compounds in commercial and authentic Java (*C. winterianus*) and Ceylon (*C. nardus*) citronella oils. The enantiomeric distributions are discussed with reference to their uses as indicators of the authenticity of these essential oils (Kreis et al. 1994).

2.4.2.2 *Cymbopogon nardus*

Citronella oil is derived from *C. nardus* (L.) Rendle and is also called "Lanabatu oil." The grass is mostly cultivated in Sri Lanka, and hence the oil obtained from it is known as Ceylon citronella oil. The main constituents of this oil have been presented in Table 2.1. The presence of phenolic derivatives (methyl eugenol and methyl isoeugenol) is the most significant difference between the Ceylon-type and Java-type oils. The wild varieties of citronella growing in Sri Lanka contain phe-nyl propanoids in abundance, whereas phenyl propanoids are present in traces in the Java-type oil, and this is the significant difference between them. The presence of elemol in the Ceylon-type has been suggested to be formed as an artifact. Wijesekara et al. (1973) isolated hedycaryol, a thermolabile precursor, by cold percolation of the macerated fresh grass. In one of the reports it was shown that the oil contained large amounts of monoterpene hydrocarbons, whereas the Java variety (Mahapengiri) contained only small amounts, mainly limonene. Both types contained com-parable amounts of geraniol, and the Java type had more quantities of citronellol and citronellal. In addition, the Ceylon type contained tricyclene, methyl eugenol, methyl isoeugenol, eugenol, and l-borneol. The GLC profiles enable the identification of the type of oil and the detection of kero-sene as a possible adulterant. The variety that grows wild in Sri Lanka (Mana) was quite different from both cultivated types (Wijesekera et al. 1973). The steam-distilled volatile oil obtained from partially dried grass (citronella grass) *C. nardus* (Linn.) Rendle (Syn. *Andropogon nardus* Linn.), cultivated in the Nilgiri Hills at Ooty, India, was analyzed by capillary GC and GC-MS. The par-tially dried grass contained 35 components, out of which 29 constituents were completely identified and comprised 92.7% of the oil. The oil contains 16 monoterpenes (79.8%), nine sesquiterpenes (11.5%), and four nonterpenic compounds (1.4%). The prominent monoterpenes were citronellal (29.7%), geraniol (24.2%), γ-terpineol (9.2%), and *cis*-sabinene hydrate (3.8%). The predominant sesquiterpenes were (*E*)-nerolidol (4.8%), β-caryophyllene (2.2%), and germacren-4-ol (1.5%) (Vijender and Mohammed 2002).

A total of 36 compounds were identified in the steam-distilled essential oil of *C. nardus* collected from Zimbabwe in 1989. The major compound was *trans*-geraniol (29.47%), followed by its ester form geraniol formate (8.79%) (Moody et al. 1995).

The Ceylon citronella oil was tested against Gram-positive bacteria and fungi, and it was found that, under in vitro conditions, the oil was as active as penicillin (Kokate and Verma 1971). *C. winterianus* oil also finds use in providing scent and good smell to low-cost products such as soaps, sprays, disinfectants, polishes, and all kinds of technical preparations. Several products and formulations have been prepared using citronella oil (Chicopharma 1970; Kichiyoshi et al. 1981) for preventing thinner sniffing and slowing the release of the rapidly evaporating substances into the atmosphere. A number of patents have been filed on the uses of citronella oil. A patent describes the use of citronella grass after the distillation in papermaking and sulfate pulping (Bhaumik and Rao 1978a, 1978b; Ciaramello et al. 1972). Another French patent by Tabakoff in 1969 and a Japanese publication by Fuji et al. in 1972 described the use of citronella oil in a composition that can be used with or without addition of water and does not corrode equipment or harm the hands. A joint Canadian and Indian Patent (Prasad and Jamwal 1971; Shaw 1971) described the use of citronella oil for the production of a formulation as mosquito repellent. Similarly, it was also found to be effective as housefly repellent (Osmani et al. 1972).

2.4.3 PALMAROSA AND GINGERGRASS OILS

Palmarosa oil is obtained (Shylaraj and Thomas 1992) from the variety *motia*, which yields an oil of better quality having more geraniol and which is commercially much more important than the oil of gingergrass derived from the variety "sofia." The oil obtained from variety *sofia* is sometimes used as an adulterant (Biaswara et al. 1976a, 1976b). The two important essential oils, palmarosa oil *C. martinii* (Roxb.) Wats. and gingergrass oil, are obtained from two of the most important grasses cultivated in India. The *sofia* variety has lower geraniol content.

2.4.3.1 *Cymbopogon martinii*

Palmarosa oil (*C. martinii* (Roxb.) Wats.), distilled from variety *motia*, has geraniol as the major component and is considered better in quality (Siddiqui et al. 1975, 1979). GC-MS analysis (Gaydou and Raudriamiharisoa 1987) of the hydrocarbons (4.75%) of the oil showed the presence of monoterpenes (45.9%), sesquiterpenes (52.2%), *n*-alkanes (1.6%), and unidentified compounds (0.4%). The main and trace constituents from the oil have been listed in Table 2.1. An interesting dihemiacetal bismonoterpenoid has been identified by Bottini et al. (1987). The x-ray diffraction method was used to establish its structure. Palmarosa oil is another that is extensively used more than any other oil in soaps, and it imparts a rose-like prominent and lasting odor (Guenther 1950; Opdyke 1974). This oil is also employed in the flavoring of tobacco and other mouth fresheners. There are several reports that the oil content increases by storing the herb prior to distillation (Singh et al. 1994). This has been discussed in another chapter by Aklandey in this book. Fungitoxic activity has also been reported from this oil (Singh et al. 1980) and, as in the case of the citronella crop, the leftover part after distillation of the plant is used in paper production (Ciaramello et al. 1972). The essential oil produced from the *sofia* variety of *C. martinii* Stapf is known as gingergrass oil. The major and minor constituents of the oil are listed in Table 2.1. The *cis* and *trans* forms of *p*-menth-2,8 diene-1-ol, *p*-menth1(7),8-dien-2-ol, carveol, and piperitol, along with limonene (20%) and monoterpene alcohols, have been reported from the wild strain of *C. martinii* var. *sofia* growing in Kumaon hills (Mathela et al. 1988; Mathela and Pant 1988; Mathela et al. 1986). One new hemiacetal bismonoterpenoid compound cymbodiacetal was characterized in the oil of *C. martinii* (Bottini et al. 1987). The oil of gingergrass is used in low-cost perfume formulations and scenting of soaps and cosmetics.

The composition of the hydrocarbon fraction of the essential oil from *C. martinii*, which represents less than 5% of the oil, has been studied. Using well-established techniques, 11 monoterpenes (ca 46%), 28 sesquiterpenes (ca 52%), and 16 *n*-alkanes (ca 1.6%) have been identified. The major constituents are limonene, α-terpinene, myrcene, β-caryophyllene, α-humulene, and β- and δ-selinenes. The study of the *n*-alkanes of *C. martinii* revealed the presence of all members of the homologous series C_{15}–C_{30} (Gaydou and Raudriamiharisoa 1986, 1987b).

[2R-(2α,4αβ,5αβ,7α,9αβ,10αβ)]-Octahydro-2,7-bis(1-methylethenyl)5α*H*, 10α*H*-4α, 9α-ethanodibenzo[*b, e*] [1,4]dioxin-5α,10α-diol (cymbodiacetal), isolated from the essential oil of *C. martinii*, was identified by means of x-ray diffraction of its 1:1 solvate with deuteriochloroform (Bottini et al. 1987). Only immature palmarosa *C. martinii* (Roxb.) Wats. var. *motia* inflorescence with unopened spikelets accumulated essential oil substantially. Geraniol and geranyl acetate together constituted about 90% of palmarosa oil. The proportion of geranyl acetate in the oil decreased significantly with a corresponding increase of geraniol during inflorescence development. An esterase enzyme activity, involved in the transformation of geranyl acetate to geraniol, was detected from the immature inflorescence using a GC procedure. The enzyme, termed as geranyl acetate-cleaving esterase (GAE), was found to be active in the alkaline pH range with the optimum at pH 8.5. The catalysis of geranyl acetate was linear up to 6 h, and after 24 h of incubation, 75% of the geranyl acetate incubated was hydrolyzed. The GAE enzymic preparation, when stored at 4°C for a week, was quite stable with only 40% loss of activity. The physiological role of GAE in the production of geraniol during palmarosa inflorescence development has been discussed (Dubey and Luthra 2001).

2.4.4 CYMBOPOGON JWARANCUSA

The roots of *C. jwarancusa* (Jones) Schult. also contain essential oil unlike other *Cymbopogon* species. The oil obtained from aerial parts of the plant is rich in monoterpene alcohols, imparting an excellent odor. Besides major and trace constituents reported in Table 2.1, several other reports are being mentioned here. *Cis-* and *trans-p*-menthenols (38%) have been reported as major components of the oil (Mathela and Pant 1988). Mathela et al. (1986) reported monoterpene alcohol (25%), piperitols (25%), and piperitinone (6.5%), along with α-thujene, camphene, *p*-cymene, piperitinone, umbellulone, and elemol in the essential oil of *C. jwarancusa*. Piperitone was reported to be the single major constituent (Saeed et al. 1978; Thapa et al. 1971), while a mixture of four isomers (3*R*,4*S*-(+)*cis-p*-menth-1-ene-3-ol-4 and 3*S*,4*S*-(–)-*trans* and two isomers *(cis* and *trans)* with same skeleton having *p*-menth-2-en-1-ol constitute 56% of oil. The oil obtained from the roots was rich in monoterpene hydrocarbon and sesquiterpene alcohol, the main sesquiterpenoid component being agarospiral (Mathela et al. 1986).

Two chemical races, *C. jwarancusa* subsp. *jwarancusa* and *C. jwarancusa* subsp. *oliveri*, have been identified from the Kumaon Himalayan region (Mathela et al. 1986); the first one being rich in piperitone and the latter having cyclic and monoterpene alcohols but low piperitone. The components of the oil of *C. jwarancusa* differed with growth conditions, particularly geographical locations. The composition of the essential oil of Khavi grass, *C. jwarancusa*, was investigated by glass capillary GC in combination with mass spectrometry. Sixty-four compounds were identified, 55 of which were reported for the first time. The oil contains a high percentage of piperitone (60%–70%), which is mainly responsible for the smell of Khavi grass (Talat et al. 1978).

The grass found in the Indian Thar desert contained citral, geraniol, geranyl acetate, and piperitone as the major components in its essential oil (Shahi 1992; Shahi and Sen 1989, 1993). In an unusual finding, paramenthenol was reported to constitute 60% of the oil (Boelens 1994; Mathela 1991). Piperitone is mainly employed as a starting material for the preparation of several valuable perfumery compounds, for example, menthol, thymol, etc. (Guenther 1950), and being the principal constituent of the essential oil and possessing a mint or camphor-like odor, this oil is used for scenting many technical preparations directly. In a study on growth performance of three cultivars of *C. jwarancusa* (Jorlab-C.j.5, Jorlab-C.j.3, and C.j.), it was found that cultivar Jorlab-C.j.5 proved to be the best among the three test cultivars in respect of herb yield (21.1 t/ha), oil content (1.6%), and major oil constituent, that is, piperitone (83%) (Singh and Pathak 1994).

2.4.5 *Cymbopogon schoenanthus*

C. schoenanthus (L.) Spreng subsp. *proximus* Hochst. is primarily native to East Africa and has been experimented under cultivation in the surrounding areas (Guenther 1950). The major and minor constituents of the oil have been discussed in Table 2.1. Cryptomeridiol, a component in the oil, has been found to be responsible for its antispasmodic activity. Diuretic and antihistaminic commercial preparations have been made for the oil. Apart from these uses, the oil finds its usual employment as a flavoring agent and in the perfumery industry.

The insecticidal activity of the crude essential oil extracted from *C. schoenanthus* and its main constituent, piperitone, was assessed in different developmental stages of *Callosobruchus maculatus* (Ketoh et al. 2006). Piperitone was more toxic to adults with an LC_{50} value of 1.6 µL/L versus 2.7 µL/L obtained with the crude extract. Piperitone inhibited the development of newly laid eggs and neonate larvae, but was less toxic than the crude extract to individuals developing inside the seeds (Guillaume et al. 2005).

2.4.6 Other *Cymbopogon* Species

The essential oils of numerous wild-growing *Cymbopogon* species have been chemically examined, and the results reveal that some of them can be used as a source of valuable essential oils.

2.4.6.1 *Cymbopogon caesius*

C. caesius (Nees) Stapf is a loosely tufted perennial grass with erect culms, sometimes stilt-rooted, to 2½ m high; of deciduous savanna bushland and wooded grassland; abundant throughout the region and, in general, over all of tropical Africa. Rao and Sudborough (1925) investigated this oil for the first time and reported limonene, geraniol, perillyl alcohol, and dipentene as the chief constituents present in it. About 50 years later, Sinha and Mehra (1977) reported that carvacrol, *d*-perillaldehyde, and *d*-nerolidol were the main constituents of the oil that were effective against *E. coli*. However, carvone (30%) is the major constituent in the Chinese oil (Liu et al. 1981). Recently, GC-MS analysis of the oil from the plants growing in the Northeast region of India have shown 35 compounds, out of which 24 have been identified and listed in Table 2.1. The essential oil of *C. caesius* is well placed to be used in perfumery. Detailed analysis by GC-MS is being undertaken, and reports are likely to be available shortly.

2.4.6.2 *Cymbopogon coloratus*

A total of 33 compounds were reported by Mallavarapu et al. (1992a) in the chemical profile of the essential oil *C. coloratus* (Nees) Stapf. The main constituents of the oil are myrcene, limonene, *trans*-β-ocimene, linalool, neral, geranial, geraniol (69.11%), geranyl acetate, and elemol. Pilley et al. (1928) described the presence of borneol, limonene, and camphene. The oil is also used for perfuming soaps and cosmetics (Gupta and Daniel 1982). Bor (1954) was skeptical about the authenticity of the species.

2.4.6.3 *Cymbopogon confertiflorus*

The synonym of *C. confertiflorus* (Steud.) Stapf is reported as *C. nardus*, which is also known as Ceylon citronella as stated earlier. The oil of this particular species is similar to the Ceylonese variety but inferior in quality with less geraniol (35%–40%) as the principal constituent (Gupta and Daniel 1982).

2.4.6.4   *Cymbopogon densiflorus*

C. densiflorus (Steud.) Stapf is a tufted perennial grass with culms 1.8 m high; of open spaces along roadsides and wooded grassland; native to central tropical Africa, Gabon to Zimbabwe, and introduced into the region (Nigeria and possibly other states) and into Brazil. It is grown in Gabon and Nigeria as an ornamental and for its aromatic oil. The leaves are avoided by browsing cattle in Zambia. The crushed leaves are used as treatment for rheumatism in Gabon. In Malawi, the flower head is smoked in a pipe as a cure for bronchial affections and, for the same complaints, the plant sap is used in the Congo (Brazzaville), where it is also given as treatment for asthma and to calm fits. It is macerated with *Ocimum basilicum* (Labiatae), and the compound is used for epilepsy in Zaïre. It is conjectured that any medicinal action is due to the camphoraceous volatile oil. The plant has also been recorded to be used as a tonic and styptic. It has fetish attributes as well. In Gabon, the inflorescence of *C. densiflorus* is burnt in fumigations required in certain rituals, for example, in incantations to chase away malignant spirits, to cleanse those who have lost their spouse, to rejuvenate and restore the efficiency of a fetish, as an amulet or talisman when the owner has violated a taboo. In Tanganyika, witch doctors smoke the flower panicle, either alone or with tobacco, to induce dreams to foretell the future. Huntsmen in Gabon use the plant as a fetish lure for game. The oil of the Brazilian plant was compared with the African oil by Koketsu et al. (1976), and olfactory analysis showed no noticeable difference. The monoterpenes found in the oil from the flowers and leaves of the Zambian plant are shown in Table 2.1. No remarkable difference in the quality of the oil from Brazilian and African plants could be recorded (Boelens 1994; Chisowa 1997).

2.4.6.5   *Cymbopogon distans*

The essential oil of *C. distans* (Nees) Wats. has been studied by GC-MS (Mathela and Joshi 1981; Mathela et al. 1989), which showed several monoterpenoids in addition to 19% sesquiterpene alcohols. Sobti et al. (1978) reported terpineol (20%) as the major constituent, whereas Thapa et al. (1971) and Liu et al. (1981) reported piperitone (30%–40%) and geraniol (10%) as the principal constituents of the oil. In another publication (Singh and Sinha 1976), limonene (29%) and methyl eugenol (13%) were reported to be the major constituents. Other major constituents have been reported in Table 2.1.

   C. distans has been reported to occur in nature in the form of several geographical races. The essential oils isolated from the leaves of *C. distans* chemotype *loharkhet* and the roots of *C. jwarancusa* (collected from India) were analyzed by GC, GC-MS, and liquid chromatography. Both oils were qualitatively very similar in sesquiterpenoid composition but contained different total concentrations of sesquiterpenoids (79.6% and 38.0% in the oils of *C. distans* and *C. jwarancusa*, respectively). The main sesquiterpenoids of the essential oil of *C. distans* were eudesmanediol (34.4%) and 5-epi-7-epi-alpha-eudesmol (11.2%). The main sesquiterpenoid in the essential oil of *C. jwarancusa* was agarospirol (9.5%) (Beauchamp et al. 1996; Dunyan et al. 1992). Exhaustive studies have been done by Mathela et al. (1988a, 1989) and Mathela and Pant (1988) on the chemical investigations of the essential oil, and they characterized *C. distans munsiyariensis*, which contained eudesmanediol (34.4%) along with geraniol (22.89%), neryl acetate (18.34%), neral (14.74%), and limonene (12.08%) as the major constituents. Mathela and Pant (1988) reported four chemotypes from the Kumaon and Garhwal regions of Uttar Pradesh (India) having marker compounds α-oxobisabolene-1 (chemotype I); citral, geraniol, and geranyl acetate (chemotype II); piperitone, limonene, and eudesmanediol (chemotype III); and sesquiterpene alcohol (chemotype IV) in their oils. One more chemotype with chemical marker *p*-menthol (66.5%) was reported later (Pande et al. 1997). A GC-MS study of the hydrocarbon fraction and the fraction containing oxygenated compounds showed the presence of 12 monoterpene hydrocarbons (28.4%), 13 sesquiterpene hydrocarbons (32.8%), 3 sesquiterpene alcohols (27.2%), 2 esters (7.2%), and 3 carbonyl compounds (4.4%) in the essential oil of *C. distans*. Of these, 27 compounds have been identified (Mathela and Joshi 1981).

2.4.6.6 *Cymbopogon khasianus*

The essential oil of *C. khasianus* (Hack.) Stapf ex Bor has been reported to contain citral (40%–60%) and geraniol (70%–80%) as the major constituents (Balyan et al. 1979; Choudhary and Leclercq 1995; Rabha et al. 1986; Rabha et al. 1989; Sobti et al. 1978a; Sobti et al. 1982; Thapa et al. 1971; Thapa et al. 1976; Verma et al. 1987). Methyl eugenol (75.82%) is the major constituent in *C. khasianus.*

2.4.6.7 *Cymbopogon ladakhensis*

Very little information is available on *C. ladakhensis.* However, Gupta and Daniel (1982) have reported that piperitone is the chief constituent of this oil.

2.4.6.8 *Cymbopogon microstachys*

The essential oil *C. microstachys* (Hook.s) Soenarke from the northeastern Indian state of Manipur was grown in Bhubaneswar in the state of Orissa in pots. The plants were found to grow well and, in 45 days, reached full growth. The essential oil thus obtained from this species was analyzed by GC and GC/MS and the results showed that the oil contained (*E*)-methyl isoeugenol (56.4%–60.7%) as the major constituent. This is the first time that an oil of *C. microstachys* has been found with (*E*)-methyl isoeugenol as the major constituent (Rout et al. 2005). The oil was found to contain about 60% phenyl propenoids with methyl eugenol (19.5%), methyl isoeugenol (4.2%), elemicin (25.3%), and isoelemicin (11.0%). The oil now reported had quite a different composition with (*E*)-methyl isoeugenol as the main constituent (56.4%–60.7%). Other constituents in significant quantities were myrcene (7.8%–12.2%), (*Z*)- and (*E*)-β-ocimene (2.4%–2.9%), *cis*-α-bergamotene (0.8%–1.7%), *trans*-α-bergamotene (0.8%–3.4%), germacrene D (0.6%–2.7%), and (*Z,E*)-α-farnesene (0.4%–2.9%), besides an unidentified compound at RT 46.1 min (0.6%–6.0%). In all, 44 components were identified, constituting 91.3%–96.2% of the oil. The study shows that the aromatic grass collected in Manipur is a chemical variant of *C. microstachys.* The easy adaptability of the plant to Bhubaneswar conditions, high oil yield, and presence of methyl isoeugenol as the major phenyl propenoid may make this a commercially important essential oil. Earlier reports have indicated that it contains citral and geraniol (Gupta and Daniel 1982). Methyl eugenol, elemicin, and isoelemicin are the major constituents (Boelens 1994; Pant et al. 1990) found in a chemical investigation of the oil, along with more than two dozen minor constituents: aromadendrene; alloaromadendrene, caryophyllene oxide, 3,4-epoxy-3,7 dimethyl-1,6-octadione, 6-7-epoxy-3,7-dimethyl-1,3-octadiene, *cis-3*-hexenol, humulene, limonene, *cis*-limonene oxide, linalool, 6-methyl-5-hepten-2-one methyl isoeugenol, myrcene, nerolidol, 4-nonanone, *trans*-ocimene, β-phellendrene, α-pinene, sabinene, terpinen-4-ol, α-terpineol, tricyclene, trimethoxy benzaldelyde, and veratraldehyde (Mathela et al. 1990b).

2.4.6.9 *Cymbopogon nervatus*

The essential oil hydrocarbons from *C. nervatus* (Hochest.) Chiov. have been investigated (Modawi et al. 1984), and sesquiterpenes β-selinene, β-elemene, β-bergamotene, and germacrene-D have been reported. The antibacterial activity of essential oil of dried inflorescence of *C. nervatus* was investigated. The essential oil remarkably inhibited the growth of tested bacteria except for *Salmonella typhi.* The maximum activity was against *Shigella dysenteriae* and *Klebsiella pneumoniae* (Kamali et al. 2005).

2.4.6.10 *Cymbopogon olivieri*

The major components of the essential oil of *C. olivieri* (Boiss.) Bor are iso-pulegol (17.2%), β-pinene (24.4%), myrcene (17.9%), piperitone (6.6%), α-pinene (7.3%), and pulegone (10.9%) (Sharma et al. 1980). The minor components include limonene, linalool, linalyl acetate, phellandrene, piperitenone, and terpinene (Bor 1954; Gupta and Daniel 1982). The oil has been tested to be fungitoxic (Singh et al. 1980).

2.4.6.11 *Cymbopogon parkeri*

The major constituents of the oil of *C. parkeri* Stapf are nerol (32%), geraniol (33%), farnesol (3.75%), geranyl acetate (8.9%), and neryl acetate (3.7%) (Rizk et al. 1985, 1983). The trace constituents reported are decane, eudesmol, geraniol, geranyl heptanoate, geranyl hexanoate, geranyl octanoate, guaiol, β-gurjunene, limonene, linalool, 14-methyl-heptadecanone, 7-methyl-4-octanone, 12-methyl-tridecanone, neral, neryl butanoate, neryl hexanoate, neryl octanoate, piperitone, α-terpineol, and xylene (Gupta and Daniel 1982; Rizk et al. 1983, 1986). A study of the antifungal activity of *C. parkeri* essential oil was done on the growth of *Rhizoctonia solani*, *Pyricularia orizea*, and *Fusarium oxysporum*, three important phytopathogenic fungi.

2.4.7 Some Lesser-Known Species

2.4.7.1 *Cymbopogon polyneuros*

The major constituents in this sweet-scented oil obtained from *C. polyneuros* (Steud.) Stapf. have been reported as limonene, perillaldehyde, and perillyl alcohol (Gupta and Daniel 1982; Thapa et al. 1971).

2.4.7.2 *Cymbopogon procerus*

Elemicin (34%) and pinene have been described by Gildemeister and Hoffmann (1956) as the major constituents of the oil from *C. procerus* (R. Br.) Domin.

2.4.7.3 *Cymbopogon rectus*

The major constituents of the oil of *C. rectus* A. Camus are geraniol (40%–60%), methylisoeugenol (30.5%), and α-pinene (Gildemeister and Hoffmann 1956).

2.4.7.4 *Cymbopogon sennarensis*

Gildemeister and Hoffmann (1956) have reported that the oil from *C. sennarensis* (Hochest.) Chiov. contains pinene and limonene (13%) and *l*-methenone-3 (45%).

2.4.7.5 *Cymbopogon stracheyi*

The essential oil of *C. stracheyi* (Hook. f.) Riaz and Jain bears a very strong aromatic note and contains geraniol, citral, geranyl acetate, citronellol, and piperitone (Gupta and Daniel 1982; Mathela 1991; Thapa et al. 1971). Some analytical studies of plants growing in the Almora region of Uttar Pradesh exhibit the presence of piperitone and car-2-ene as major components, along with geraniol, α-copaene, β-elemene, caryophyllene, and calarene. Lohani et al. (1986) carried out detailed chemical investigation of its oil and revealed the presence of car-2-ene (29.4%) and piperitone (47.8%), along with several minor constituents such as acoradiene, β-bisabolene, α-cadinene, camphene, *trans*-caryophyllene, α-copaene, *p*-cymene, dihydro-α-capaene-8-ol, geraniol, β-elemene, α-himachalene, intermediol, juniper camphor, and nerolidol.

2.4.7.6 *Cymbopogon tortilis*

A group from China (Liu et al. 1981) reported that methyl eugenol (55%) is the principal component of the oil *C. tortilis* (Presl.) Hitche.

2.4.7.7 *Cymbopogon travancorensis*

Elimicin is the chief constituent of the oil from *C. travancorensis* Bor. Other constituents reported are borneol, elemol, camphene, limonene, and citral (Gupta and Daniel 1982; Mallavarapu et al. 1992b; Menon 1956).

2.4.7.8 *Cymbopogon goeringii*

The essential oil obtained from *C. goeringii* (Steud.) A. Camus exhibits antiarrhythmic action on isolated guinea pig heart. The results from chemical analysis of the oil are not available (Liu and Feng 1989).

2.4.7.9 *Cymbopogon asmastonii*

The compounds reported from the essential oil of *C. asmastonii* are *d*-limonene (5.5%), carveol (69.5%), and a complex mixture of ketones (17.5%) (Manjoor-i-khuda et al. 1986).

2.4.7.10 *Cymbopogon giganteus*

The composition of the essential oil isolated by hydrodistillation from the leaves of *C. giganteus* Chiov. growing wild in Ivory Coast was determined by GC-RI, GC-MS, and ^{13}C-NMR after fractionation on silica gel. The oil was characterized by high contents of *trans*- and *cis-p*-mentha-2,8-dien-1-ols (18.4% and 8.7%, respectively), *cis*- and *trans-p*-mentha-1(7),8-dien-2-ols (16.0% and 15.7%, respectively), and limonene (12.5%). A total of 46 components were identified, including 25 compounds reported for the first time in the oils of this species (Boti et al. 2006). The essential oils of fresh flowers (2 samples), leaves, and stems of *Cymbopogon giganteus* (Hochest.) Chiov. from the Cameroon were investigated by GC and GC-MS. More than 55 components have been identified in the samples 1 (flowers sample 1), 2 (leaves), 3 (stems), and 4 (flowers sample 2) with main compounds possessing the *p*-menthane skeleton as follows: *cis-p*-mentha-1(7),8-then-2-ol (1: 22.8%, 2: 27.7%, 3: 29.1%, 4: 20.5%), *trans-p*-mentha-1(7),8-dien-2-ol (1: 24.9%, 2: 21.6%, 3: 28.1%, 4: 26.5%), *trans-p*-mentha-2,8-then-1-ol (1: 17.3%, 2: 22.1%, 3: 21.4%, 4: 16.3%), and *cis-p*-mentha-2,8-dien-1-ol (1: 8.3%, 2: 5.4%, 3: 4.6%, 4: 9.7%).

The oils of several other species have also been distilled out, but their chemistry is still not well studied. These could well be potentially valuable aromatic crops. Further research is needed to reveal the chemistry of these uncommon grasses that can be brought under cultivation on a larger scale. The most interesting species are *C. caesius*, *C. coloratus*, *C. distans*, *C. khasianus*, *C. microstachys*, *C. parkeri*, and others. On the other hand, all species that have been chemically investigated or not need to be cultivated in a particular phytogeographical region and reinvestigated using modern spectroscopic methods such as GC-MS, IR, NMR, ^{13}C-NMR, etc. Popielas et al. (1991) and Vole et al. (1997) studied the oil from its inflorescence and identified 18 compounds, most of them belonging to oxygenated monoterpenes with *p*-menthadiene skeleton. It is expected for this species that chemotaxonomists will be enabled to classify all the species in a perfect and systematic manner so that the prevailing confusion regarding the correct chemotaxonomy of this large genus may be removed. Additional components in higher concentrations, responsible for the characteristic aroma impressions of the samples from *C. giganteus*, are especially limonene, *trans*-verbenol, and carvone as well as some other mono- and sesquiterpenes. Antimicrobial activities of the oils from the leaf, stem, and flowers were found against Gram-positive and Gram-negative bacteria as well as the yeast *Candida albicans*, and these results were discussed with the compositions of each sample (Leopold et al. 2007).

2.4.8 BIOSYNTHESIS OF TERPENES IN *CYMBOPOGON* SPECIES

Terpenoids are a class of compounds derived from the universal precursor isopentenyl diphosphate (IPP) and its allylic isomer dimethylallyldiphosphate (DMAPP), also called isoprene units (Scheme 2.1). Terpenoid building blocks are then formed through condensation of additional IPP moieties (C_5) via prenyltransferases. Monoterpenoids are derived from geranyl pyrophosphate (GPP, C_{10}), sesquiterpenoids are derived from farnesyl pyrophosphate (FPP, C_{15}), and diterpenoids are

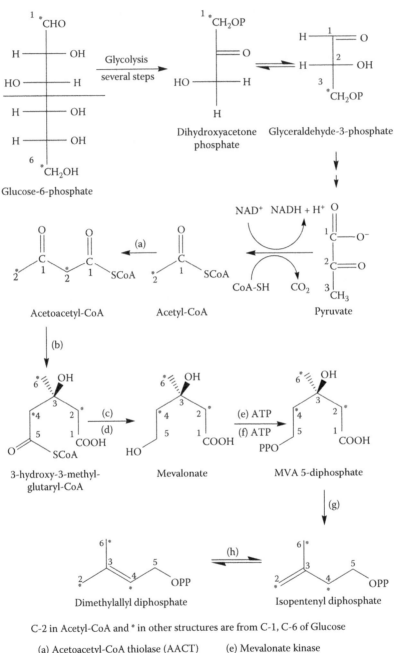

C-2 in Acetyl-CoA and * in other structures are from C-1, C-6 of Glucose

(a) Acetoacetyl-CoA thiolase (AACT)   (e) Mevalonate kinase

(b) HMG CoA Synthase   (f) Phosphomevalonate kinase

(c) HMG-CoA reductase   (g) IPP synthase

(d) NADPH   (h) IPP Isomerase

SCHEME 2.1   MVA-pathway from glucose to IPP and DMAPP.

derived from geranylgeranyl pyrophosphate (GGPP, C_{20}). Even higher-order terpenoids are possible through condensation of these intermediates to larger precursor moieties. For example, sterols are derived from the triterpenoid squalene (C_{30}), which contains six isoprene units through condensation of two molecules of FPP, and carotenoids (C_{40}) are largely formed through condensation of two molecules of GGPP to yield eight-isoprene-unit compounds. After the formation of the acyclic terpenoid structural building blocks (e.g., GPP, FPP, GGPP), terpene synthases act to generate the main terpene carbon skeleton. Additional transformations often involving oxidation, reduction, isomerization, and conjugation enzymes decorate or alter the main skeleton with varied functional groups to yield the tremendously diverse terpenoid family of compounds. The essential oils obtained from aroma-bearing plants mainly possess mono- and sesquiterpenoids in high percentage besides a few nonterpenoidal compounds. The essential oil obtained from various *Cymbopogon* species also contains a large number of mono- and sesquiterpenoids as discussed in this chapter.

It has now been unequivocally proven that two distinct and independent biosynthetic routes exist to IPP and its allylic isomer DMAPP, the two building blocks for isoprenoids in plants. The cytosolic pathway is triggered by acetyl coenzyme A (Scheme 2.1) where the classical intermediate mevalonic acid is formed, which in turn converts into IPP and DMAPP. These further combine to elongate into sesquiterpenes (C_{15}) and triterpenes (C_{30}) (Newman and Chappell 1999). In contrast, the plastidial pathway (Eisenreich et al. 1998, 2001; Lichtenthaler 1999; Rohmer 1999) provides precursors for the biosynthesis of isoprene (C_5), mono- (C_{10}), di- (C_{20}), and tetraterpenes (C_{40}) (Lichtenthaler 1999; Eisenreich et al. 1997, Scheme 2.2). The pathway (Scheme 2.2) is initiated by the transketolase-type condensation of pyruvate (C-2 and C-3) and glyceraldehyde-3-phosphate to 1-deoxy-d-xylulose-5-phosphate (DXP), followed by the isomerization and reduction of this intermediate to 2-*C*-methyl-d-erythritol-4-phosphate, formation of the cytidine 5′-diphosphate derivative, phosphorylation at C-2, and cyclization to 2-*C*-methyl-d-erythritol-2,4-cyclodiphosphate (MECDP) as the last defined step. This is further converted to 1-hydroxy-2-methyl-2-(*E*)-butenyl-4-diphosphate (HMBDP). HMBDP is converted to two C-5 units for further condensation (Scheme 2.3). Later, the genes encoding each enzyme of the plastid pathway up to formation of the cyclic diphosphate are isolated from plants and from eubacteria where the pathway exists (Takahashi et al. 1998; Bouvier et al. 1998; Rohdich et al. 1999, 2000; Sprenger et al. 1997; Lange and Croteau 1999; Little and Croteau 1999; Lange et al. 1998; Schwender et al. 1999; Kuzuyama et al. 2000a, 2000b; Lüttgen et al. 2000; Herz et al. 2000).

The cymbopogons have been reported to possess about 100 monoterpenes and around 50 sesquiterpenes in all, and thus the MVA and DXP pathways are very likely to be present during the biosynthesis of these compounds in the cymbopogons. Not much biosynthetic studies have been carried out on these species. However, Akhila (1985) has conducted studies on the biosynthetic relationship of citral-*trans* and citral-*cis* using doubly labeled [^{14}C, ^{3}H] precursors. The results revealed that the leaf blades of *C. flexuosus* converted geraniol into citral-*trans* with the loss of pro-(*1S*)-hydrogen whereas nerol lost the pro-(*1R*)-hydrogen while being converted into citral-*cis*. Secondly, the citral-*trans* is converted into citral-*cis* and vice versa, and there is no separate route for the biosynthesis of either of the two aldehydes. The mechanism for interconversion has been shown in Scheme 2.4.

The biosynthesis of three major components in *C. winterianus* has been studied by Akhila (1986) using ^{3}H- and ^{14}C-labeled precursors. Geraniol, citronellol, and citronellal formed in the blades of *C. winterianus* from doubly labeled mevalonic acid predominantly labeled only that C_5 moiety that was derived from IPP. This was believed to be due to the presence of a metabolic pool of DMAPP. These results later on support the newly discovered theory of a nonmevalonate pathway (DXP pathway) in which mevalonic acid is believed to be converted into IPP in cytosol. The very low incorporation of radioactivity into monoterpenes (C_{10}) may be the result of seepage of IPP through the plastidial membrane. Monoterpenes are believed to be biosynthesized in plastids. According to the reports, geraniol is converted to citronellol, which in turn is transformed into citronellal as per the mechanism shown in Scheme 2.5.

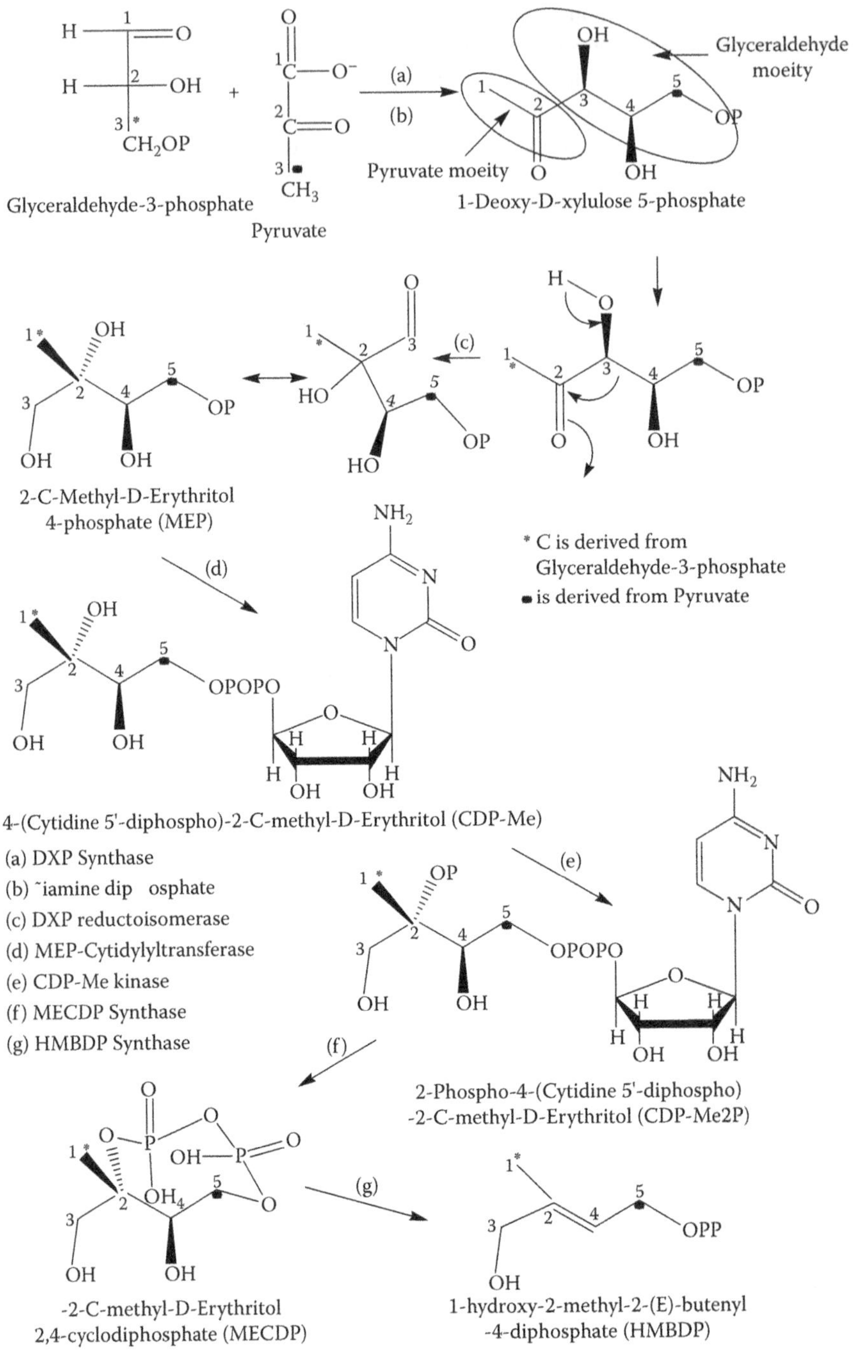

SCHEME 2.2 Nonmevalonate pathway (DXP) to monoterpenes.

Biosynthesis of several mono- and sesquiterpene skeletons and compounds has been worked out in various plant species (Akhila et al. 1980a, 1980b, 1980c, 1988a, 1988b, 1987a, 1987b, 1985, 1986, 1990; Croteau et al. 1981; Crotaeau and Davis 2005). Though about 150 mono- and sesquiterpenes are present in all the *Cymbopogon* species collectively, biosynthetic experiments have not been carried out. Based on the literature, two comprehensive schemes (Schemes 2.6 and 2.7) have been made showing biosynthetic pathways to most of the mono- and sesquiterpene skeletons and compounds

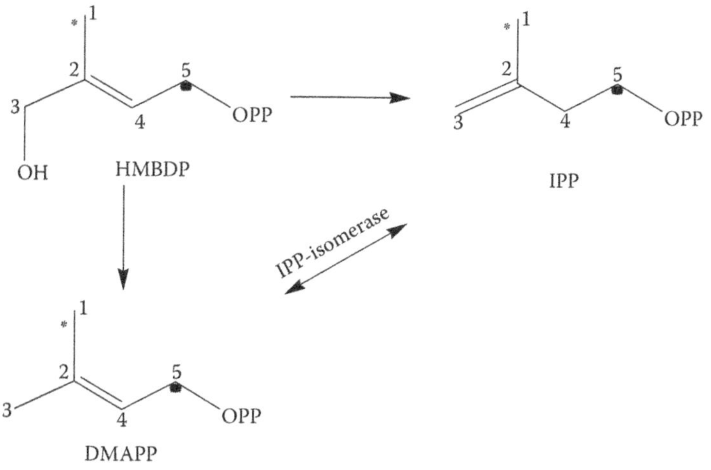

SCHEME 2.3   1-hydroxy-2-methyl-2-(*E*)-butenyl-4-diphosphate (HMBDP) to IPP and DMAPP.

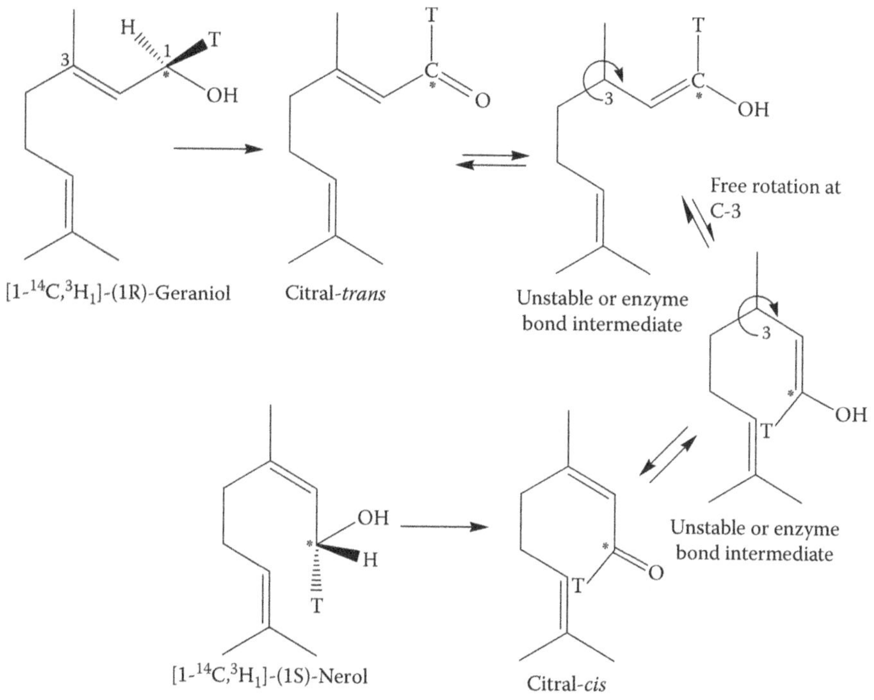

SCHEME 2.4   Biosynthesis of citral-*trans* and citral-*cis* from doubly labeled geraniol and nerol.

derived therefrom. Possible biosynthetic pathways for cadinane, bisabolane, eudesmane, and gurju-nane compounds have been shown in Schemes 2.8, 2.9, 2.10, and 2.11.

   cis-Farnesyl pyrophosphate (FPP), also known as 2,3-(*Z*)-farnesyl pyrophosphate (Scheme 2.12), has been considered as a universal intermediate starter for many sesquiterpenes. However, it isomer-izes to its *trans*-isomer 2,3-(*E*)-FPP through enzymatic conversion, possibly via nerolidol. FPP also cyclizes to a major group of sesquiterpenes such as caryophyllenes and humulenes. For convenience and according to generally adopted nomenclature, the numbering of carbon atoms (1 to 15) has been

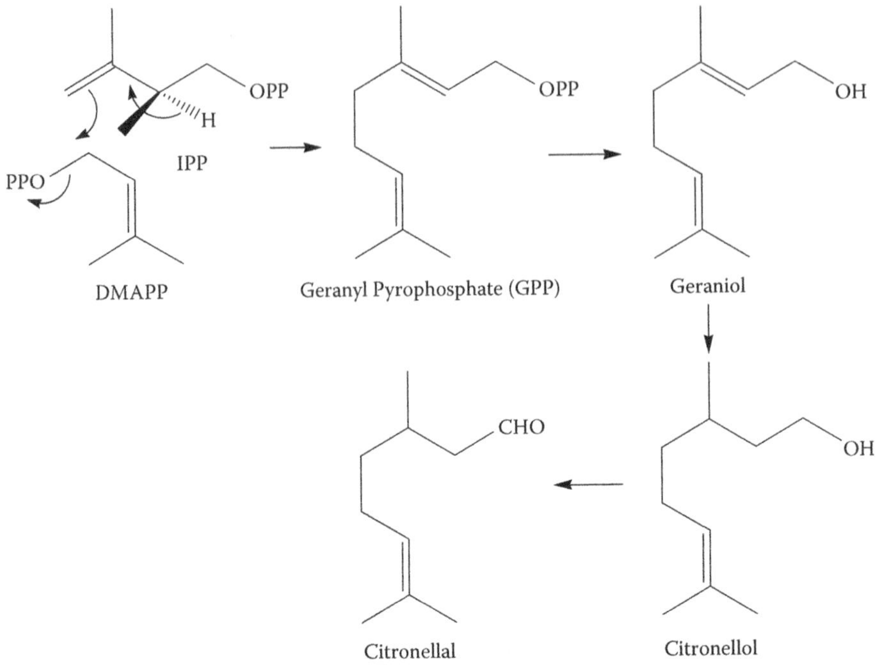

SCHEME 2.5 Biosynthesis of geraniol, citronellol, and citronellal from IPP + DMAPP in *Cymbopogon winterianus*.

shown in FPP in Scheme 2.12. A nucleophilic attack by an enzyme at C-3 of *trans*-FPP triggers the cyclization process that forms the nine-member cyclic skeleton along with the cyclobutane ring with elimination of pyrophosphate from C-1. This is the penultimate precursor of β-caryophyllene and β-caryophyllene epoxide. Alternatively, an enzymatic attack at C-10 will facilitate the formation of the C11–C1 bond with the elimination of pyrophosphate from C-1 and formation of C_{11}-ring skeleton of α- and β-humulene.

Alloaromadendrene and aromadendrene are a different group of sesquiterpenes present in the essential oils of cymbopogons. A biogenetic route has been proposed from the *cis*-FPP in Scheme 2.13. There are three double bonds in FPP inviting electrophiles (enzymes) to attack them at suitable positions depending on the accessibility of the enzyme to the site of the attack. This depends upon the stereochemistry and spatial arrangement of the hydrogens and other attached atoms such as oxygen and phosphorus. In this case, an enzyme is attracted at C-7 of $\Delta^{6,7}$, which attaches to C-2 resulting in the formation of a cyclopentane ring, whereas the other enzyme attacks at C-10, enabling $\Delta^{10,11}$ to attack at C-1 and release FPP from there. This forms the aromadendrene skeleton. The *re*- and *si*-face attacks by $\Delta^{6,7}$ on C-2 generates the two isomers alloaromadendrene and aromadendrene. Another mechanism for the biosynthetic route to butenol and acoradienes has been illustrated in Scheme 2.14. These schemes are biogenetic speculations based on chemical considerations and mechanisms, and are very good experimental models to be verified by radiotracer techniques.

2.4.9 BIOLOGICAL ACTIVITIES

The essential oil obtained from various *Cymbopogon* species is widely used in the perfumery, cosmetic, food, and flavor industries. Besides, the oil possess several biological activities (Dikshit and Hussain 1984), which have been discussed in detail in Section (2.4.9), but some highlights are

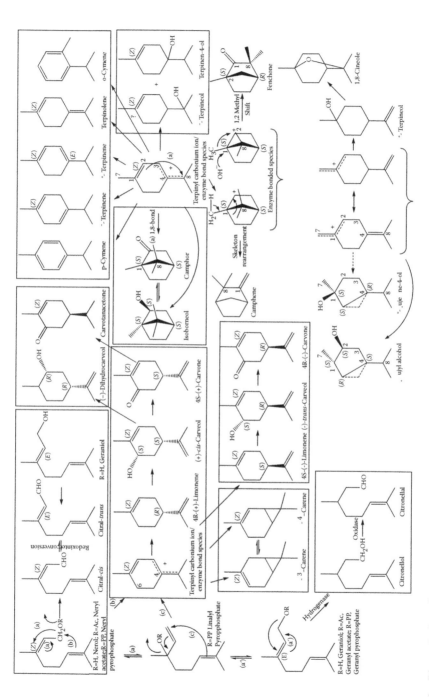

SCHEME 2.6 Biosynthetic pathways to monoterpene hydrocarbons and oxygenated monoterpenes commonly found in various *Cymbopogon* species. The most widely occurring three possible precursors (C$_{10}$) GPP, NPP, and LPP have been shown.

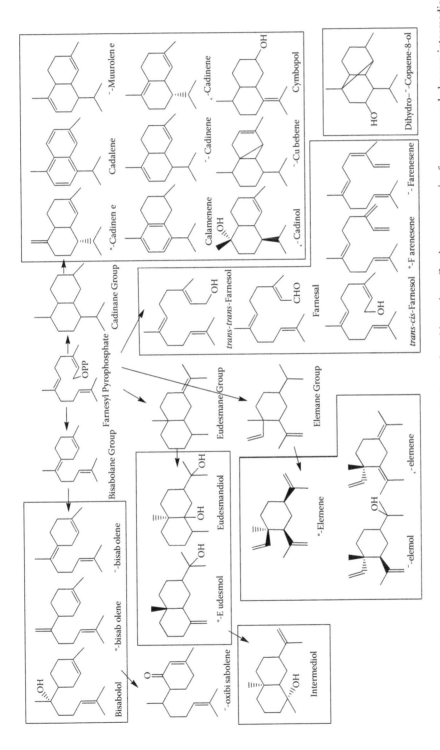

SCHEME 2.7 Diversity in biosynthetic pathways to various sesquiterpene skeletons found in various *Cymbopogon* species from commonly known intermediate *cis*-farnesylpyrophosphate (FPP).

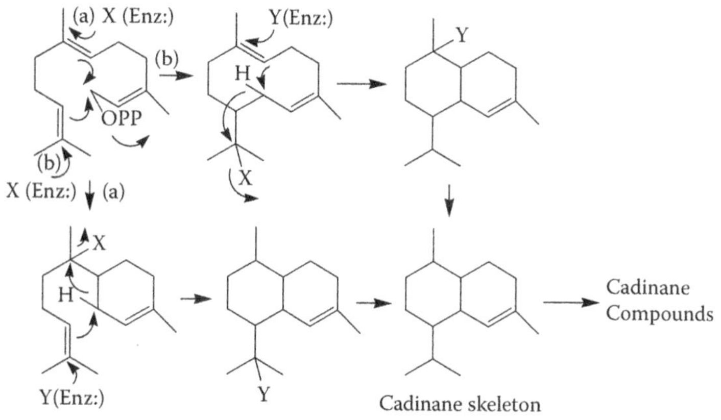

SCHEME 2.8 Cyclization of FPP to cadinane skeleton and compounds of cadinane group. X and Y denote attacking enzymes involved in biosynthesis.

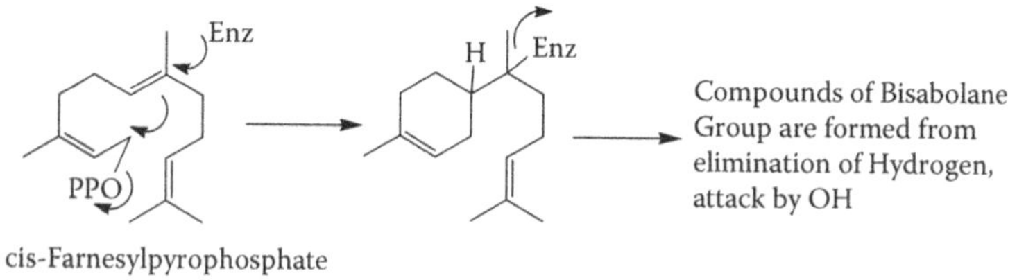

SCHEME 2.9 Cyclization of *cis*-FPP to monocyclic bisabolane compounds.

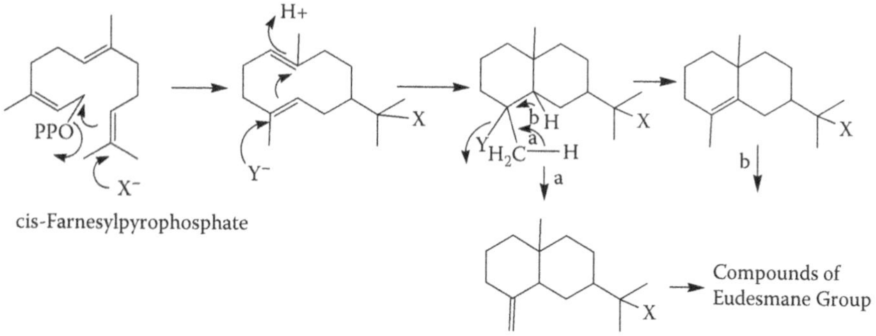

SCHEME 2.10 Possible biogenesis of compounds of eudesmane group in cymbopogons. X and Y denote enzymes or their biogenetic equivalents that make a nucleophilic attack on electron-deficient sites of the skeleton.

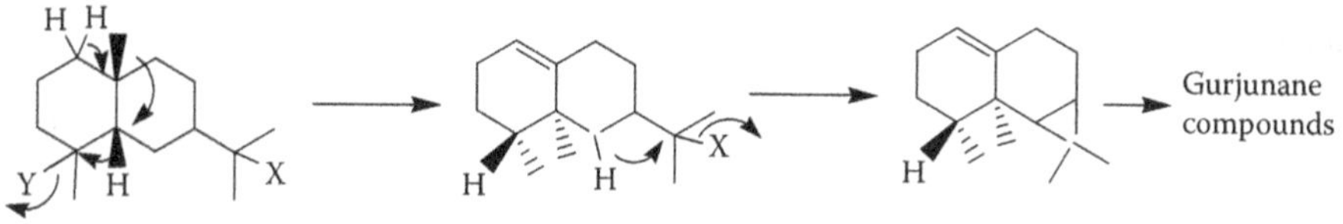

SCHEME 2.11 Biogenetic pathways to Gurjunane compounds after 1,2 methyl shifts and hydrogen loss while forming a cyclopropane ring.

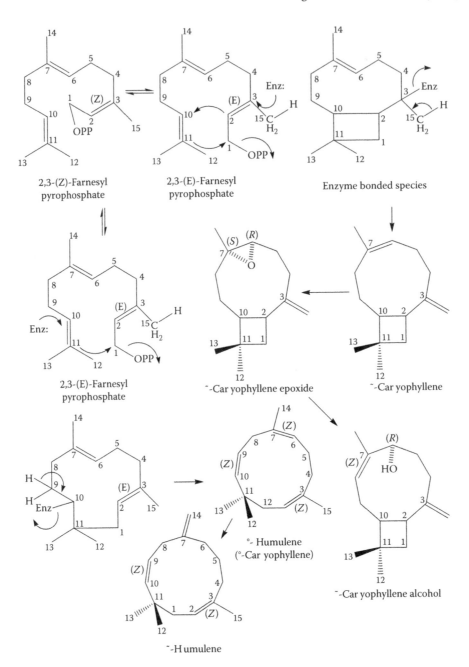

SCHEME 2.12 Biosynthesis of caryophyllenes and humulenes.

provided here to give the reader an overall view of the pharmacological properties of this oil. The important activities include antibacterial, antifungal, anticancer, pesticidal, anthelmintic, mosquito repellant, mosquito larvicidal, antiinflammatory, analgesic, and hypoglycemic. The main components such as monoterpenes are nonnutritive dietary components found in the essential oils of citrus fruits and other plants. A number of these dietary monoterpenes have antitumor activity. For example, *d*-limonene, which comprises >90% of orange peel oil, has chemopreventive activity against rodent mammary, skin, liver, lung mammary, lung and forestomach cancers when fed during the

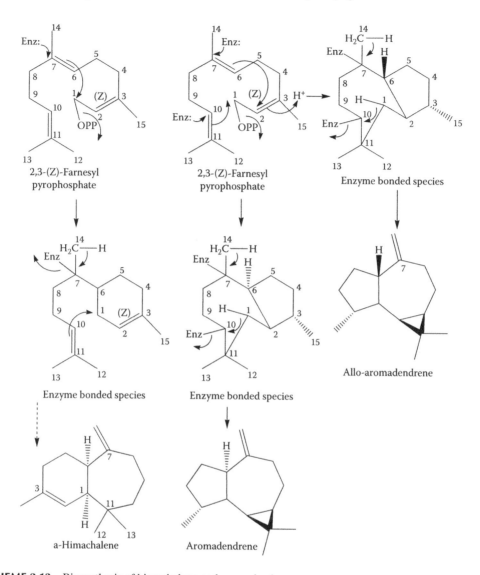

SCHEME 2.13   Biosynthesis of himachalene and aromadendrene.

initiation phase. In addition, perillyl alcohol has promotion phase chemopreventive activity against rat liver cancer, and germaniol has in vivo antitumor activity against murine leukemia cells. Perillyl alcohol and *d*-limonene also have chemotherapeutic activity against rodent mammary and pancreatic tumors. As a result, their cancer chemotherapeutic activities are under evaluation in Phase I clinical trials. Several mechanisms of action may account for the antitumor activities of monoterpenes (Crowell 1999).

2.4.9.1 Pain Reliever

Cymbopogon winterianus (Poaceae) is used for its analgesic, anxiolytic, and anticonvulsant properties in Brazilian folk medicine. The cited report aimed to perform phytochemical screening and investigate the possible anticonvulsant effects of the essential oil from fresh leaves of *C. winterianus* in different models of epilepsy (Quintans-Júnior et al. 2008). Myrcene was identified as the active constituent responsible for the activity. The peripheral analgesic effect of myrcene was confirmed

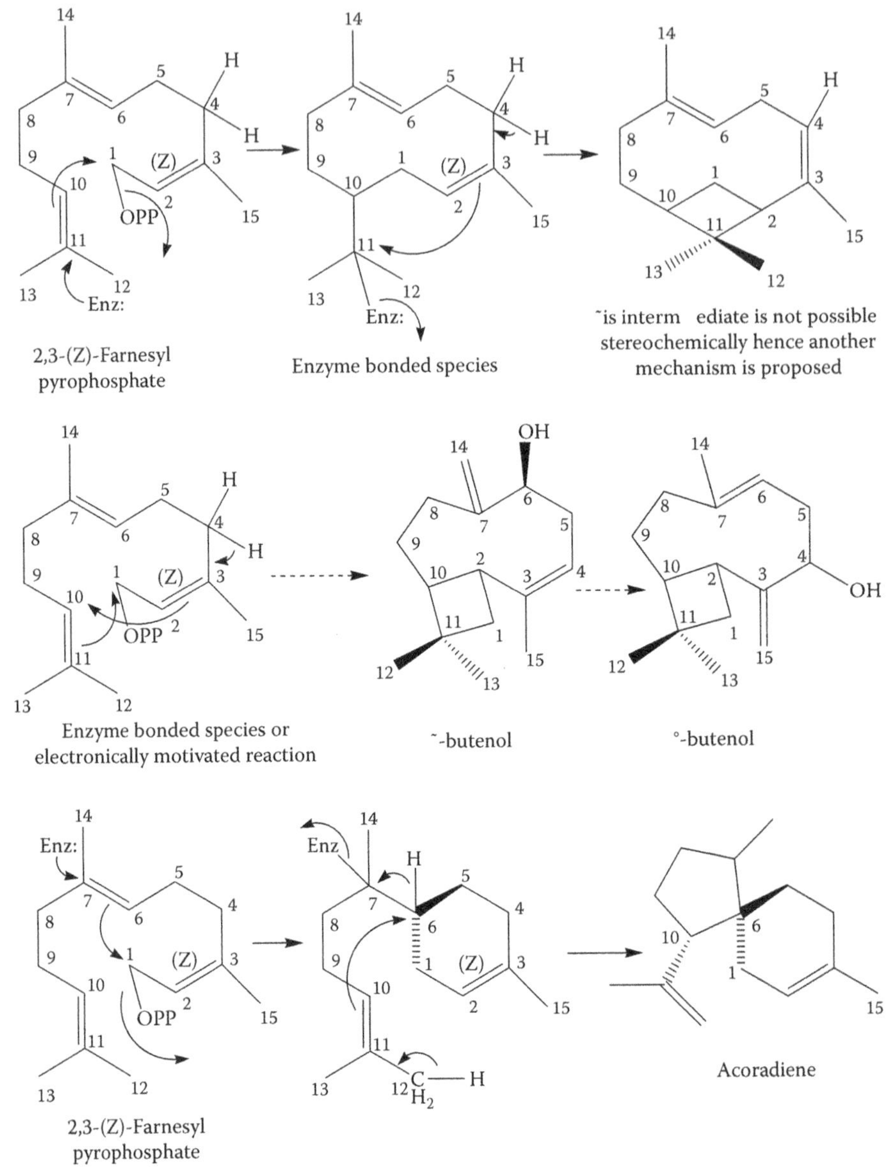

SCHEME 2.14 Biosynthetic routes to butenol and acoradiene.

by testing a saturated compound preparation on the hyperalgesia induced by prostaglandin in the rat paw test, and upon the contortions induced by intraperitoneal (ip) infections of iloprost in mice (Lorenzetti et al. 1991). Oral administration of an infusion of C. *citratus* fresh leaves to rats produced a dose-dependent analgesia for the hyperalgesia induced by subplanter infections of either caragenin or prostaglandin E2, but did not affect that induced by dibutyryl cyclic AMP. These results indicated a peripheral site of action. In contrast to the central analgesic effect of morphine, myrcene did not cause tolerance on repeated infection in rats. This analgesic property supports the use of lemongrass tea as a sedative in folk medicine. It is suggested that terpenes such as myrcene may constitute a lead for the development of new peripheral analgesics with a profile of action different from that of aspirin-like drugs.

2.4.9.2 Activity against Leukemia and Malignancy

The essential oil from a lemongrass variety of *Cymbopogon flexuosus* (CFO) and its major chemical constituent sesquiterpene isointermedeol (ISO) were investigated for their ability to induce apoptosis in human leukemia HL-60 cells because dysregulation of apoptosis is the hallmark of cancer cells. CFO and ISO inhibited cell proliferation with 48 h IC_{50} of ~30 and 20 μg/mL, respectively (Kumar et al. 2008). Two active compounds, *d*-limonene and geraniol, were isolated by glutathione-*S*-transferase (GST) assay and fractionation of lemongrass (*C. citratus*) oil. These were tested for their capacity to induce activity of the detoxifying enzyme GST in several tissues of the female A/J mice. *d*-Limonene increased GST activity two- to threefold than controls in the mouse liver and the mucosa of the small and large intestines. Geraniol showed high GST-inducing activity in the mucosa of the small and large intestines, which was about 2.5-fold greater than controls. Induction of increased GST activity, which is believed to be a major mechanism for chemical carcinogen detoxification, has been recognized as one of the characteristics of the action of anticarcinogens (Zheng et al. 1993).

The essential oil from *C. citratus* and its isolated principal citral have been tested for cytotoxicity against P388 leukemia cells. The cytotoxicity of citral, IC_{50} against P388 mouse leukemia cells was 71 μg/mL (Dubey et al. 1997). In another experiment, IC_{50} of *C. citratus* oil in P388 leukemia cells was found to be 5.7 μg/mL (Dubey et al. 1997).

2.4.9.3 Activation of Male Hormones

The antimale sex hormone agent is a 5-reductase inhibitor that converts testosterone to active dihydrotestosterone. This agent is extracted from the leaves, stems, rhizomes, roots, or whole plant of *C. flexuosus*. The composition containing the antimale sex hormone agent is especially useful as hair growth stimulants (Kisaki et al. 1998).

2.4.9.4 Activity against Worms

The essential oil from *C. martinii* var. *motia*, in varying concentrations (0–0.4%), have shown to have good to excellent anthelmintic activity against tapeworms, round worms, and earthworms in in vitro tests (Sangwan et al. 1985). The activity exceeded that of the drug piperazine phosphate (Siddiqui and Garg 1990). Helminthiasis is one of the most important groups of parasitic diseases in the Indo–Pakistan subcontinent, resulting in heavy production losses in livestock. A wide variety of anthelmintics is used for the treatment of helminths in animals. However, the development of resistance in helminths against commonly used anthelmintics has always been a challenge faced by the animal health care professionals. Therefore, exploitation of the anthelmintic potential of plants indigenous to the Indo–Pakistan subcontinent is an area of research interest (Akhtar et al. 2000).

2.4.9.5 Lowering Blood Sugar

A study was designed to investigate the hypoglycemic and hypolipidemic effects of the single daily oral dosing of 125–500 mg/kg of fresh leaf aqueous extract of *Cymbopogon citratus* Stapf (CCi) in normal male Wistar rats for 42 days (Adeneye and Agbaje 2007). In another report, *Cymbopogon proximus* herb was assessed by Eskandar and Won Jun (1995) for hypoglycemic and hyperinsulinemic action on alloxan diabetic rats. A dose of 1.5 mL of herb suspension/100 g B weight was orally administered to the rats for intervals of 4, 8, and 16 days. The results revealed that considerable hypoglycemic effect was exerted after 16 days. The level of serum insulin was also increased in diabetic rats.

2.4.9.6 Potential to Repel Mosquitoes and Kill the Larvae

Ointment and cream formulations of lemongrass oil in different classes of base and the oil in liquid paraffin solution have been evaluated for mosquito repellency in a topical application. Mosquito repellency was tested by determining the bite deterrence of product samples applied on an experimental bird's skin against a 2-day-starved culture of *Aedes aegypti* L. mosquitoes. The 1% v/v solution and 15% v/w cream and ointment preparations of the oil exhibited ≥50% repellency lasting 2–3 h, which may be attributed to citral, a major oil constituent. This activity was comparable to that of a commercial mosquito repellent. Base properties of the lemongrass oil formulations influenced their effectiveness. The oil demonstrated efficacy from the different bases in the order of hydrophilic base > emulsion base > oleaginous base (Oyedele et al. 2002). Ansari and Razdan (1995) studied various essential oils for their mosquito-repellent activity. Essential oils from *C. martinii* var. *sofia*, *C. citratus*, and *C. nardus* were found as effective as chemical base oil. The percent protection against *Culex quinquefasciatus* ranged third 95%–96%. Fractional distillation of Ceylon citronella (*C. nardus*) oil yielded 13 fractions. Monoterpene hydrocarbon fractions were highly lethal to late third instar *Culex quinquefasciatus* larvae. The results suggested that myrcene was responsible for this activity. Elemol and/or methyl iso-eugenol were identified as active larvicidal principles in the latter fractions. The residue after the fractional distillation also possessed considerable larvicidal activity (Ranaweera and Dayananda 1997).

2.4.9.7 Activity to Reduce Edema

The species of *Cymbopogon giganteus* is widely used in traditional medicine against several diseases. This study reports the inhibitory effect produced by the chemical constituents of the essential oil from leaves of *C. giganteus* of Benin in vitro on 5-lipoxygenase, and has been found useful as an antiinflammatory agent. The scientists assayed the antiradical scavenging activity of the sample by the 1,1-diphenyl-2-picrylhydrazyl (DPPH) method (Alitonou et al. 2006). Earlier, the antiinflammatory activity of the oil was attributed to the inhibition of the prostaglandin pathway (Krishnamoorthy et al. 1998). Oral administration of essential oils extracted from *C. martinii* leaves produced dose-dependent inhibition of carrageenan-induced paw edema in experimental male albino rats.

2.4.9.8 Potential to Control Aging Process

Leaves from cultivated *Cymbopogon citratus* were extracted with methanol, 80% aqueous ethanol, and water (infusion and decoction), and the extracts were assessed for their antiradical capacity by 2,2-diphenyl-1-picrylhydrazyl (DPPH') assay; the infusion extract exhibited the strongest activity. Tannins, phenolic acids (caffeic and *p*-coumaric acid derivatives), and flavone glycosides (apigenin and luteolin derivatives) were identified in three different fractions obtained from an essential-oil-free infusion, and a correlation with their scavenger capacity for reactive oxygen species was studied. The tannin and flavonoid fractions were the most active against species involved in oxidative damage processes. In the flavonoid fraction, representing 6.1% of the extract, 13 compounds (*O*- and *C*-glycosylflavones) were tentatively identified by high-performance liquid chromatography, coupled to photodiode-array and electrospray ionization mass spectrometry detectors (HPLC–PDA–ESI/MS), nine of which were identified for the first time in this plant, all of them being *C*-glycosylflavones (mono-*C*-, di-*C*- and *O*,*C*-diglycosylflavones). The potential beneficial and protective value of the identified polyphenols for human health is discussed (Figueirinha et al. 2008). Hyalurodinase inhibitors are extracted from *C. nardus* and some other plants for the preparation of cosmetics. Hyalurodinase inhibitors prevented the degradation of aging-related hyaluronic acid (Namba et al. 1995). Antioxidant activity of *C. schoenanthus* was measured by DPPH assay. The results ranged from 36.0% to 73.8% (2 μL of essential oil per milliliter of test solution). The antioxidant activity was also assayed using the β-carotene–linoleic acid bleaching method. The best results ($IC_{50} = 0.47 \pm 0.04$ mg mL^{-1}) were obtained with the fresh leaves of plants collected in the desert region. The greatest acetylcholinesterase inhibitory activity ($IC_{50} = 0.26 \pm 0.03$ mg mL^{-1}) was exhibited by the essential oil of the fresh leaves from the mountain region (Khadri et al. 2008).

2.4.9.9 Activity against Pests

The susceptibility of *Spodoptera litura* larvae to different concentrations (0.2%–0.8%) of the essential oil of *C. citratus* has been studied in relation to host plant resistance in peanut. Field trials indicated that larvae developing on the most susceptible variety had the lowest mortality due to biopesticide lemongrass oil. The larvae treated with the oil before feeding showed significant higher mortality on the diet containing resistant pods than on that containing susceptible pods (Rajapakse and Jayasena 1991). The essential oils from *C. martinii* var. *motia*, *C. flexuosus*, and *C. winterianus* are reported to possess insect-repellant, nematicidal, and insect-attractant properties (Ahmad et al. 1993).

A number of essential oils including citronella (*C. winterianus*) and palmarosa (*C. martinii*) showed pesticidal activity against the stored grain insect *Tribolium castaneum* (Naik et al. 1995). Within a storage period of 10 days, samples of maize and cowpea treated with lemongrass powder and essential oil showed no physical deterioration (Adegoke and Odesola 1996).

2.4.9.10 Activity against Microbes

The essential oil of *C. martinii* var. *motia* and its different dilutions have shown significant antibacterial activity against *Staphylococcus aureus*, *S. pyagens*, *E. coli*, and *Corynebacterium ovis* (Gangrade et al. 1990).

The essential oils of lemongrass (*C. citratus*), palmarosa (*C. martinii* var. *motia*), and khavi grass (*C. jwarancusa*) were tested for antibacterial property against *E. coli*, *S. aureus*, *Shigella flexneri*, and *Salmonella typhi*. Lemongrass oil was the most active and caused complete inhibition of *S. aureus* at less than 400 ppm. Palmarosa was more active against *S. flexneri* and *S. typhi*, whereas khavi grass showed less activity than lemongrass and palmarosa. The activity of the oil might be attributed to the components citral, geraniol, and piperitone (Syeed et al. 1990).

The essential oil from the leaves of *C. martinii* was tested for toxicity against *Fusarium oxysporum*. Toxicity was the strongest as the mycelial growth of the pathogen was inhibited. Fungitoxicity remained unchanged in temperature treatment after a long storage period. It had no effect on the *Cajanus cajan* plant (Shrivastava et al. 1990). The essential oil of *C. nardus* exhibited a very good order of antifungal activity (Lemos et al. 1994).

The effect of auto-oxidation of lemongrass (*C. citratus*) oil on its antibacterial activity was studied. Using the active oxygen method, the oil was found to undergo rapid oxidation under accelerated test conditions. The oxidized oil samples were found to have reduced activity against bacteria. The activity was completely lost in extensively oxidized oil samples. Inclusion of antioxidants in the oil samples reduced the rate of oxidation and enhanced the antibacterial activity of the oil. The effects of the antioxidants were concentration-dependent, and at their effective concentration oxidation was completely prevented for the period of the test (Orafidiya 1993).

The inhibitory effect of lemongrass (*C. flexuosus*) essential oil isolated from local and Thai cultivars against pathogenic fungi were reviewed. No significant difference was found. The oil completely inhibited the growth of *Monilia sitophilia*, *Penicillium digilotum*, *Aspergillus parasiticus*, *A. niger*, and *A. fungis* (Shadab-Qamar et al. 1992).

The minimum inhibitory concentration (MIC) and minimum lethal concentration (MLC) of the oil obtained from lemongrass *(C. citratus),* and citral against 35 clinical isolates of four dermatophytes were determined by the agar dilution method. The MIC and MLC of lemongrass oil were found to be higher than those of citral. The mode of action of lemongrass oil and citral were proven to be fungicidal. A comparative study of efficacy of cream containing four different concentrations of lemongrass oil was performed in vitro by hole diffusion assay. The 2.5% lemongrass oil was demonstrated to be the minimum concentration for the preparation of an antifungal cream for subsequent clinical study (Wannissorn et al. 1996).

The essential oil from several plants, including lemongrass (*C. citratus*), were tested for antimicrobial activities against *Paenibacillus* larvae, the causal agent of American Foul Broad (AFB) disease of honeybees. Trials for determining the MIC of the oil revealed that lemongrass and thyme were most effective. The results indicated that lemongrass and thyme oils could be used as effective inhibitors of AFB in honeybee colonies (Allpi et al. 1996).

Thirteen essential oils of African origin (mostly from the Cameroon) were correlated with the antimicrobial activities of the oil toward six microbial strains: *S. aureus*, *B. coli*, *Proteus mirabilis*, *Klebsiella pneumoniae*, *Candida albicans*, and *Pseudomonas aeruginosa*. The oil of *C. citratus* displayed noteworthy antifungal and antibacterial properties (Chalchat et al. 1997). Fresh oil as well as 2-, 7-, and 12-year-old oils of a local variety of lemongrass (*C. citratus*) were distilled and redistilled, and tested against *B. coli*, *S. aureus*, *Shigella flexneri*, *Salmonella typhi*, Para-A, and *Klebsiella pneumonae*. The oil that was kept for two years exhibited, after redistillation, maximum activity due to its high citral content. *S. flexneri* and *S. typhi* were inhibited effectively at low doses of the oil. The inhibition appeared to be mostly by the citral content of the oil (Syeed et al. 1990).

During the screening of some aromatic plants for fungitoxicity of their volatile oils, *C. pendulus* var. *Praman* exhibited the strongest activity, completely inhibiting the mycelial growth of the test organisms *Microsporuin gypseum* and *Trichophyton mentagrophytes*. The volatile oil distilled from fresh leaves was found to be fungicidal at its MIC of 200 μg/mL, inhibiting heavy inocula of the test fungi. During the testing of its fungitoxic spectrum, it also inhibited mycelial growth of three other fungi and was found to be more active than some commercial drugs tested (Pandey et al. 1996). Essential oils obtained from the leaves of 29 medicinal plants commonly used in Brazil were screened against 13 different *E. coli* serotypes. The oils were obtained by water distillation using a Clevenger-type system, and their MIC was determined by the microdilution method. Essential oil from *C. martinii* exhibited a broad inhibition spectrum, presenting strong activity (MIC between 100 and 500 μg/mL) against 10 out of 13 *E. coli* serotypes: three enterotoxigenic, two enteropatho-genic, three enteroinvasive, and two shiga-toxin producers. *C. winterianus* strongly inhibited two enterotoxigenic, one enteropathogenic, one enteroinvasive, and one shiga-toxin producer serotypes. *Aloysia triphylla* also shows good potential to kill *E. coli* with moderate-to-strong inhibition. Other essential oils showed antimicrobial properties, although with a more restricted action against the serotypes studied. Chemical analysis of *C. martinii* essential oil performed by GC and GC–MS showed the presence of compounds with known antimicrobial activity, including geraniol, geranyl acetate, and *trans*-cariophyllene, which, tested separately, indicated geraniol as antimicrobial active compound. The significant antibacterial activity of *C. martinii* oil suggests that it could serve as a source for compounds with therapeutic potential (Duarte et al. 2006).

An essential oil from a lemongrass variety of *C. flexuosus* (CFO) and its major chemical con-stituent sesquiterpene isointermedeol (ISO) were investigated for their ability to induce apoptosis in human leukemia HL-60 cells because dysregulation of apoptosis is the hallmark of cancer cells. CFO and ISO inhibited cell proliferation with 48 h IC of ~30 and 20 μg/mL, respectively. Both induced concentration-dependent strong and early apoptosis as measured by various endpoints, for example, annexin V binding, DNA laddering, apoptotic bodies formation, and an increase in hypodiploid sub-G0 DNA content during the early 6 h period of study. This could be because of early surge in reactive oxygen species (ROS) formation with concurrent loss of mitochondrial membrane potential observed. Both CFO and ISO activated apical death receptors TNFR1, DR4, and caspase-8 activity. Simultaneously, both increased the expression of mitochondrial cytochrome *c* protein with its concomitant release to cytosol leading to caspase-9 activation, suggesting thereby the involvement of both the intrinsic and extrinsic pathways of apoptosis. Further, Bax transloca-tion and decrease in nuclear NF-κB expression predict multitarget effects of the essential oil and ISO while both appeared to follow similar signaling apoptosis pathways. The easy and abundant availability of the oil combined with its suggested mechanism of cytotoxicity makes CFO highly useful in the development of anticancer therapeutics (Kumar et al. 2008). The essential oil from a lemon grass variety of *Cymbopogon flexuosus* was studied for its in vitro cytotoxicity against twelve human cancer cell lines. The in vivo anticancer activity of the oil was also studied using both solid and ascitic Ehrlich and Sarcoma-180 tumor models in mice. In addition, the morphological changes in tumor cells were studied to ascertain the mechanism of cell death. The in vitro cytotoxicity studies showed dose-dependent effects against various human cancer cell lines (Sharma et al. 2009).

C. winterianus (Poaceae) is used for its analgesic, anxiolytic, and anticonvulsant properties in Brazilian folk medicine and these reports are aimed to perform phytochemical screening and to investigate the possible anticonvulsant effects of the essential oil from fresh leaves of *C. winterianus* in different models of epilepsy. The phytochemical analysis of the oil showed the presence of geraniol (40.06%), citronellal (27.44%), and citronellol (10.45%) as the main compounds. A behavioral screening demonstrated that the essential oil (100, 200, and 400 mg/kg, ip) caused depressant activity on CNS. When administered concurrently (200 and 400 mg/kg, ip) it significantly reduced the number of animals that exhibited PTZ- and PIC-induced seizures in 50% of the experimental animals ($p < 0.05$). Additionally, EO (100, 200, and 400 mg/kg, ip) significantly increased ($p < 0.05$) the latencies of clonic seizures induced by STR. Our results demonstrated a possible anticonvulsant activity of the essential oil (Quintans-Júnior et al. 2008).

REFERENCES

Abdullah MA, Foda YH, Saleh M, Zaki MSA, Mostafa MM. 1975. Identification of the volatile constituents of the Egyptian lemongrass oil I. Gas chromatographic analysis. *Nahrung* **19:** 195–200 (C.A., 83, 48089 u).

Abegaz B, Yohannes PG, Deiter RK. 1983. Constituents of essential oil of Ethiopian *Cymbopogon citratus* Stapf. *J Nat Prod* **46:** 424–426.

Adegoke GO, Odesola BA. 1996. Storage of maize and cowpea and inhibition of microbial agents of biodeterioration using the powder and essential oil of lemongrass *(C. citratus)*. *Int Biodeterioration Biodegrad* **37:** 81–84.

Adeleke AK, Adebola OO, Adeolu OA. 2001. Volatile leaf oil constituents of *Cymbopogon citratus* (DC) Stapf. *Flav Frag J* **16(5):** 377, 378.

Adeneye AA, Agbaje EO. 2007. Hypoglycemic and hypolipidemic effects of fresh leaf aqueous extract of *Cymbopogon citratus Stapf.* in rats. *J Ethnopharmacol* **112(3):** 440–444.

Ahmad R, Sheikh V, Ahmad A, Ahmad M. 1993. Essential oils as insect and animal attractant and repellants. *Hamdard Med* **36:** 99–105.

Ahmed ZF, Rizk AM, Hammouda FM. 1970. Postep Dziedzinie Leku Rosi. *Proc Ref Dosw Wygloszone Symp* 20–23 (C.A.78: 94856-u).

Akhila A, Francis MJO, Banthorpe DV. 1980a. Biosynthesis of carvone in *Mentha spicata. Phytochemistry* **19:** 1433–1437.

Akhila A, Rani K, Thakur RS. 1990. Biosynthesis of artemisinic acid in *Artemisia annua. Phytochemistry* **29(7):** 2129–2132.

Akhila A, Sharma PK, Thakur RS. 1988a. 1,2-Hydrogen shifts in the formation of Sesquithujane skeleton in ginger rhizomes. *Phytochemistry* **27(11):** 3471–3473.

Akhila A, Sharma PK, Thakur RS. 1987a. 1,2-Hydrogen shifts during the biosynthesis of Patchoulenes in *Pogostemon cablin. Phytochemistry* **26(10):** 2705–2707.

Akhila A, Sharma PK, Thakur RS. 1988b. Biosynthetic relationship in Patchouli alcohol, Seychellene and Cycloseychellene in *Pogostemon cablin. Phytochemistry* **27(7):** 2105–2108.

Akhila A. 1980b. Banthorpe DV: Biosynthetic origin of *gem*-methyls of geraniol. *Phytochemistry* **19:** 1429–1430.

Akhila A. 1980c. Biosynthesis of the skeleton of Pulegone in *Mentha pulegium. Pflanzenphysiol Zeitschrift fur* **99(3):** 277–282.

Akhila A. 1985. Biosynthetic relationship of citral-*trans* and citral-*cis* in *Cymbopogon flexuosus* (Lemongrass). *Phytochemistry* **24(11):** 2585–2587.

Akhila A. 1986. Biosynthesis of monoterpenes in *Cymbopogon winterianus. Phytochemistry* **25(2):** 421–424.

Akhila A. 1987b. Biosynthesis of menthol and related monoterpenes in *Mentha arvensis. J Plant Physiol* **126(4/5):** 379–386.

Akhtar MS, Iqbal Z, Khan MN, Lateef M. 2000. Anthelmintic activity of medicinal plants with particular reference to their use in animals in the Indo–Pakistan subcontinent. *Small Ruminant Res* **38(2):** 99–107.

Alam K, Agua T, Manen H, Taie R, Rao KS, Burrows I, Huber ME, Rali T. 1994. Preliminary screening of sea weeds, sea grass for antimicrobial and antifungal activity. *J Pharmacognosy* **32:** 169–399.

Alitonou GA, Avlessi F, Sohounhloue DK, Agnaniet H, Bessiere JM, Menut C. 2006. Investigations on the essential oil of *Cymbopogon giganteus* from Benin for its potential use as an anti-inflammatory agent. *Int J Aromatherapy* **16(1):** 37–41.

Allpi AM, Ringuilet JA, Ceremele El, Re MS, Henning CP. 1996. Antimicrobial activity of some essential oil against *Paenibacillus* larvae, the causal agent of American foulbrood disease. *J Herbs Spice Med Plants* **4:** 9–16.

Anonymous. 1950. *The Wealth of India,* Vol. II, pp. 411–419, CSIR, New Delhi.

Anonymous. 1952. Indian Standard Specifications, Indian Standards Institutions, New Delhi.

Anonymous. 1958. Essential Oils and Aromatic Chemicals, Symposium at Dehradun, CSIR, New Delhi.

Anonymous. 1973. Application of gas liquid chromatography to the analysis of essential oil III. Determination of geraniol in oils of citronella. *Analyst* **98:** 823–829.

Anonymous. 1978. Insect repellent complex. *Pat Ger Offen* 2, 752, 140, 24 May 1978 (C.A. **89:** 175043 f).

Anonymous. 1980. Terpenoids from palmarosa grass *(Cymbopogon martinii* var. *motia)* R.A K. Riechst. *Aromen Kosmet* **30:** 20–22 (C.A. 92: 160599e).

Ansari MA, Razdan RK. 1995. Relative efficacy of various oils in repelling mosquitoes. *Indian J Mal* **32:** 104–111.

Ansari SH, Ali M, Siddiqui AA. 1996. Evaluation of chemical constituents and trade potential of *Cymbopogon citratus* (Lemongrass). *Hamdard Med* **39:** 55–59.

Ansari SH, Quadry JS. 1987. TLC and GLC studies on Khawi grass oil. *Indian J Nat Prod* **3:** 10–12.

Arctander S. 1960. *Perfume and Flavour Materials of Natural Origin.* Allured Publishing, Elizabeth, New Jersey.

Arora R, Pandey GN. 1977. The application of essential oils and their isolates for blue mold decay control in *Citrus reticulata* Blanco. *J Food Sci Technol* **14:** 14–16.

Atal CK, Bradu BL. 1976a. Search for aroma chemicals of industrial value from genus *Cymbopogon*. Part II. *C. pendulus* (Nees ex Steud.) Wats. (Jammu lemongrass)—a new superior source of citral. *Indian J Pharmacol* **38:** 61–63.

Atal CK, Bradu BL. 1976b. Search for aroma chemicals of industrial value from genus *Cymbopogon* Pt. III: *Cymbopogon pendulus* (Nees ex Steud.) Wats. (Jammu lemongrass), a new superior source of citral. *Indian Perfumer* **20:** 29–33.

Atal CK, Bradu BL. 1976c. Search for aroma chemicals of industrial value from genus *Cymbopogon* Pt. IV. Chandi and Kolar grasses as source of methyl eugenol. *Indian Perfumer* **20:** 35–38.

Atal CK, Bradu BL. 1976d. Search for aroma chemicals of industrial value from *Cymbopogon* Part V. Chandi and Kolar grasses as source of methyl eugenol. *Indian J Pharm* **38:** 63, 64.

Baiswara RB, Nair KNG, Mathew TV. 1976a. Detection of adulteration of palmarosa oil in ginger grass oil by thin layer chromatography. *Res Ind* **21:** 37–39.

Baiswara RB, Nair KNG, Mathew TV. 1976b. Detection of ginger grass oils in lemongrass oil by thin layer chromatography. *Res Ind* **21:** 39, 40.

Balyan SS, Shahi AK, Choudhury SN, Singh A, Atal CK. 1979. Performance of five strains of genus *Cymbopogon* at Jammu. *Indian Perfumer* **23:** 121–124.

Baruah P, Mishra BP, Pathak MG, Ghosh AC. 1995. Dynamics of essential oil of *Cymbopogon citrates* (D.C.) Stapf. under rust disease indices. *J Essen Oil Res* **7:** 337–338.

Baslas RK, Baslas KK. 1968. Essential oil potentialities of Kumaon. *Indian Perfumer* **12:** 3–6.

Baslas RK. 1970. Chemistry of Indian essential oils. Part IX. *Flav Industry* **1:** 475–478.

Beauchamp PS, Dev V, Docter DR, Ehsani R, Vita G, Melkani AB, Mathela CS, Bottini AT. 1996. Comparative investigation of the sesquiterpenoids present in the leaf oil of *Cymbopogon distans* (Steud.) Wats. var. *loharkhet* and the root oil of *Cymbopogon jwarancusa* (Jones) Schult. *J Essen Oil Res* **8:** 117–121.

Beech DF. 1977. Growth and oil production of lemongrass (*C. citratus*) in the Ord Irrigation Area, Western Australia. *Aust J Exp Agric Anim Husb* **17:** 301–307.

Bhattacharya AK, Kaul PN, Rao BRR, Mallavarapu GR, Ramesh S. 1997. Interspecific and inter cultivar variations in the essential oil profiles of lemongrass. *J Essen Oil Res* **9:** 361–364.

Bhattacharya SC. 1970. Perfumery chemicals from indigenous raw materials. *J Indian Chem Soc* **47:** 307–313.

Bhaumik SK, Rao VS. 1978a. Studies on prehydrolysis sulfate pulp from *Cymbopogon* grass (oil extracted). *Indian Pulp Pap* **33:** 3–5.

Bhaumik SK, Rao VS. 1978b. Evaluation of citronella grass (waste) as a raw material for paper making. *Indian Pulp Pap* **33:** 17–19.

Blake O, Booth R, Carrigon D. 1993. The tanin content of herbal teas. *Br J Phytotherapy* **3:** 124–127.

Blanco MM, Costa C, Freire AO, Santos Jr. JG, Costa M. 2007. Neurobehavioral effect of essential oil of *Cymbopogon citratus* in mice. *Phytomedicine* **16**: 265–270.

Boelens MH. 1994. Sensory and chemical evaluation of tropical grass oils. *Perfum Flavor* **19**: 29–45.

Bor NL. 1954. The genus *Cymbopogon* Spreng. In India, Burma and Ceylon Part II. *J Bombay Nat Hist Soc* **52**: 149–183.

Borovik VN, Kuravskaya IM. 1977. Flammability and explosiveness of some synthetic perfumes and inter-mediate products. *Maslo-Zhir Prom-st* **10**: 34, 35 (C.A.89: 48794-b).

Boti JB, Muselli A, Tomi F, Koukoua G, N'Guessan TY, Costa J, Casanova J. 2006. Combined analysis of *Cymbopogon giganteus* Chiov. leaf oil from Ivory Coast by GC/RI, GC/MS and ^{13}C-NMR *C.R. Chimie* **9**: 164–168.

Bottini AT, Dev V, Garfagnoli DJ, David J, Hope H, Joshi P, Lohani H, Mathela CS, Nelson TB. 1987. Isolation and crystal structure of a novel dihemiacetal bismonoterpenoid from *Cymbopogon martinii*. *Phytochemistry* **26**: 2301, 2302.

Bouvier F, d'Harlingue A, Suire C, Backhaus RA, Camara B. 1998. Dedicated roles of plastid transketolases during the early onset of isoprenoid biogenesis in pepper fruits. *Plant Physiol* **117**: 1423–1431.

Brazil S, Gilberto A de A, Bauer L. 1971. Essential oil of *Cymbopogon citratus*. *Rio Grande do Sul* **52**: 193–196 (C.A., **76**, 131371-p).

Bruns K, Heinrich B, Pagel I. 1981. Citronellol. *Untersuchung von Handels und Hybridolen Verschiedener Provenienz in Vorkommen und Analytic Atherischer Ole, Band 2* (Kubeczkaeds KH, Ed.), G. Thieme Verlag, Stuttgart, Germany.

Carlini EA, Contar JDP, Silva-Filho AR, Da Silveira-Filho NG, Frochtengarten ML, Bueno OFA. 1986. Pharmacology of lemongrass (*Cymbopogon citratus* Stapf.). I. Effects of teas prepared from the leaves on laboratory animals. *J Ethnopharmacol* **17**(1): 37–64.

Chakrabarti MM, Ghosh SK. 1974. Essential oils from plant grown in West Bengal. *Indian Perfumer* **18**: 35–39.

Chalchat JC, Garry RP, Menut C, Lonaly G, Malhuret R, Chopineau J. 1997. Correlation between chemical composition and antimicrobial activity. VI. Activity of some African essential oils. *J Essen Oil Res* **9**: 67–75.

Chauhan YS, Singh KK, Ganguly D. 1976. Improvement of Java citronella *(C. winterianus* Jowitt) by chemical mutagenesis. *Indian Perfumer* **20**: 73–77.

Cherian S, Chattattu GJ, Viswanathan TV. 1993. Chemical composition of lemongrass varieties. *Indian Perfumer* **37**: 77–80.

Chiang HC, Ma YC, Huang KF. 1981. Quantitative analysis of citronella oil and lemongrass oil by hydrogen NMR spectrometry. *Taiwan Yao Huesch Ta Chin* **33**: 95–103 (C.A. 97, 11630-j).

Chicopharma NV. 1970. Pat. Neth. No. 6909123 (C.A. 7479503-a).

Chisowa BH. 1997. Chemical composition of flower and leaf oils of *Cymbopogon densiflorus* Stapf from Zambia. *J Essen Oil Res* **9**: 469, 470.

Chopra RN, Nayar SL, Chopra IC. 1956. Glossary of Indian Medicinal Plants. CSIR, New Delhi, p. 87.

Choudhary DK, Kaul BL. 1979. Radiation-induced methyl eugenol deficient mutant of *Cymbopogon flexuosus* (Nees ex Steud.) Wats. *Proc Indian Acad Sci* **Sect 88 B**: 225–228.

Choudhary SN, Leclercq PA. 1995. Essential oil of *Cymbopogon khasianus* (Munro ex Hack.) Bor from north eastern India. *J Essen Oil Res* **7**: 555, 556.

Ciaramello D, Azzini A, Pinto AJD, Guilherme M, Donalisio R. 1972. Use of citronella, lemongrass, palmarosa and vetiver leaves for the production of cellulose and paper. *An Acad Bras Cienc* **44**(Suppl.): 430–441. (C.A. **83**, 8 1640-x).

Crawford M, Hanson AW, Koker MES. 1975. Structure of *Cymbopogon*, a novel triterpenoid from lemongrass. *Tetrahedron Lett* **35**: 3099–3102.

Cronquits A. 1980. Chemistry in plant taxonomy, In *Chemosystematics Principles and Practice*. (Bisby FA, Vaughan JC, and Wright CA, Eds), Academic Press, London. pp. 1–27.

Croteau R, Felton M, Karp F, Kjonaas R. 1981. Relationship of camphor biosynthesis to leaf development in sage (*Salvia officinalis*). *Plant Physiol* **67**(4): 820–824.

Croteau R, Davis EM. 2005. (–)-Menthol biosynthesis and molecular genetics. *Naturwissenschaften* **92**: 562–567.

Crowell PL. 1999. Prevention and therapy of cancer by dietary monoterpenes. *J Nutr* **129**: 775–778.

Dawidar AM, Esmiraly ST, Abdel MM. 1990. Sequiterpenes from *Cymbopogon schoenanthus*. *Pharmazie* **45**: 296, 297.

De Martinez MV. 1977. Adaptation of the determination of citral with hydroxyl aminehydrochloride to lemongrass oil. *Arch Bioquim Quim Farm* **20:** 51–56 (C.A., 91,181225-j).

De Sylva MG. 1959. Lemongrass oil from Ceylon. *Mfg Chemist* **30:** 415, 416.

Dev Kumar C, Narayana MR, Khan MNA. 1977. Synthetic products from oil of citronella. *Indian Perfumer* **22:** 139–145.

Dev V, Melander D, Yee JT. 1988. Terpenoid composition of the essential oil from *Cymbopogon jwarancusa*. *Dev Food Sci Flav Frag* **18:** 317–321.

Dhar AK, Lattoo S. 1985. Essential oil content in *Cymbopogon jwarancusa* after cutting. *Pafai J* **7:** 28–31.

Dhar AK, Thapa RK, Atal CK. 1981. Variability in yield and composition of essential oil in *Cymbopogon jwarancusa*. *Planta Med* **41:** 386–388.

Dhar RS, Dhar AK, 1997. Ontogenetic variation in essential oil concentration and its constituents in five genotypes of *Cymbopogon jwarancusa* (Jones) Schult. *J Essen Oil Res* **9:** 433–439.

Dikshit A, Hussain A. 1984. Anti-fungal action of some essential oils against animal pathogens. *Fitoterapia* **55:** 171–176.

Duarte MCT, Leme EE, Delarmelina C, Soares AA, Figueira GM, Sartoratto A. 2006. Activity of essential oils from Brazilian medicinal plants on *Escherichia coli. J Ethnopharmacol* **111(2):** 197–201.

Dubey NK, Kishore N, Varma J, Lee SY 1997. Cytotoxicity of the essential oil of *Cymbopogon citratus* and *Ocimum gratissimum. Indian J Pharm Sci* **59:** 263–264.

Dubey NK, Takeya K, Itakawa H. 1997. Citral: A cytotoxic principle isolated from the essential oil of *Cymbopogon citratus* against 1388 leukaemia cells. *Curr Sci* **73:** 22–24.

Dunyan Xue, Maosen Song, Ning Chen, Yaozu Chen. 1992. Chemical ingredients of the essential oil of *Cymbopogon distans. Gaoden Xoxiao Huaxue X. Xuebao* **13:** 1551, 1552.

Eisenreich W, Rohdich F, Bacher A. 2001. Deoxyxylulose phosphate pathway to terpenoids. *Trends Plant Sci* **6:** 78–84.

Eisenreich W, Sagner S, Zenk MH, Bacher A. 1997. Monoterpenoid essential oils are not of mevalonoid origin. *Tetrahedron Lett* **38:** 3889–3892.

Eisenreich W, Schwarz M, Cartayrade A, Arigoni D, Zenk MH, Bacher A. 1998. The deoxyxylulose phosphate pathway of terpenoid biosynthesis in plants and microorganisms. *Chem Biol* **5:** R221–R223.

Elagamal MH, Wolff P. 1987. A further contribution to the sesquiterpenoid constituents of *Cymbopogon proximus. Planta Med* **53:** 293, 294.

El Tawil BAH, El Beih FK. 1982. Study of volatile oil of Saudi *Cymbopogon citratus* D.C. and *Cymbopogon schoenanthus* (L). *J Chin Chem Soc* (Taipei) **30:** 281, 282 (C.A. 100,39433).

Eskandar EF, Won Jun H. 1995. Hypoglycaemic and hyperinsulinenic effect of some Egyptian herbs used for the treatment of diabetes mellitus (type II) in rats. *Egypt J Pharm Sci* **36:** 334–341.

Esmort H, David RH, Dudley IF. 1998. Volatile constituents of the essential oil of *Cymbopogon citratus* Stapf. grown in Zambia. *Flav Frag J* **13(1):** 29, 30.

Evans FE, Miller DW, Cairus T, Baddeley GV, Wen KB. 1982. ^{13}C NMR of naturally occurring substances, Part 74. Structure analysis of proximadiol (Cryptomeridiol) by ^{13}C NMR spectroscopy. *Phytochemistry* **21:** 855–858.

Figueirinha A, Paranhos A, Pérez-Alonso JJ, Santos-Buelga C, Batista MT. 2008. *Cymbopogon citratus* leaves: Characterization of flavonoids by HPLC-PDA-ESI/MS/MS and an approach to their potential as a source of bioactive polyphenols. *Food Chem* **110(3):** 718–728.

Foda YH, Abdullah MA, Zaki MS, Mostafa MM. 1975. Identification of the volatile constituents of the Egyptian lemongrass oil. II. IR Spectroscopy. *Nahrung* **19:** 395–400 (C.A.83, 1683 17-w).

Formacek K, Kubeczka KH. 1982. *Essential Oils Analysis by Capillary Chromatography and ^{13}C NMR Spectroscopy.* John Wiley & Sons, New York.

Fuji T, Furukawa S, Suzuki S. 1972. Compounded perfumes for toilet goods. Non-irritative compounded perfumes for soaps. *Yukagaku* **21:** 904–908 (C.A. 78, 75786-e).

Gagrade SK, Sri Vastava RD, Sharma OP, Maghe MN, Trivedi KC. 1990. Evaluation of some essential oil for antibacterial properties. *Indian Perfumer* **34:** 204–208.

Ganguly P, Singh KK, Bhagat SD, Upadhyay DN, Chauhan YS, Gupta NK, Singh HS. 1979. "RRLJOR-3-1970" an improved strain of Java citronella *(Cymbopogon martinii). Phytochemistry* **26:** 183–185.

Gaydou BM, Raudriamiharisoa RP. 1987. Hydrocarbon from the essential oil of *Cymbopogon winterianus* Jowitt. *Indian Perfumer* **23:** 107–111.

Gaydou BM, Raudriamiharisoa RP. 1987. Composition of palmarosa *(cymbopogon martinii)* essential from Madagascar. *J Agric Food Chem* **35(1):** 62–66.

Gaydou EM, Raudriamiharisoa RP. 1986. Hydrocarbons from the essential oil of *Cymbopogon martinii*. *Phytochemistry* **26**(1): 183–185.

Gildemeister B, Hoffmann F. 1956. *Die Artherischen Ole*, Vol. IV. Akademie Verlag, Berlin, pp. 307–419.

Gonzalo VF, Villarrubia MM. 1974. Essential oil of lemongrass in Tucuman, *Int Congr Essen Oils,* Allured Publishing, Oak Park, Ilinois. p. 110 (C.A.,84,49730-m).

Gonzalo VF, Villarrubia MM. 1973. Essential oil of lemongrass in Tucuman. *Arch Bioquim Quim Farm* **18**: 51–57.

Grace O, Yisak W, Ogunlana EO. 1984. Antibacterial constituents in the essential of *cymbopogon citratus*. (D.C) Sapf. *Journal of Ethnopharmacology* **12**: 279–286.

Guenther E. 1948. *The Essential Oils*. Vol. 1, Van Nostrand, London. pp. 232–236.

Guenther E. 1950. *The Essential Oils*. Vol. IV, Van Nostrand, London. pp. 3–153.

Guillaume KK, Honore KK, Isabella AG, Jacques H. 2006. Comparative effects of *cymbopogon shoenanthus* essential oil and piperitone on *collasabuchus maculatur* development. *Fitoterapia* 77(7–8): 506–510.

Gulati BC, Sadgopal 1972. Chemical examination of oil of *Cymbopogon nardus*. *Indian Oil Soap J* **37**: 305–310.

Gulati BC, Garg SN. 1976. Essential oil of *Cymbopogon pendulus* Wats. under the Tarai climate of Uttar Pradesh. *Indian J Pharm* **38**: 78, 79.

Gupta BK, Daniel P. 1982. Aromatic grasses of India and their utilization—a plea for further research. *Pafai J* Jan–March, 13–27.

Gupta BK, Jam N. 1978. Cultivation and utilization of genus *Cymbopogon* in India. *Indian Perfumer* **21**: 55–68.

Gupta BK. 1969. Studies in the genus *Cymbopogons*. *Perfum Essen Oil Res* **70B**: 241–247.

Gupta PC, Singh R, Singh K, Pradhan K. 1975. Chemical composition and *in vitro* matter digestibility of some important grasses. *Ann Arid Zone* **14**: 245–250 (C.A. 84, 178699-g).

Gupta YN, Chauhan PNS. 1970. Effect of ultraviolet radiation on essential oils. *Indian Perfumer* **14**: 63–66.

Gyane DD. 1976. Preservation of shea butter. *Drug Cosmet Ind* **18**: 36–38 and 138–140.

Handique AK, Singh HB. 1990. Antifungal action of lemongrass oil on some soil-borne plant pathogens. *Indian Perfumer* **34**: 232–234.

Han IK, Kim KL, Park SH, Kim BH, Ahn BH. 1971. Nutritive values of the native grasses and legumes in Korea V. Chemical determination of the mineral content of native herbage plants. *Han'guk Chuksanak'hoe Chi* **13**: 329–334 (C.A.81, 74938-f).

Hanson SW, Crawford M, Koker MES, Menezes FA. 1976. Cymbopogonol: A new triterpenoid from *C. citratus*. *Phytochemistry* **15**: 1074, 1075.

Herath HMW, Iruthayathas EE, Ormrod DP. 1979. Temperature effects on essential oil composition of citronella selection. *Econ Bot* **33**: 425–430.

Herz S, Wungsintaweekul J, Schuhr CA, Hecht S, Lüttgen H, Sagner S, Fellermeier M, Eisenreich W, Zenk MH, Bacher A, Rohdich F. 2000. Biosynthesis of terpenoids: YgbB protein converts 4-diphosphocytidyl-2C-methyl-d-erythritol 2-phosphate to 2C-methyl-d-erythritol 2,4-cyclodiphosphate. *Proc Natl Acad Sci USA* **97**: 2486–2490.

Idrissi A, Bellakhdar J, Canigueral S, Iglesian J, Vila R. 1993 Composition of the essential oil of lemongrass *(C. citratus* D.C.) Stapf. cultivated in Morocco. *Plantes Medicinales et Phytotherapie* **24**: 274–277.

Ille K. 1986. Gas chromatographic analysis of some essential oils used in perfumery. Mezhdunar 4th Kongar Efirnym Maslam. Naumenko PV (Ed.), Moscow. pp. 112–123.

Iruthayathas EE, Herath HMW, Wijesekara ROB, Jayawardene AL. 1997. Variations in the composition of oil in citronella. *J Nat Sci Coun Sri Lanka* **5**: 133–146.

Jirovetz L, Buchbauer G, Eller G, Ngassoum, MB, Maponmetsem PM. 2007. Composition and antimicrobial activity of *Cymbopogon giganteus* (Hochst.) Chiov. Essential flower, leaf and stem oils from Cameroon. *J Essen Oil Res* **19**: 485–489.

Jyrkit T. 1983. Composition of the essential oil of *C. flexuosus*. *J Chrom* **262**: 364–366.

Kalia NK, Sood RP, Padha CD, Jamwal RK, Chopra MM. 1980. Utilization of wild growing *Cymbopogon martinii* (var. *sofia*). *Indian Perfumer* **24**: 182–184.

Kamali HHEL, Hamza MA, Amir MYEL. 2005. Antibacterial activity of the essential oil from *Cymbopogon nervatus* inflorescence. *Fitoterapia* **76**(5): 446–449.

Kanjilal PB, Pathak MG, Singh RS, Ghosh AC. 1995. Volatile constituents of the essential oil of *Cymbopogon caesius* (Nees ex Hook. et Am) Stapf. *J Essen Oil Res* **7**: 437–449.

Kaul BL, Choudhary DK, Atal CK. 1977. Introduction and chemical evaluation of Burmese citronella in Jammu. *Indian J Pharmacol* **39**: 42, 43.

Ketoh GK, Koumaglo HK, Glitho IA, Huignard J. 2006. Comparative effects of *Cymbopogon schoenanthus* essential oil and piperitone on *Callosobruchus maculatus* development. *Fitoterapia* **77(7–8):** 506–510.

Khadri A, Serralheiro MLM, Nogueira JMF, Neffati M, Smiti S, Araújo MEM. 2008. Antioxidant and antiacetylcholinesterase activities of essential oils from *Cymbopogon schoenanthus* L. Spreng. Determination of chemical composition by GC-mass spectrometry and ¹³C NMR. *Food Chem* **109(3):** 630–637.

Kichiyoshi Koryo KK, Akira Sangy KK. 1981. An unpleasant odour composition for prevention of thinner sniffing. *Jpn Tokyo Koho JP Pat* 48538 (C.A., **96,** 117345-q).

Kisaki A, Matsuyama Y, Sakano T, Fujiwara N, Nakaguchi O. 1998. Antimale sex hormone agent and composition. *Jpn Kokai Tokyo JP* **17:** 486.

Kokate CK, Verma KC. 1971. Antimicrobial activity of volatile oils of *Cymbopogon nardus* and *Cymbopogon citratus*. *Sci Cult* **37:** 196–198.

Koketsu M, Moura LL, Magalhaes MT. 1976. Essential oil of *Cymbopogon densiflorus* Stapf. and *Tagetes minuta* L. grown in Brazil. *An Acad Bras Cienc* **48:** 743–746 (C.A.88, 141480-k).

Kreis P, Masandl A. 1994. Chiral compounds of essential oil. Part XVII. Simultaneous stereoanalysis of *Cymbopogon* oil constituents. *Flav Frag J* **9:** 257–260.

Krishnamoorthy G, Kavimani S, Loganthan C. 1998. Antiinflammatory activity of the essential oil of *Cymbopogon martinii*. *Indian J Pharm Sci* **60:** 114–116.

Krishnarajah SR, Ganesalingam VK, Senanayake UM. 1985. Repellancy and toxicity of some plant oils and their terpene components to *Stotroga cereallelia* (Olivier). *Trop Sci* **25:** 249–252.

Kulkarni RN, Mallavarapu GR, Bhaskaran K, Ramesh S, Kumar S. 1997. Essential oil composition of citronella-like variant of lemongrass. *J Essen Oil Res* **9:** 393–395.

Kumar A, Malik F, Bhushan S, Sethi VK, Shahi AK, Kaur J, Taneja SC, Ghulam N, Qazi GN, Singh J. 2007. An essential oil and its major constituent isointermedeol induce apoptosis by increased expression of mitochondrial cytochrome *c* and apical death receptors in human leukaemia HL-60 cells. *Chem-Biol Interact* **171(3):** 332–347.

Kuriakose KP. 1995. Genetic variability in East Indian lemongrass *(C. flexuosus* Stapf.). *Indian Perfumer* **39:** 76–83.

Kurian A, Nair BVG, Rajan KC. 1984. Effect of antioxidants on the preservation of citral content of lemongrass oil. *Indian Perfumer* **28:** 28–32.

Kusumov FY, Babaev RI. 1983. Components of the essential oil of *Cymbopogon citratus*. *Stapf Khim Prir Soedin* **1:** 108–109 (C.A., **98** 204188-a).

Kuzuyama T, Takagi M, Kaneda K, Dairi T, Seto H. 2000. Formation of 4-(cytidine-5′-diphospho)-2-*C*-methyl-*d*-erythritol from 2-*C*-methyl-*d*-erythritol 4-phosphate by 2-*C*-methyl-*d*-erythritol 4-phosphate cytidylyl-transferase, a new enzyme in the non-mevalonate pathway. *Tetrahedron Lett* **41:** 703–706.

Kuzuyama T, Takagi M, Kaneda K, Watanabe H, Dairi T, Seto H. 2000 Studies on the nonmevalonate pathway: conversion of 4-(cytidine-5′-diphospho)-2-*C*-methyl-*d*-erythritol to its 2-phospho derivative by 4-(cytidine-5′-diphospho)-2-*C*-methyl-*d*-erythritol kinase *Tetrahedron Lett* **41:** 2925–2928.

Lange BM, Croteau R. 1999. Isoprenoid biosynthesis via a mevalonate-independent pathway in plants: cloning and heterologous expression of 1-deoxy-d-xylulose-5-phosphate reductoisomerase from peppermint. *Arch Biochem Biophys* **365:** 170–174.

Lange BM, Wildung MR, McCaskill DG, Croteau R. 1998 A family of transketolases that directs isoprenoid biosynthesis via a mevalonate-independent pathway. *Proc Natl Acad Sci USA* **95:** 2100–2104.

Lawrence BM. 1991. Citronella oil: A monograph. In *Lawrence Review of Natural Products*, p. I.

Le KT, Chu PNS. 1976. Chemical composition of lemongrass essence (*Cymbopogon flexuosus* Stapf.) from Vietnam. *Tap Clii Hoc* **14:** 31–36.

Lemos TLG, Monte EJO, Matos FJA, Alencer JW, Craveiro AA, Barobasa RCSB, Lima BO. 1994. Chemical composition and antimicrobial activity of essential oils from Brazilian plants. *Fitoterapia* **63:** 226–228.

Lichtenthaler HK. 1999. *Annu Rev Plant Physiol Plant Mol Biol* **50:** 47–66.

Little DB, Croteau R. 1999. Biochemistry of essential oil terpenes: A thirty year overview. In *Flavour Chemistry*: Thirty Years of Progress (Hornstein I et al., Eds). Kluwer Academic/Plenum Press, New York. pp. 239–253.

Liu C, Zhang J, Yiao R, Gan L. 1981. Chemical studies on ten essential oils of *Cymbopogon* genus. *Huaxue Xuebao* 241–247 (C.A.98, 104358-in).

Liu M, Feng G. 1989. Effect of *Cymbopogon goeringii* (Steud.) A. Camus volatile oil on physiological properties of isolated guinea pig myocardium. *Zlionguo Zliongyao Zazlii* **14:** 160–622.

Lohani H, Bisht JC, Melkani AB, Mathela DK, Mathela CS, Dev V. 1986. Chemical composition of essential oil of *Cymbopogon stracheyi* (Hook f.) Riaz and Jani. *Indian Perfumer* **30:** 447–452.

Lorenzetti BB, Souza GEP, Sarti SJ, Santos ED, Frreira SH. 1991. Myrcene mimics the peripheral analgesic activity of lemongrass tea. *J Ethnopharmacol* **34:** 43–48.

Lorenzo D, Dellacassa E, Atti-Serafini L, Santos AC, Frizzo C, Paroul N, Moyna P, Mondello L, Dugo G. 2000. Composition and stereoanalysis of *Cymbopogon winterianus* Jowitt. oil from Southern Brazil. *Flav Frag J* **15(3):** 177–181.

Lucius G, Adler B. 1971. *Mezhdunar Kangr. Effirnym Maslam (Mater)* 4th, **1:** 211–214. Naumenko PV, Ed., "*Pishcheuaya, Promyshlennot,*" Moscow, USSR (C.A. 78,15 1557-y).

Lüttgen H, Rohdich F, Herz S, Wungsintaweekul J, Hecht S, Schuhr CA, Fellermeier M, Sagner S, Zenk MH, Bacher A, Eisereich W. 2000. Biosynthesis of terpenoids: YchB protein of *Escherichia coli* phosphorylates the 2-hydroxy group of 4-diphosphocytidyl-2C-methyl-d-erythritol. *Proc Natl Acad Sci USA* **97:** 1062–1067.

Maheshwari ML, Chandel KPS, Chien MJ. 1988. The composition of essential oil from Hyssopus officinalis and *Cymbopogon jwarancusa* (Jones) Schult. collected in the cold desert of Himalaya. *Dev Food Sci Flav Frag* **18:** 171–176.

Maheshwari ML, Mohan J. 1985. Geranyl formate and other esters in palmarosa oil. *Pafai J* **7:** 21–26.

Mallavarapu GR, Kulkami RN, Ramesh S. 1992a. Composition of the essential oil of *Cymbopogon coloratus*. *Planta Med* **58:** 479, 480.

Mallavarapu GR, Ramesh S, Kulkami RN, Shyamsunder KV. 1992b. Composition of essential oil of *Cymbopogon travencorensis*. *Planta Med* **58:** 219–222.

Mallavarapu GR, Rajeshwara Rao, Kaul PN, Ramesh S, Bhattacharya AK. 1998. Volatile constituents of the essential oil of the seeds and the herb of palmarosa (*C.martinii* (Roxb.) Wats. var. *motia* Burk. *Flav Frag J* **13:** 161–169.

Manjoor-i-Khuda M, Rahman M, Yusuf M, Choudhary J. 1984. Essential oils of *Cymbopogon* species of Bangladesh. *J Bangladesh Acad Sci* **8:** 77–80.

Manjoor-i-Khuda M, Rahman M, Yusuf M, Choudhary J. 1986. Studies on essential oil bearing plants of Bangladesh. Part II. Five species of *Cymbopogon* from Bangladesh and the chemical constituents of their essential oils. *Bangladesh J Sci Ind Res* **21:** 70–82.

Mathela CS, Joshi P. 1981. Terpenes from the essential oil of *Cymbopogon distans*. *Phytochemistry* **20:** 2770, 2771.

Mathela CS, Melkani AB, Pant AK, Pande C. 1988a. Chemical variations in *Cymbopogon distans* and their chemosystematic implications. *Biochem Syst Ecol* **16:**161–165.

Mathela CS, Chittattu GI, Thomas J. 1996. OD-468. A Lemongrass chemotype rich in geranyl acetate. *Indian Perfumer* **40:** 9–12.

Mathela CS, Lohani H, Pande C, Mathela DK. 1988. Chemosystematics of terpenoids in *Cymbopogon martini*. *Biochem Syst Ecol* **16:** 167–169.

Mathela CS, Melkani AB, Pant AK, Vasu Dev, Nelson TE, Hope H, Bottini AT. 1989. Eudesmanediol from *Cymbopogon distans*. *Phytochemistry* **28:** 1936–1938.

Mathela CS, Pande C, Pant AK, Singh AK. 1990a. Terpenoid composition of *Cymbopogon distans f. munsiyariensis*. *Herb Hung* **27:** 117–121.

Mathela CS, Pande C, Pant AK, Singh AK. 1990b. Phenylpropanoid constituents of *Cymbopogon microstachys*. *J Indian Chem Soc* **67:** 526–528.

Mathela CS, Pant AK, Melkani AB, Pant A. 1986. Aromatic grasses of U.P Himalaya: A new wild species as a source of aroma chemicals. *Sci Cult* **52:** 342–344.

Mathela CS, Pant AK. 1988. Production of essential oils from some new Himalayan *Cymbopogon* species. *Indian Perfumer* **32:** 40–50.

Mathela CS, Sinha GK. 1978. Study of the essential oil of *C. nardus* var. *stracheyi*. *J Indian Chem Soc* **55:** 621, 622.

Mathela CS. 1991. Flavour constituents of Himalayan aromatic grasses. International Symposium on Newer Trends in Essential Oils and Flavours, RRL Jammu, India, p. 27.

Mehmood Z, Ahmad S, Mohammed F. 1997. Antifungal activity of some essential oils and their major constituents. *Indian J Nat Products* **13:** 10–13.

Melkani AB, Joshi P, Pant PK, Mathela CS, Dev V. 1985. Constituents of the essential oil from two varieties of *Cymbopogon distans*. *J Nat Prod* **46:** 1995–1997.

Menon TCK. 1956. The essential oil of *Cymbopogon travancorensis*. *J Bombay Nat Hist Soc* **53:** 742, 743.

Modawi BM, Magar HRY, Satti AM, Duprey RJH. 1984. Chemistry of Sudanese flora. Part III. *Cymbopogon nervatus. J Nat Prod* **47**: 167–169.

Mohammad F, Nigam MC, Rehman W. 1981a. Detection of new trace constituents in the essential oils of *Cymbopogon flexuosus. Pafai* **13**: 22–24.

Mohammad F, Nigam MC, Rehman W. 1981b. Essential oil of *Cymbopogon martinii* var. *motia*: detection of new trace constituents. *Pafai J* **3**: 11–13.

Moody JO, Adeleye SA, Gundidza MG, Wyllie G, Ajayi-Obe OO. 1995. Analysis of the essential oil of *Cymbopogon nardus* (L.) Rendle growing in Zimbabwe. *Pharmazie* **50(1)**: 74, 75.

Moris JA, Khettry A, Seitz EW. 1979. Antimicrobial activity of aroma chemicals and essential oils. *J Am Oil Chem Soc* **56**: 595–603.

Muthuswami S, Sayed S. 1980. Preliminary note on the extraction of lemongrass oil. *Indian Perfumer* **24**: 226, 227.

Naik SN, Kumar A, Maheshwari RC, Kumar B, Chandra R, Guddewar MB. 1995. Evaluation of toxicity and growth-inhibiting activity of *Tripolium castaneum* (Herbst). *Indian Perfumer* **39**: 1–8.

Nair BVG, Chinnamma NP, Pushpakumari R. 1980a. Influence of varieties on the grass, oil yield and quality of oil of palmarosa (*C. martinii* Stapf. var. *motia*). *Indian Perfumer* **24**: 22–24.

Nair BVG, Chinnamma NP, Pushpakumari R. 1980b. Investigations on some types of lemongrass (*C. flexuosus* Stapf.) *Indian Perfumer* **24**: 20, 21.

Nair EVG, Mariam KA. 1978. The new promising aromatic plant of Kerala. *Indian Perfumer* **22**: 300, 301.

Nair RV, Siddiqui MS, Sen T, Nigam MC. 1982. Identification of three new species of *Cymbopogon* in perfumery value and their G.C. evaluation. *Pafai J* **4**: 21–23.

Namba T, Hatsutori Y, Shimomoura K, Nakamura M. 1995. Extraction of hyalurodinase inhibitors from *Azadirachta indica* or other plants for manufacturing cosmetics or for therapeutic use. *Japan Kokai Tokai Tokyo Koho* JP 07, 138,180 (95,138,180).

Nath SC, Saha BN, Bordoloi DN, Mathur RK, Leclercq PA. 1994. The chemical composition of the essential oil of *Cymbopogon flexuosus* Steud. Wats. growing in northeast India. *J Essen Oil Res* **6**: 85–87.

Naves YR. 1970. Palmarosa oil compounds. *Parfum Cosmet Savons* **13**: 354–357 (C.A.73, 59212–6).

Naves YR. 1971. Components of palmarosa essential oil. In *Meijhdunar Kongr. Efirnym Maslam (Mater)*, 4th, **1**: 239–244. Naumenko PV, Ed., *"Pishchevaya Promyshlennost"*, Moscow, USSR (C.A. 78, 16393 1-d).

Newman JD, Chappell J. 1999 Isoprenoid biosynthesis in plants: carbon partitioning within the cytoplasmic pathway. *Crit Rev Biochem Mol Biol* **34**: 95–100.

Neyberg AG. 1953. Some essential oil from the east of the colony. *Bull Agr Conogo* **44**: 1–38 and 319–364 (C.A. 48, 957-d).

Ng TT. 1972. Growth performance and production potential of some aromatic grasses in Sarawak, Preliminary assessment. *Trop Sci.* **14**: 47–58.

Nigam MC, Datta SC, Bhattacharya AK, Duhan SPS. 1975. Utilization of citral rich oils for production of geraniol. *Indian Perfumer* **18**: 55, 56.

Nigam MC, Nigam IC, Levi L. 1965. Essential oil and their constituents XXVII: Composition of gingergrass. *Can J Chem* **43**: 521–525.

Nigam MC, Srivastava HK, Siddiqui MS. 1987. Chemistry of *Cymbopogon* and their essential oils. *Pafai* **19**: 13–20.

Olaniyi AA, Sofowora EA, Oguntimehin BO. 1975. Phytochemical investigation of some Nigerian plants used against fevers II. *Cymbopogon citratus. Planta Med* **28**: 186–189.

Oliveros-Belardo L, Aureus B. 1978. Essential oil from *C. citratus* (D.C.) Stapf. growing in the Philippines. *Asian J Pharm* **3**: 14–17 (C.A.90, 15698 1-q).

Oliveros-Belardo L, Aureus E. 1979. Essential oil from *Cymbopogon citrates* (D.C.) Stapf. growing wild in the Phillippines. *Int Congr Essential Oils.* **7**: 166–168 (C.A. 92, 37768-g).

Oliveros-Belardo L, Smith RM, Gupta ZB, Abadiano MJP. 1989. Leaf essential oil of wild *Cymbopogon martinii* (Roxb.) Wats. from Sorsogon, Philippines. *Philippine J Sci* **118**: 31–58.

Onawunmi GO, Yisak W, Ogunlana EO. 1984. Antibacterial constituents in the essential oil of *Cymbopogon citratus* (DC.) Stapf. *J Ethnopharmacol* **12(3)**: 279–286.

Opdyke DLJ. 1973. Monographs on fragrance raw materials: Caryophyllene. *Food Cosmet Toxicol* **11**: 1059–1060.

Opdyke DLJ. 1974. Monographs on fragrance raw materials: Palmarosa oil. *Food Cosmet Toxicol* **12** (Suppl.): 947.

Opdyke DLJ. 1976. Inhibition of sensitization reactions induced by certain aldehydes. *Food Cosmet Toxicol* **14**: 197, 198.

Opdyke DLJ. 1976. Monographs on fragrance raw materials: Lemongrass oil: East Indian. *Food Cosmet Toxicol* **14:** 455.

Opdyke DLJ. 1976. Monographs on fragrance raw materials: Lemongrass oil: West Indian. *Food Cosmet Toxicol* **14:** 457.

Orafidiya LO. 1993. The effect of autoxidation of lemongrass oil on its antibacterial activity. *Phytotherapy Res* **7:** 269–271.

Osmani S, Anees I, Naidu MB. 1972. Insect repellent creams from essential oils. *Pesticides* **6:** 19–21.

Oyedele AO, Gbolade AA, Sosan MB, Adewoyin FB, Soyelu OL, Orafidiya OO. 2002. Formulation of an effective mosquito-repellent topical product from lemongrass oil. *Phytomedicine* **9(3):** 259–262.

Pande C, Melkani AB, Mathela CS. 1997. *Cymbopogon distans*: report on new chemotype. *Indian Perfumer* **41:** 35, 36.

Pandey MC, Sharma IR, Dikshit A. 1996. Antifungal evaluation of the essential oil of *Cymbogon pendulus* (Nees ex Steud) *Flav Frag J* **11:** 257–260..

Pandey TC, Gupta YN. 1973. Effect of γ-radiations on citronella oils (Ceylon type). *Indian Perfumer* **17:** 18, 19.

Pant AK, Mathela CS, Pande N, Pant AK. 1990. Phenyl propenoid constituents of *Cymbopogon microstachys* *J Indian Chem Soc* **67:** 526, 527.

Patra NK, Singh HP, Kalra A, Singh HB, Mengi N, Singh VR, Naqvi AA, Kumar S. 1997. Isolation and development of geraniol-rich cultivar of citronella *(C. winterianus)*. *J Med Arom Plant Sci* **19:** 672–676.

Patra NK, Srivastava RK, Chauhan SP, Ahmad A, Misra LN. 1990. Chemical features and productivity of a geraniol-rich variety (GRL-1) of *Cymbopogon flexuosus*. *Planta Med* **56:** 239, 240.

Patra P, Dutta PK. 1986. Evaluation of improved lemongrass strains for herb and oil yield and citral content under Bhubaneswar condition. *Res Ind* **31:** 358–360.

Peyron L. 1972. Essence produced in Mata Grasse. *An Acad Bras Cienc* **44** (Suppl.): 332–340 (C.A. 83, 136686-z).

Peyron L. 1973. Essential oil from Mato Grosso. *Perftimes Cosmet Savons* **3:** 371–378.

Pilley PP, Rao P, Sanjiva B, Simonson JL. 1928. Constituents of some Indian essential oils, Part XXII. The essential oil from flower heads of *C. coloratus* Stapf. *J Indian Inst Sci* **IIA:** 181–186.

Pino IA, Rosada A, Correa MT. 1996. Chemical composition of the essential oil of *Cymbopogon winterianus* Jowitt from Cuba. *J Essen Oil Res* **8:** 693, 694.

Popielas L, Moulis C, Keita A, Fouraste I, Bessiere IM. 1991. The essential oil of *Cymbopogon giganteus*. *Planta Med* **57:** 586, 587.

Prasad S, Jamwal R. 1971. Insecticidal composition. *Indian Pat.* 122,504 (C.A. 77, 30347-a).

Quintans-Júnior LJ, Souza TT, Leite BS, Lessa NMN, Bonjardim LR, Santos MRV, Alves PB, Blank AF, Antoniolli AR. 2008. Phythochemical screening and anticonvulsant activity of *Cymbopogon winterianus* Jowitt. (Poaceae) leaf essential oil in rodents. *Phytomedicine* **15(8):** 619–624.

Rabha LC, Baruah AKS, Bordoloi DN. 1979. Search for aroma chemicals of commercial value from plant resources of North East India. *Indian Perfumer* **23:** 178.

Rabha LC, Hagarika AK, Bordoli DN. 1986. *Cymbopogon khasianus,* a new rich source of methyl eugenol. *Indian Perfumer* **30:** 339–344.

Rabha LC, Hazarika AK, Bordoloi. 1989. A chemotype of *Cymbopogon khasianus* (Hack.) Stapf. ex Bor: An additional information for a source of higher oil and geraniol from North India. *Indian Perfumer* **33:** 261–265.

Rajapakse RHS, Jayasena W. 1991. Plant resistance and biopesticides from lemongrass oil for suppressing *Spodoptera litura* in peanut. *Agri Int* **43:** 166, 167.

Rajendrudu G, Rama Das VS. 1983. Interspecific differences in the constituents of essential oil of *Cymbopogon*. *Proc Indian Acad Sci* **92:** 331–334.

Ramdan FM, EI-Zanfaly HT, Allian AM, EI-Wakeil FA. 1972a. Antibacterial effects of some essential oils I. *Chem Microbiol Technol Lebensm* **1:** 96–102 (C.A. 77, 122532-k).

Ramdan FM, El-Zanfaly HT, Allian AM, El-Wakeil FA. 1972b. Antibacterial effects of some essential oils II. *Chem. Microbiol Technol Lebensm* **2:** 51–55 (C.A. 77. 122533-m).

Ramirez AR, Bressani R, Elias LG. 1977. Utilization of lemongrass *Cymbopogon* species begasse in ruminant nutrition. *Int. Symp Feed Compas Anim Nutr Requir Comput Diets* 198–203 (C.A. **91,** 90167-r).

Ranaweera SS, Dayananda KR. 1997. Mosquito-larvicidal activity of Ceylon citronella *(C. nardus* L.) Rendle oil fractions. *J Nat Sci SriLanka* **24:** 247–252.

Rao BGV, Narsimha JPL. 1971. Activity of some essential oils towards phytopathogenic fungi. *Riechst. Aromen Koerperflegen* **21:** 405, 406 and 408–410 (C.A. 76, 81621-x).

Rao BL, Kala S, Dhar KL, Kaul BS. 1992. New aromatic chemicals in *Cymbopogon* for future. *Indian Perfumer* **36:** 241–245.

Rao BN, Pandita S, Kaul BL. 1995. Gamma radiation as a mean of inducing variation in *Cymbopogon flexuosus* a chemotype for α-bisabolol *Indian Perfumer* **39:** 84–87.

Rao BS, Sudborough JJ. 1925. Kachi grass oil. *J Indian Inst Sci* **8A:** 9–27.

Rao JJ, Grover GS, Saxena VD. 1989. Mutagenic property of *Cymbopogon martinii* essential oil. *Parfum Kosmet* **70:** 488.

Rao JS. 1957. Grass flora of Coimbatore district, South India, *J Bombay Nat Hist Soc* **54:** 674–690.

Rauber CS, Guterres SS, Schapoval EES. 2005. LC determination of citral in *Cymbopogon citratus* volatile oil. *J Pharm Biomed Anal* **37(3):** 597–601.

Razdan TK, Koul GL. 1973. Chromatographic studies of Indian essential oils: Oil of Citronella Java type III. *Indian J Chem* **8:** 27–32.

Razdan TK.1984. Calorimetric analysis of the constituents of essential oils. *Parfum Kosmet* **65:** 190–195.

Rizk AM, Heiba HI, Mashaly M, Sandra P. 1985. Constituents of plants growing in Qatar. X . Seasonal variations of the volatile oil of *Cymbopogon parkeri*. *Qatar Univ Sci Bull* **5:** 71–76 (C.A. 106, 2908h).

Rizk AM, Heiba HI, Sandra P, Mashaly M, Bicchi C. 1983. Constituents of the volatile oil of *Cymbopogon parkeri*. *J Chrom* **279:** 145–150.

Rizk AM, Rimpler H, Ghaleb H, Heiba HI. 1986. Constituents of plants growing in Qatar. IX. The antispasmodic components of *Cymbopogon parkeri* Stapf. *Int J Crude Drug Res* **24:** 69–74.

Rohdich F, Wungsintaweekul J, Fellermeier M, Sagner S, Herz S, Kis K, Eisenreich W, Bacher A, Zenk MH. 1999. Cytidine 5′-triphosphate-dependent biosynthesis of isoprenoids: YgbP protein of *Escherichia coli* catalyzes the formation of 4-diphosphocytidyl-2-*C*-methylerythritol. *Proc Natl Acad Sci USA* **96:** 11758–11763.

Rohdich F, Wungsintaweekul J, Lüttgen H, Fischer M, Eisenreich W, Schuhr CA, Fellermeier, M, Schramek N, Zenk MH, Bacher A. 2000. Biosynthesis of terpenoids: 4-Diphosphocytidyl-2-*C*-methyl-d-erythritol kinase from tomato. *Proc Natl Acad Sci USA* **97:** 8251–8256 (First Published July 4; 10.1073/pnas.140209197).

Rohmer M. 1999. The discovery of a mevalonate-independent pathway for isoprenoid biosynthesis in bacteria, algae and higher plants. *Nat Prod Rep* **16:** 565–574.

Rouesti P, Voriate G. 1960. Lemongrass extract in Italian Somali Land. *Fr Parf* **3:** 39–44 (C.A. **55**, 7765-b).

Rout PK, Sahoo S, Rao YR. 2005. Essential oil composition of *Cymbopogon microstachys* (Hook.) Soenarke occurring in Manipur. *J Essen Oil Res* **17:** 358–360.

Saeed T, Sandra PJ, Verzele MJE. 1978. Constituents of the essential oil of *Cymbopogon jwarancusa*. *Phytochemistry* **17(8):** 1433, 1434.

Sangwan NK, Verma KK, Verma BS, Malik MS, Dhinsa KS. 1985. Nematicidal activity of essential oil of *Cymbopogon* grasses. *Nematologica* **31:** 93–99.

Sarer E, Scheffer JJC, Svendsen AB. 1983. Composition of the essential oil of *Cymbopogon citratus* Stapf. cultivated in Turkey. *Sci Pharm* **51:** 58–63 (C.A. **99**, 10685-k).

Sargenti SR, Lancas FM. 1997. Supercritical fluid extraction of *Cymbopogon citratus*. *Chromatographia* **46:** 285–290.

Schwander J, Müller C, Zeidler J, Lichtenthaler HK. 1999. Cloning and heterologous expression of a cDNA encoding 1-deoxy-D-xylulose-5-phosphate reductoisomerase of *Arabidopsis thaliana*. *FEBS Lett* **455:** 140–144.

Selvag E, Holm IO, Thune P. 1995. Allergic contact dermatitis in an aroma therapist with multiple sensitization to essential oils. *Contact dermatitis* **33:** 354, 355.

Shadab-Qamar E, Hanif M, Choudhary FM. 1992. Antifungal activity of lemongrass essential oils. *Pakistan J Sci Ind Res* **35:** 246–249.

Shahi AK, Sen DN. 1993. Essential oil yielding grasses of the Indian Thar desert. *Agric Equip Int* **45:** 62–65.

Shahi AK, Sen DN, Mathela CS. 1990. Chemical studies on *Cymbopogon schoenanthus* (Linn). Spreng a grass of Indian Thar desert. *Indian Drugs* **28:** 37, 38.

Shahi AK, Sen DN. 1989. Note on *Cymbopogon jwarancusa* (Jones) Schult: Source of piperitone in Thar desert. *Curr Agric* **13:** 99–101.

Shahi AK, Sharma SN, Tava A. 1997. Composition of *Cymbopogon pendulus* (Nees ex Steud.) Wats., an elemicin rich oil grass grown in Jammu region of India. *J Essen Oil Res* **9:** 561–563.

Shahi AK, Tava A. 1993. Essential oil compositions of three *Cymbopogon* species of Indian Thar desert. *J Essen Oil Res* **5:** 639–343.

Shahi AK. 1992. A search of *Cymbopogon jwarancusa* (Jones) Schult. subsp. *olivieri* (Boiss.) Soenakro. A citral yielding grass from Indian Thar desert. *Indian Perfumer* **36:** 182–184.

Sharma ML, Pandey MB, Khanna KR, Kapoor LD. 1972. Essential oils from plants raised on alkaline soil. *Indian Perfumer* **16:** 27–30.

Sharma ML, Srivaastava GS, Singh A. 1980. Essential oil from *C. olivieri* (Boiss.) Bor.). *Indian Perfumer* **24:** 17–19.

Sharma PR, Mondhe DM, Muthiah S, Pal HC, Shahi AK, Saxena AK, Qazi GN. 2009. Anticancer activity of an essential oil from *cymbopogon flexosus Chemica Biological Interactions* **179(2–3):** 160–168.

Shaw DW. 1971. Roll-on insect repellent. *Can Pat.* 865069 (C.A.74: 140098-d).

Shrivastava S, Rothria A, Misra N. 1990. Evaluation of essential oil of *Cymbopogon martinii* (Roxb.) Wats. for the control of *Fusarium oxysporurn* f. sp. *udum* casual organism of wilt of Arhar *(Cajnus cajon). J Nat Acad Sci India* **60B:** 291, 292.

Shylaraj KS, Thomas J. 1992. Effect of irradiation on the oil content of palmarosa *(C. martinii* var. *motia). Indian Perfumer* **36:** 168–170.

Siddique-Ullah M, Haque MM, Kalimuddin M. 1979. Dissolving pulp from lemongrass stem *(C. flexuosus* Linn.) by prehydrolysis Kraft process. *Bangladesh J Sci* **14:** 359–361.

Siddiqui MS, Mishra LN, Nigam MC, Abu-AI-Futuh, IM. 1980. Chemotaxonomie von *Cymbopogon*: Gas chromatographische untersuchung des etherischem oil von *Cymbopogon proximus. Parfum Kosmet* **61:** 419, 420.

Siddiqui MS, Mohammad F, Srivastava SK, Gupta RK. 1979. A thin layer chromatographic method for estimation of geraniol in oils of palmarosa. *Perfum Flavorist* **4:** 19, 20.

Siddiqui MS, Sen T, Nigam MC, Datta SC. 1975. Isolation of alcoholic constituents of the oil of citronella (Java) by sodium complex method. *Parfum Kosmet* **56:** 193, 194.

Siddiqui N, Garg SC. 1990. *In vitro* anthelmintic activity of some essential oils. *Pak J Sci Ind Res* **33:** 536, 537.

Singh A, Balyan SS, Shahi AK. 1978. Harvest management studies and yield potentiality of Jammu lemongrass. *Indian Perfumer* **22:** 189–191.

Singh AK, Dikshit A, Sharma ML, Dixit SN. 1980. Fungitoxic activity of some essential oils. *Econ Bot* **34:** 186–190.

Singh AK, Naqvi AA, Ram G, Singh K. 1994. Effect of bay storage on oil yield and quality of three *Cymbopogon* species (*C. winterianus*, *C. martinii*, and *C. flexuosus*) during different harvesting seasons. *J Essen Oil Res* **6:** 289–294.

Singh AK, Ram G, Sharma S. 1996. Accumulation pattern of important monoterpenes in the essential oil of citronella Java *(C. winterianus)* during one year of crop growth. *J Med Arom Plant Sci* **18:** 883–887.

Singh AP, Sinha GK. 1976. Study of essential oil of *C. distans* Linn. Wats. *Indian Perfumer* **20:** 67–70.

Singh RS, Pathak MG. 1994. Herb yield and volatile constituents of *cymbopogon jwaranusa* (Jones) Schult cultivars. *Industrial Crops and Products* **2(3):** 197–199.

Singh RS, Pathak MG, Singh KK. 1970. Dynamics and diurnal changes in oil of Java citronella (*C. winterianus* Jowitt). *Indian Perfumer* **23:** 116–120.

Sinha AK, Mehra MS. 1977. Study of some aromatic medicinal plants. *Indian Perfumer* **22:** 129–131.

Sobti SN, Bradu BL, Rao BL, Verma V, Jamwal PS. 1978a. Search for aroma chemicals of industrial value from the genus *Cymbopogon* Pt. VI. *C. khasianus* from Arunanchal Pradesh—a new source of citral. *Indian Perfumer* **22:** 222, 223.

Sobti SN, Rao BL, Jani BB. 1978b. Genetic variability in oils of *Cymbopogon*: Scope for evolving new strains. *Indian Perfumer* **22:** 281–283.

Sobti SN, Rao BL, Pushpangadan P, Verma V. 1978c. Investigations in some newly introduced *Cymbopogon* species. *Indian Perfumer* **22:** 278–280.

Sobti SN, Rao BL, Sharma SN, Atal CK. 1981. Jamrosa, a new geraniol rich *Cymbopogon. Indian Perfumer* **25:** 116–118.

Sobti SN, Verma V, Rao BL. 1982. Scope for development of new cultivars of cymbopogons as a source of terpene chemicals. In *Cultivation and Utilization of Aromatic Plants* (Atal CK, Kapur BM, Eds), Regional Reasearch Laboratory, Jammu, India.

Soulari M, Fanghaenel E. 1971. Essential oil of *C. winterianus* (citronella oil) produced in Cuba. *Rev Cenic Cienc Fis.* **3:** 79–91.

Sprenger GA, Schörken U, Wiegert T, Grolle S, De Graaf AA, Taylor SV, Begley TP, Bringer-Meyer S, Sahm H. 1997. Identification of a thiamin-dependent synthase in *Escherichia coli* required for the formation of the 1-deoxy-d-xylulose 5-phosphate precursor to isoprenoids, thiamin, and pyridoxol. *Proc Natl Acad Sci USA* **94:** 12847–12862.

Srikulvandhana S, Jennings WG, Derafols W. 1976. Lemongrass oil from mutant clones of *Cymbopogon flexuosus. Chem Mikrobiol Technol Lebensm* **4:** 129–131.

Sukari MA, Rahmani M, Manas A, Takahashi S. 1992. Toxicity studies of plants extracts on insects and fish *Partanika* **15:** 41–44.

Syeed M, Khalid MR, Choudhary FM. 1990. Essential oil of Gramineae family having antibacterial activity. Part I. *C. citratus, C. martinii, C. jwarancusa* oils. *Pak J Sci Ind Res* **33:** 529–531.

Takahashi S, Kuzuyama T, Watanabe H, Seto H. 1998. A 1-deoxy-d-xylulose 5-phosphate reductoisomerase catalyzing the formation of 2-*C*-methyl-d-erythritol 4-phosphate in an alternative non-mevalonate pathway for terpenoid biosynthesis. *Proc Natl Acad Sci USA* **95:** 9879–9884.

Talat S, Patrick JS, Maurice JEV. 1978. Constituent of the essential oil of *cymbopogon jwarancusa*. *Phytochemistry* **17(8):** 1433–1434.

Taskinen J, Mathela DK, Mathela CS. 1983. Composition of the essential oil of *Cymbopogon flexuosus*. *J Chromatogr* **262:** 364–366.

Thapa RK, Agarwal SG, Dhar KL, Atal CK. 1981. Citral containing *Cymbopogon* species. *Indian Perfumer* **25:** 15–18.

Thapa RK, Agarwal SK. 1989. *Cymbopogon flexuosus*: A rich source for (±) α-bisabolol. *J Essen Oil Res* 107–110.

Thapa RK, Bradu BL, Vashist VN, Atal CK. 1971. Screening of *Cymbopogon* species for useful constituents. *Flav Industry* **2:** 49–51.

Thapa RK, Dhar KL, Atal CK. 1976. The essential oil of *Cymbopogon flexuosus* RRL-57 and RRL-59, Part II. *Indian Perfumer* **20:** 39–44.

Thieme H, Benecke R, Brotka J. 1980. Application of gas chromatography in determining the constituents of Oleam citronella. *Zentralbl Pharm Pharmakother Laboratoriums Diagn* **119:** 953–957 (C.A. 94, 36417-w).

Torres RC, Ragadio AG. 1992. Chemical composition of the essential oil of Philippines *Cymbopogon citratus* D.C Stapf. The Asian symposium of Medicinal Plants, Spices and other Natural Products (ASOMAS VII), Manila, 2–7 February.

Torres RC, Ragodio AG. 1996. Chemical composition of the essential oil of Philippine *Cymbopogon citratus* (D.C.) Stapf. *Philippine J Sci* **125:** 147–156.

Torres RC. 1993. Citral from *Cymbopogon citratus* (D.C.) Stapf. lemongrass oil. *Philippine J Sci* **122:** 269–278.

Verma V, Sobti SN, Atal CK.1987. Chemical composition and inheritance pattern of five *Cymbopogon* species. *Indian Perfumer* **31:** 295–305.

Vijender SM, Mohammed A. 2002. Volatile constituents of *Cymbopogon nardus* (Linn.) Rendle. *Flav Frag J* **18(1):** 73–76.

Virmani OP, Datta SC. 1973. Scope of production of some essential oil in the Lucknow district. *Indian Perfumer* **17:** 35–41.

Virmani OP, Datta SC. 1971. Essential oil of *Cymbopogon winterianus* (oil of citronella). *Flav Ind* **2:** 595–602.

Vole CD, Baeta J, Antonio P da Cunha. 1997. Comparative chemical study of inflorescence of *Cymbopogon gigariteum*. *Bot Esc Farm Univ Coimbra Ed Cient* **27:** 17–52.

Wannissom B, Jarikasen S, Soontorn TT. 1996. Antifungal activity of lemongrass oil and lemongrass oil cream. *Phytother Res* **10:** 551–554.

Wijesekera ROB, Jayewardene AL, Foneska BD. 1973a. Varietal difference in the constituents of citronella oil. *Phytochemistry* **12:** 2697–2704.

Wijesekera ROB, Jayewardene AL, Foneska BD. 1973b. The chemical composition and analysis of citronella oil. *J Nat Sci Coun Sri Lanka* **1:** 67–81 (C.A. 81.111382-t).

Yadav P, Dubey NK. 1994. Screening of some essential oils against ring worm fungi. *Pharma Sci* **56:** 227–230.

Zaki MSA, Foda YH, Mustafa MM, Abd Allah MA. 1975. Identification of the volatile constituents of the Egyptian lemongrass oil II. Thin layer chromatography. *Nahrung* **19:** 201–205.

Zamureenka VA, Klyuev NA, Grandberg IH, Dmitriev LB, Esvandghya GA. 1981. Composition of essential oil of lemongrass *(C. citratus* D.C.). *Izv Timiryazensk S-Kh Akad* **2:** 167–169 (C.A. 94, 145166-j).

Zheng GQ, Kenney PM, Tam TDT. 1993. Potential anticarcinogenic natural products isolated from lemongrass oil and galanga root oil. *J Agric Food Chem* **41:** 153–156.

3 The Cymbopogons
Harvest and Postharvest Management

A. K. Pandey

CONTENTS

3.1 INTRODUCTION

Cymbopogon (Poaceae) represents an important genus of about 120 species that grow in tropical and subtropical regions around the world. On account of their diverse uses in pharmaceutical, cosmetics, food and flavor, and agriculture industries, *Cymbopogon* grasses are cultivated (medicultured) on a large scale, especially in the tropics and subtropics. There is a large worldwide demand for the essential oils of *Cymbopogon* species (Dutta 1982; Gunther 1956). They are well known as a source of commercially valuable compounds such as geraniol, geranyl acetate, citral (neral and geranial), citronellal, piperitone, eugenol, etc., which are either used as such in perfumery and allied industries or as starting materials for the synthesis of other products commonly used in perfumery (Shahi and Tava 1993). Distillation of the grass produces an essential oil and a hydrosol (distillate water) that have powerful antibiotic, antiviral, and antifungal properties which are used effectively against infectious and inflammatory symptoms. Several *Cymbopogon* species are being cultivated in different parts of world. Lemongrass, palmarosa, and citronella essential oils are the main raw material products of the cultivated cymbopogons. However, other *Cymbopogon* species are also grown in other parts of the world (Oyen and Dung 1999).

Volatile constituents of the essential oil of *C. caesius* were studied by Kanjilal et al. (1995). The main constituents were perillyl alcohol (25.61%), geraniol (19.80%), and limonene (7.26%) along with 21 other compounds. Choudhury and Leclercq (1995) also studied the essential oil composition of *C. khasianus* (Munro ex Hack.) Bor from northeastern India. *C. pendulus*, an elemicin-rich aromatic grass of the Meghalaya region of India, grew well under the subtropical climatic conditions at Jammu, India. The essential oil obtained from this plant was rich in elemicin (53.7%), a starting material for developing the antibacterial drug trimethoxyprim (Shahi et al. 1997). Chowdhury et al. (1998) studied the essential oil of *Cymbopogon* species growing in Bangladesh. Essential oil of *C. nardus* (L.) Rendle growing in Zimbabwe was studied by Moody et al. (1995). Comparative investigation of the sesquiterpenoids present in the leaf oil of *C. distans* (Steud.) Wats. var. *loharkhet* and the root oil of *C. jwarancusa* (Jones) Schult. was performed by Beauchamp et al. (1996). Chisowa (1997) studied the chemical composition of flower and leaf oils of *C. densiflorus* Stapf from Zambia.

A wide range of variation has been observed in the oil content of *Cymbopogon* species, and this is influenced by genetic, agronomic, and geoclimatic factors (Rao et al. 1980; Patra et al. 1990; Pandey and Chowdhury 2000). It is reported that oil content is lower during the month of heavy rainfall compared to the dry months (Guenther 1961). Similarly, monthly variation in the oil content in lemongrass over a year has been studied (Handique et al. 1984). It has also been reported that in some aromatic crops, the factors photoperiod, intensity of light, temperature (Voirin et al. 1990; Lincoln and Langenhein 1978; Clark and Menary 1980), and seasons or months of harvesting (Rudloff and Underhill 1965; Adams 1970; Singh et al. 2000) exert a profound influence on the essential oil content and terpenoid composition of these crops. Diseases such as iron chlorosis significantly reduced biomass, essential oil yields, and total chlorophyll content of the leaves of Java citronella (*C. winterianus*), lemongrass (*C. flexuosus* var. *flexuosus*), and palmarosa (*C. martinii* var. *motia*) plants.

The presence, yield, and composition of essential oils have been affected in a number of ways by various factors, from their formation in plants to their final isolation. Several of the factors of influence have been studied, particularly for commercially important crops, to optimize the cultivation conditions and time of harvest and to obtain higher yields of high-quality essential oils that fit market requirements. Knowledge of factors that determine the oil yield and chemical variability of aromatic plants species are thus very important. These include (1) physiological variations, (2) environmental conditions, (3) geographic variations, (4) genetic factors and evolution, (5) political/social conditions, and also (6) harvest time and technique (Figueiredo et al. 2008; Singh et al. 1996; Dhar et al. 1996b, 1996c; Costa et al. 2005). This chapter gives an account of the harvesting and distillation practices that will optimize yield of quality essential oil from *Cymbopogon* species (Cassel and Vergas Rubem 2006).

3.2 LEMONGRASS

Lemongrass, a perennial herb widely cultivated in the tropics, encompasses three different species: *C. flexuosus* (Steud.) Wats. (East Indian), *C. citratus* Stapf (West Indian), and *C. pendulus* (North Indian). The common name lemongrass has been given to these species because of the typical strong lemongrass-like odor of the essential oil present in the leaves. Two species, *C. flexuosus* and *C. pendulus*, are cultivated in India, whereas *C. citratus* is cultivated in the West Indies, Guatemala, Brazil, etc. *C. citratus* is a tufted perennial grass with numerous stiff leafy stems arising from short rhizomatous rootstocks. The aboveground parts, which contain the oil, grow to 2 m in height. A number of cultivars are acknowledged, which differ considerably in yield and citral content (Nair et al. 1979).

Lemongrass has been used fresh, dried, or powdered. The fresh stalks are found in Asian markets and now in many health food markets. Lemongrass is widely used in Thai and Vietnamese cooking. This aromatic herb is used in Caribbean and many types of Asian cooking, and has become very popular in the United States. Lemongrass has been used for centuries in Indonesia and Malaysia by herbalists, and it is also used in Ayurvedic herbalism. It is used in teas to combat depression and bad moods, and to fight fever and combat nervous and digestive disorders. Studies show that lemongrass has antibacterial (Burt 2004; Hussain 1994) and antifungal properties (Dikshit and Hussain 1984; Wannissorn et al. 1996; Pandey et al. 1996; Paranagama et al. 2003; Mahanta et al. 2007). The oil is used to cleanse oily skin, and in aromatherapy, it is used as a relaxant. Valued for its exotic citrus fragrance, it is commercially used in soaps, perfumes, cosmetics, and disinfectants, and is a raw material for manufacturing ionones and vitamin A.

The leaves yield aromatic oil, containing 70%–90% citral (the aldehyde responsible for the lemon odor). The quality of lemongrass is generally determined by its citral content (Chisowa et al. 1998). Citral consists of the cis-isomer geranial and the trans-isomer neral. These two are normally present in the ratio of about 2 to 1. *C. flexuosus* has higher citral content than *C. citratus* (Weiss 1997; Taskinen et al. 1983).

Lemongrass grows wild across the tropics, and the content and quality of the oil varies among provenances. It prefers a warm climate with well-distributed rainfall and well-drained soil. Usually, it grows on poor, gravelly soils. Lemongrass is a perennial grass mainly cultivated on hill slopes as a rainfed crop (Pandey et al. 2001). The crop provides maximum yield from the second to fourth year of planting and economic yield up to the fifth year. Thereafter the yield declines considerably. Different cultivars of lemongrass, for example, CKP-25, OD 19, Cauvery, Pragati, and Praman, were evaluated for herbage and oil yield. Brief information about various cultivars is given in Table 3.1. Cauvery recorded the highest oil yield (Singh 1997; Rao and Lala, 1992).

3.2.1 HARVESTING

Harvesting time is one of the most important factors influencing optimum superior-quality oil yield. Depending on soil and climatic conditions, a lemongrass plantation lasts, on average, for 3–5 years only. The yield of biomass and oil is less during the first year, but it increases in the second year and reaches a maximum in the third year; after this, the yield declines. Correct harvesting procedure is very important. The essential oil content varies considerably during the development of the plant. If the plant is harvested at the wrong time, the oil yield can be severely reduced. The oil is usually contained in oil glands, veins, or hairs that are often very fragile. Handling will break these structures and release the oils. This is why a strong smell is given off when these plants are handled. Therefore,

TABLE 3.1
Currently Grown Varieties of Lemongrass and Their Description

Variety	Description
Sugandhi (OD 19)	It is adapted to a wide range of soil and climatic conditions.
	A red-stemmed variety with a plant height of 1 to 1.75 m and profuse tillering.
	The oil yield ranges from 80 to 100 kg per hectare with 85%–88% of total citral produced under rainfed conditions (with life-saving irrigation).
Pragati	It is a tall-growing variety with dark purple leaf sheath suitable for North Indian Plains and the Tarai belt of subtropical and tropical climate.
	Average oil content is 0.63% with 75%–82% citral.
Praman	It has evolved through clonal selection and belongs to the species *C. pendulus*.
	It is a medium-sized variety with erect leaves and profuse tillering.
	The oil yield is high with 82% citral.
Jama rosa	It is very hardy, with vigorous growth.
	The variety yields about 35 t of herbage per hectare, containing 0.4% oil (FWB).
	The variety yields up to 300 kg oil in 4–5 cuts in 16–18 months of growing period.
RRL 16	Average herbage yield of this variety is 15–20 t/ha/annum, giving 100–110 kg oil.
	The oil content varies from 0.6% to 0.8% (fresh weight basis) with 80% citral.
CKP 25	This is a hybrid between *C. khasianum* × *C. pendulus*.
	It gives 60 t/ha herbage in North Indian plains under irrigation.
	The oil contains 82.85% citral.
Other varieties	OD-408, Cauvery (OD-408 is a white-stemmed selection of OD-19 and is an improvement in yield in terms of oil and citral content. Cauvery needs high soil moisture for luxuriant growth and was evolved for river valley tracts.)

these plants have to be handled very carefully to prevent valuable oils from being lost. Harvesting is done with the help of sickles; the plants are cut 10 cm above ground level and are allowed to wilt in the field before being transported to the distillation site. Seasonal influence on lemongrass has been reported (Handique et al. 1984; Thomas et al. 1980).

During the first year of planting, three cuttings are obtained and, subsequently, five to six cuttings per year are taken, subject to weather conditions (Rana et al. 1996). The harvesting season begins in May and continues until the end of January. The first harvest is generally obtained after 4–6 months of transplanting seedlings. Subsequent harvests are done at intervals of 60–70 days, depending on the fertility of the soil and other seasonal factors. Under normal conditions, three harvests are possible during the first year and three to four in subsequent years, depending on the management practices followed. The optimum interval between harvests to obtain maximum quantity of oil is 40–45 days for local types of lemongrass. For OD-19, the optimum interval was found to be 60–65 days when grown in hilltops and 45–55 days in valleys and lower areas (Jha et al. 2004). Singh et al. (2000) have reported that the first harvest should be taken 90 days after planting to boost the development of tillerings in lemongrass. Under Jammu (India) conditions, more tillers with fewer harvests of lemongrass was also reported (Pal et al. 1990; Singh et al. 1978).

Rao et al. (2005) conducted an experiment in Bengaluru (formerly Bangalore), Karnataka, India, during 2001–2003 to study the effect of harvest intervals on oil and citral accumulation in *C. flexuosus* cv. Krishna. The highest percentages of oil (4.8) and citral (87.1) were obtained in lemongrass harvested on February 17, 2002, whereas the highest percentages of geraniol (9.3) and geranyl acetate (2.5) were obtained with July 21, 2003, harvest. Variations in major chemical constituents in the oil of *C. flexuosus* were found in different seasons under Brahmaputra valley agroclimatic conditions (Sarma et al. 2003). Effect of leaf position and age on quality of oil has been discussed (Singh et al. 1989) whereas sucrose metabolism to components of essential oil have been studied by Singh and Luthra in 1988.

TABLE 3.2
Effect of Different Harvesting Intervals on Herb Yield, Oil Content, and Oil Yield of Lemongrass

Cutting	Fresh Herb Yield (q/ha)			Oil Content (%)			Oil Yield (L/ha)		
	75[a]	100	125	75[a]	100	125	75[a]	100	125
First	69.4	136.0	147.6	0.50	0.53	0.56	35.1	72.2	82.7
Second	152.3	141.6	50.0	0.56	0.61	1.02	85.9	86.8	51.7
Third	77.4	—	—	0.77	—	—	58.0	—	—
CD 5%	62.0	NS	35.4	0.16	NS	0.12	NS	NS	19.2

Note: NS = Not significant, quintal.
[a] Harvesting intervals (in days).
Source: Gill B S et al. 2007. *India Perfumer* 51: 23–27.

TABLE 3.3
Biomass Yield and Essential Oil Concentration in Different Varieties of Lemongrass at Different Harvests

Varieties	Biomass Yield (t/ha)				Essential Oil Concentration (%)		
	First Harvest	Second Harvest	Third Harvest	Total Harvest	First Harvest	Second Harvest	Third Harvest
OD-19	7.9	4.3	16.0	28.2	0.62	0.75	0.66
Cauvery	6.3	3.5	25.1	34.9	0.88	1.05	1.02
Pragati	7.7	3.4	22.5	33.6	0.71	0.80	0.70
SHK-7	12.5	4.0	24.4	40.9	0.42	0.51	0.57
Praman	2.2	3.8	26.6	32.6	0.63	0.71	0.61
LSD ($P = 0.05$)	4.5	NS	7.2	11.9	0.18	0.04	0.14

Note: NS = Not significant.
Source: Rajeswara Rao B R et al. 1998. *Journal of Medicinal and Aromatic Plant Sciences* 20: 407–412.

Harvesting intervals are determined by infrastructure, management, climate, and cultivar. If harvested too often, the productivity of the plant will be reduced, and the plant may even die. If the plant is allowed to grow too large, the oil yield will be reduced. It should be 1.2 m high with four to five leaves. The grass should be harvested early in the morning if it is not raining. Manual harvesting is common, but harvesting methods adopted depend on infrastructure. The bush is cut from 7 cm (manually) to 25 cm (mechanically) above ground. Weiss (1997) reports that each worker in Sri Lanka harvests 2 t of fresh grass daily.

An experiment was conducted in Punjab, India, to determine harvest intervals in lemongrass (Gill et al. 2007). The results given in Table 3.2 reveal that maximum herb and oil yields were obtained when the cutting was taken 125 days after general harvest (in the first week of March) and for subsequent cuttings taken after 75-day harvesting intervals. The oil content was maximum 75 days after general harvesting, and the oil content decreased with delay in harvesting, that is, from 75 to 125 days in the first as well as in the second and third cuttings. The decrease in oil content with delay in harvesting could be attributed to the fact that as the age of the crop increases, the plant becomes woodier and lower leaves become dry.

The effects of plant age (60, 67, 74, 81, 88, 95, 102, 109, and 116 days) on *C. citratus* biomass production and essential oil yield were studied in Campos dos Goytacazes, Rio de Janeiro, Brazil.

The essential oil yield decreased as plant age increased. Nevertheless, the increase in dry-matter production with plant age resulted in an increase in the total essential oil production (Leal et al. 2003). Essential oil from different plant parts of lemongrass grown in Brazil was studied by Ming et al. (1995). The study revealed that leaf blade contains 0.42% essential oil, whereas leaf sheath has 0.13% essential oil.

A study was conducted in the Western Cape, which has a Mediterranean climate. The plot was first harvested manually when the plants were 6 months old and every month thereafter. Leaf growth was significantly slower in winter; therefore, a monthly cut during the colder periods was not possible. Studies suggest that oil yield and citral content increase in hot dry seasons (Weiss 1997). Frequent cutting of *C. flexuosus* can increase total oil yield (Weiss 1997); a similar scenario exists for *C. citratus*.

Experiments with irrigated lemongrass (*C. citratus*) in Western Australia have shown that the highest oil yield of 419 L/ha over a 360-day period was obtained when the plants were cut at 60-day intervals and at a height of 20 cm. Longer intervals and higher cutting heights gave lower oil yields, although in some cases, fresh and dry-matter yields were increased. Studies on the effect of water stress showed that time between irrigations in the dry season should not be more than 10 days if oil yields are to be maintained. Wilting of cut lemongrass in the dry season was shown to result in a loss of oil, with losses increasing with the duration of wilting up to 11 h (Beech 1997). The effects of seasonal variation and harvest time on the foliar content of essential oil from lemongrass *C. citratus* were also studied by Beech (1990) and Leal et al. (2001).

The effects of harvesting time (07.00, 09.00, 11.00, 13.00, 15.00, or 17.00 h) on the essential oil content of lemongrass (*C. citratus*) and on the citral and myrcene contents of the essential oil were studied. The highest essential oil content was obtained at 09.00 h (5.59 mL/kg dry matter) and 11.00 h (5.31 mL/kg dry matter). The average citral (61.23%) and myrcene (24.14%) contents suggested that the essential oil was probably of the West Indian type. The results indicated that lemongrass may be harvested between 09.00 and 11.00 h for optimum essential oil, citral, and myrcene yields.

Outbreaks of *Puccinia nakanishikii* on commercial plantations of *C. citratus* caused reductions in the yield of essential oils. Healthy, uninfected leaves yielded 0.80% oil, while leaves with a disease index of 60%–75% yielded 0.50% oil. This reduction in essential oil yield was also accompanied by a decrease in the content of geraniol, and increases in the contents of neral and myrcene (Boruah et al. 1995).

3.2.2 YIELD

The grass yield during the first year was about 10 t/ha, which gives about 28 kg of oil. From the second year onward, the grass yield was about 25 t/ha, giving about 75 kg of oil. On an average, 25–30 t of fresh herbage are harvested per hectare per annum from four to six cuttings, which yields about 80 kg of oil. Under irrigated conditions, from newly bred varieties, an oil yield of 100–150 kg/ha is obtained. The average recovery of oil is 0.30%–0.35% with 70% citral for local types of lemongrass, while OD-19 variety gives 0.40%–0.45% oil recovery and 85%–90% citral content. *C. citratus* yields 30–50 t/ha and continues around this level for its 4–6-year plantation life. Oil yield from fresh herbage is 0.25%–0.5% and even 0.4%–0.6% with good management. This data is from Weiss (1997), but a wide range of variation in herbage and oil yield has been reported with infrastructure and management playing a significant role.

3.2.3 POSTHARVEST MANAGEMENT

3.2.3.1 Predistillation Handling

The cut grass may be distilled fresh, but some natural reduction in the moisture content by withering in the sun allows greater still vessel packing and oil recoveries per batch distillation. Wilting is done in the field where grass lies after cutting, rather than in heaps, but it should not exceed 24 h or oil losses may occur through brittleness and evaporation. Around 250 kg of partially wilted grass can be packed into 1000 L of a still vessel. The crop is chopped into small pieces before filling the stills.

3.2.3.2 Drying

Drying techniques influence the essential oil yield and composition. The grass is allowed to wilt for 24 h before distillation, as it reduces the moisture content by 30% and improves oil yield. In Brazil, a study was conducted to analyze the influence of drying on yield and composition of essential oil. The temperature varied from 40°C to 60°C, and the air velocities investigated were 0.2, 0.5, and 0.8 m/s. The highest yield was obtained at 60°C, the highest temperature investigated, and at 0.8 m/s. At air velocities near 0.2 m/s, the lowest masses of essential oil were extracted. The differences between air velocities did not influence the composition of the essential oils, but did influence the quantity of the components in the fractions and the time of drying. The essential oil extracted from the wet plants present fewer components than others, and it can be explained by the fact that water molecules solvate the components (Peisíno et al. 2005).

Experiments were also conducted to study the effects of drying temperature on the amount and quality of essential oils extracted from *C. citratus* (Buggle et al. 1997). Leaf blades were cut into small parts (about 1–1.5 cm in length) and dried for several days at 30°C, 50°C, 70°C, or 90°C, until a constant weight was achieved. A higher amount of oil was collected at lower drying temperatures; but at 30°C, leaves were affected by fungal (*Aspergillus* sp., *Penicillium* sp., *Rhizopus* sp., *Cladosporium* sp., *Trichoderma* sp., and *Alternaria* sp.) growth. The analysis of the oils by gas chromatography-mass spectrometry (GC-MS) showed variations in citral concentration (86.1%–95.2%). The best results were obtained at a drying temperature of 50°C (1.43% oil content).

The effects of drying method (oven drying at 40°C or drying at room temperature using a moisture drier) and fragment size (powder obtained from the mill, and 1.0 cm or 20.0 cm fragments) on the yield and composition of the essential oil of lemongrass (*C. citratus*) were studied. The essential oil was extracted using Clevenger's modified apparatus for 2 h. Higher essential oil and citral contents were obtained when the leaves were dried at room temperature. The fragment size had no significant effect on the evaluated parameters.

3.2.3.3 Storage

The safe limit of herb storage varied according to the species and storage conditions. Storage of *C. flexuosus* herbage always caused a reduction in oil content except during the summer, when it was not affected by 3 days of storage under shade. Little variation in the geranial and neral contents of the essential oils of *C. flexuosus* leaves was observed during storage for 15 days. Temperature and humidity were found to play a vital role in biosynthesis accumulation of essential oils in stored herbs (Singh et al. 1994).

3.2.3.4 Distillation

Distillation represents a dynamic part of a whole process in which the ethereal oils contained within a plant's aromatic sac or glands are liberated through heat and pressure and transformed into a liquid essence of sublime beauty. The recovery of essential oils (the value-added product) from the raw botanical starting material is very important since the quality of the oil is greatly influenced during this step. There are different methods for obtaining volatile oils from plants, such as steam distillation, aqueous infusion, solvent extraction, cold or hot expression, and supercritical fluid extraction (SFE) with carbon dioxide. The chemical composition of the oil, both quantitative and qualitative, differs according to the technique used to obtain it from the plant parts (Sandra and Bicchi 1987).

A comprehensive review of various techniques employed to obtain essential oil from the material in which it occurs was prepared by Weurman (1969). Harvesting the plant at the appropriate time and endeavoring to distill its essence is the art and craft of the distiller. The grass is either distilled afresh or is allowed to wilt for 24 h. Wilting reduces the moisture content and allows a larger quantity of grass to be packed into the still, thereby economizing the fuel use. The current method of distillation adopted in different parts of India is primitive and gives oil of poor quality, as it is based on hydrodistillation or direct-fired still. It is being distilled in the same distilleries used for Japanese mint in India. Variation in the essential oil composition of rose-scented geranium (*Pelargonium* sp.) distilled by different distillation techniques, for example, water distillation, water steam distillation, and steam distillation, was studied by Kiran et al. (2004). For good-quality oil, it is advisable to adopt steam distillation. The equipment for distillation consists of a boiler to produce steam, a distillation tub, a condenser, and one to three separators.

3.2.3.4.1 Steam Boiler
A steam boiler is made of steel or galvanized sheets for steam distillation of essential oil. In developed countries, an oil- or gas-fired package boiler is used. Biomass-fired boilers are used in India; they are also quite common in remote third world countries. However, there are some distilleries that purchase wood, as it is a cheap local fuel, or burn fuel oil, while the spent biomass of the process is discarded, primarily for being too wet.

3.2.3.4.2 Stills
Stills come in all sizes, shapes, and materials of construction. A still is basically a tank with some means of injecting steam at the bottom in a way that allows its uniform distribution, such as perforated crosses or plates, false bottoms, manifolds, etc. This method is known as *hydrodiffusion*, as opposed to *hydrodistillation*. In the latter, the still is filled with the material submersed in water, and the oil is "boiled" out of the aromatic raw material. The opening of the still can be a simple manhole cover or a full-sized lid with the same diameter as the tank, depending on the unloading method. The steam/oil vapors exit at the upper ends of the still or through an opening in the lid, which is sometimes fitted with a coarse filter.

Most stills operate at atmospheric pressure, but some are designed to withstand higher pressures, usually in the 2 bar range. Stills that operate under pressure are sometimes unloaded under pressure through a large opening at the end of a cone-shaped bottom. Materials that require frequent loading and unloading are processed in stills mounted on pivots that allow the still to swivel to the upside-down position and dump the entire contents. The distillation tub is made of mild steel or copper and has a perforated bottom on which the grass rests. The tub has a steam inlet pipe at the bottom. A removable lid is fitted to the top. Charging and discharging can be done in perforated cages with iron chains, which can be lowered in the tub with the help of a chain-pulley block.

3.2.3.4.3 Condensers
Different types of condensers are available, but tubular condensers are better than others. The condenser is provided with an inlet and an outlet through which cold water is made to flow through the chamber to cool the pipes when the distillate flows through them. Condensers are of various types; they range from truck radiators to copper coils, shell and tube heat exchangers, pipes submersed in river-fed canals, air-cooled condensers, tube condensers inside sprinkler towers, etc., depending on the location, climate, available space, and resources.

3.2.3.4.4 Oil Separators
The oil separator is the one component that is most critical to overall product recovery and profitability of the plant, whether conventional or continuous. Except in modern facilities, the separator often seems to get the least engineering attention from distillery operators in the field. The separator, too, comes in a wide range of homemade designs, although the main idea is that of a continuous decanter, sometimes referred to as a *Florentine flask*. Its efficiency is governed by a number of well-known variables such as oil and water specific gravity differential at various temperatures; phase viscosities versus ascending and descending cross-sectional velocities at various distillate flow rates and tank diameters; coalescing effects of different packing materials; emulsification effects;

oil solubility at various temperatures; chemical composition and polarity of the oil and its effect on solubility; etc.

The stills in a typical distillery, usually two in number, are 6–9 ft high and 3–6 ft in diameter. After the still is tightly filled with grass, the lid is fastened on and steam let in at the bottom. The steam passes up through the grass, carrying off the oil through a water-cooled coil. The distillate consisting of water and oil is collected in steel or copper tanks about 3 ft in diameter and 18 in. deep. When the tank is nearly full, a siphon attachment begins to discharge the water in the lower level of the tank. The oil, which is lighter, floats and, when a quantity has been collected, is drawn out into bottles or drums.

To obtain the maximum yield of oil and to facilitate its release, the grass is chopped into shorter lengths. Chopping the grass has further advantages in that more grass can be charged into the still and even packing is facilitated. It can be stored for up to 3 days under shade without any adverse effects on yield or quality of oil. Oil is obtained through steam distillation. The grass should be packed firmly as this prevents the formation of steam channels. The steam is allowed to pass into the still with a steam pressure of 18–32 kg in the boiler. The mixture of vapors of water and lemongrass oil passes into the condenser. As the distillation proceeds, the distillate collects in the separator. The oil, being lighter than water and insoluble, floats on the top of the separator and is continuously drawn off. It is then decanted and filtered. Small cultivators can use direct-fire stills, but in such cases, properly designed stills should be used. These stills are provided with a boiler at the bottom of the tub. This is separated by a false bottom from the rest of the tub. Water is poured at the bottom of the tub, and grass is charged in the top portion. In the still, the water does not come in contact with the grass. Providing a perforated disk just above the water level in the copper still will be helpful in producing oil of better quality. This method is known as *water and steam method*. The quality of oil suffers because of the crude method of production. To get a maximum yield of good-quality oil, it is advisable to use steam distillation.

Thick stems should be removed before distillation as these are devoid of oil. Time required for one distillation is about 4 h, including the time required for charging and discharging, provided the firewood is well dried and of good quality. For one distillation, about 40 kg of firewood is required. A light yellow, lemon-scented volatile oil is obtained. When crop area is large enough, the steam method is found to be more economical. Coal can also be used as fuel. The oil has a strong lemon-like odor. The oil is yellowish in color, having 75%–85% citral and small amounts of other minor aromatic compounds. The oil content recovered from the grass ranges from 0.5%–0.8%.

3.2.3.5 Supercritical Fluid Extraction (SFE)

SFE of *C. citratus* yielded oil at par with steam distillation but with more compounds (Sargenti and Lancas 1997). A comparative analysis of the oil and supercritical CO_2 extract of *C. citratus* was done by Marongiu et al. (2006). Dried and ground lemongrass leaves were used as a matrix for supercritical extraction of essential oil with CO_2. The objective of this study was to analyze the influence of pressure on the supercritical extraction. A series of experiments were carried out for 360 min, at 50°C, and at different pressures: 90, 100, 110, and 120 bar. Extraction conditions were chosen so as to maximize the citral content in the extracted oil. The collected extracts were analyzed by GC-MS, and their composition was compared with that of the essential oil isolated by hydrodistillation and steam distillation. At higher solvent densities, the extract's aspect changes, passing from a characteristic yellow-colored essential oil to a yellowish semisolid mass, because of the extraction of high-molecular-mass compounds. The optimum conditions for citral extraction were found to be 90 bar and 50°C. At these conditions, citral represents more than 68% of the essential oil and the extraction yield was 0.65%, while the yield obtained from hydrodistillation was 0.43% with a citral content of 73%. Lemongrass oil was also extracted by the SFE method. The essential oil yield did not increase as expected, but decreased (Rozzi et al. 2002).

Lemongrass essential oil was extracted with dense carbon dioxide at 23°C–50°C and 85–120 bar. Liquid carbon dioxide extracts had a larger quantity of coextracted waxes than the supercritical extracts. The process condition of 120 bar and 40°C was considered ideal for the extraction of lemongrass essential oil, as a good-quality product was obtained together with good extraction rate and yield (Carlson et al. 2001).

3.2.3.6 Treatment of Oil Prior to Storage

Oil recovered from each distillation run is added to a settling tank, where it should be left for 1–2 days to allow occluded water to sink to the bottom and be run off. The oil is then filtered to remove any solid debris and is next transferred to a bulking or storage tank. Bulking is an important operation since this allows for variation in oil composition between individual distillations and the supply to a buyer of an oil of reasonably consistent quality. Clean, new containers (steel/aluminum drums; UN standard) of 200 L capacity are used for shipment. Since the specific gravity of the oil is around 0.9, each drum will contain approximately 180 kg. For international shipment by land, sea, or air, drums must be labeled with appropriate identification and hazard codes.

3.2.3.6.1 Purification of Oil

The insoluble particles present in the oil are removed by the simple filtration method after mixing it with anhydrous sodium sulfate and keeping it overnight or for 4–5 h. In case the color of the oil changes because of rusting, then it should be cleaned by the steam rectification process.

3.2.3.6.2 Storage and Packing of Oil

The oil should be stored in glass bottles or containers made of stainless steel or aluminum or galvanized iron, depending on the quantity of oil to be stored. The oil should be filled up to the brim, and the containers should be kept away from direct heat and sunlight in cool or shaded places. The oil should be stored in well-sealed glass bottles, at 5°C–25°C, and in a dry, well-ventilated area away from direct heat and sunlight. Lemongrass oil can be stored for up to 3 years without affecting the quality of oil, if kept in aluminum containers sealed airtight using wax. Containers should be completely filled to exclude any air and protect the oil from sunlight as air and sunlight affect the citral content.

In a study conducted at Lucknow, India, the freshly distilled, light-yellow-colored oils of palmarosa (*C. martinii*), citronella (*C. nardus*), and lemongrass (*C. flexuosus*) were stored separately in 1 L containers made of amber glass, plain glass, aluminum, iron, and high-density polyethylene (HDPE). The containers were placed in a dark room at room temperature for up to 26 months. The oil samples were drawn from the storage containers periodically and analyzed for the color and percentage composition of the major constituents of the oil. The oil changed color only when stored in iron containers. There was little, if any, change in the percentage composition of major constituents in the three oils stored in other containers (Raina et al. 1998). The following points are taken into consideration while storing oils:

- Care is taken to ensure that the essential oil does not contain any water before storage. Amber-colored bottles are convenient for small quantities of oil. For large quantities, steel or aluminum drums are widely used.
- The oils are left to stand for some time so that water can settle down. If the oil is still turbid, a small amount of sodium sulfate is added and the oil is filtered. The containers are filled up to the brim, tightly capped, and stored in a cool, dry, and dark place.
- Exposure to air, light, and water causes deterioration of the quality of essential oil.

3.2.4 QUALITY ANALYSIS

Identification and estimation of various constituents of essential oils are carried out by gas chromatography.

3.2.4.1 Standard Specifications

Oils Association (BEOA) in its Chemical Hazard Information and Packaging (CHIP) Regulations list of 1999 gives the following details:

Hazard symbol	l:	Xn
Risk phrase	:	R65
H/C %	:	15
Safety phrase	:	S62

In the EU, all member countries today follow the standards published by the International Organization for Standardization (ISO 3217-1974). The main physiochemical requirements of this standard for lemongrass oil are the following:

Relative density at 20°C/20°C	:	0.872–0.897
Optical rotation at 20°C	:	−3° to +1°
Refractive index at 20°C	:	1.483–1.489
Carbonyl compounds as citral mininum	:	75%
Solubility in ethanol (70% v/v) at 20°C	:	soluble

In the United States, the Fragrance Manufacturers' Association (FMA) has published a standard (CAS # 08007-02-1) with very similar requirements. Both the ISO and FMA standards include gas chromatography analysis fingerprints for West Indian type lemongrass oil, and this analytical technique is the first of its kind used on a sample received by a buyer. The older physicochemical analyzes are used when adulteration or other quality deficiencies are suspected. It is important to recognize that the published standard specifications are the minimum requirements of buyers and users. More demanding in-house quality criteria may be set by end users, and these will include subjectively assessed odor characteristics.

3.3 PALMAROSA (*CYMBOPOGON MARTINII* VAR. *MOTIA* (ROXB.) WATS.)

Palmarosa (*Cymbopogon martinii*) is a widely distributed plant in India that yields a sweet, fragrant, aromatic oil. In India, palmarosa oil is mostly obtained from wild-growing grass in the states of Madhya Pradesh, Maharashtra, Andhra Pradesh, and Karnataka. A plantation of *motia* variety was started in Punjab in 1924. The late Prof. Puran Singh, Chief Chemist, Forest Research Institute and Colleges, Dehra Dun, succeeded in cultivating the grass at Jaranwala (Lyallpur) over an area of 93 ha in a short period of 4–5 years. He put up a steam distillation plant, and 1350–1600 kg of oil was produced annually. It was later cultivated near Dehra Dun by Purandad Essential Oil Plantation (Industry), and oil of good quality has been produced. Palmarosa is adapted to marginal areas and poor soils and can be grown under dense canopies of trees and used for soil conservation.

Sahoo (1994) selected cultivars, namely, RRL(b)69, RRL(B)77, IW31245, IW3630, CI8041, and HR89 for commercial production of high-quality palmarosa oil. Recently, its cultivation has also been taken up in the states of Karnataka, Maharashtra, Madhya Pradesh, and Uttar Pradesh. Palmarosa is also flourishing on a red sandy loam soil in the semiarid tropical climate of South India under rainfed conditions (Rajeswara Rao 2001). The use of farmyard manure (FYM) and nitrogen fertilizer had a positive impact on biomass and essential oil yield. Palmarosa grass has also been cultivated as an intercrop with pigeon pea. The oil content and quality, in terms of total geraniol, of palmarosa were not adversely affected by intercropping (Maheshwari et al. 1995). The flowering tops and foliage contain a sweet-smelling oil that emits a rose-like odor and is widely used in soaps, cosmetics, and perfumery industries. The oil is also used as a raw material for producing geraniol, which is extensively used in the perfumery industry.

Mainly, oil of palmarosa is obtained from flowering shoots and aboveground parts of the motia variety of *C. martinii*. The variety is also referred to as *rosha grass* or *russa grass* and yields an oil of high geraniol content (75%–90%), which is also called *East Indian geranium oil* or *rosha oil*. Another variety, *sofia*, is also found growing wild in India, which yields an oil of lower geraniol content known as gingergrass oil. The oil is of inferior grade and fetches much less price than palmarosa oil. Oil of palmarosa is one of the most important essential oils of India that is exported, and once, India was the principal supplier of this oil to the world. The market for export has fallen because of the deterioration of the quality of oil, competition with other countries, and appearance of synthetic geraniol in the market.

Essential oil composition of palmarosa grass from different places of India was studied by Raina et al. (2003). Geraniol (67.6%–83.6%) was the major constituent and, although the composition of the three oils was similar, quantitative differences in the concentrations of some constituents were observed.

3.3.1 USES

Oil of palmarosa is used in perfumery, particularly for flavoring tobacco and for the blending of soaps, owing to the lasting rose note it imparts to the blend. In soap perfumes, it has a special importance because geraniol remains stable in contact with alkali. It also serves as a source for very high grade geraniol. Geraniol is highly valued as a perfume and as a starting material for a large number of aromatic chemicals, such as geranyl esters, that have a permanent rose-like odor.

The essential oil is distributed in all parts of the grass, such as flower heads, leaves, and stems, with the flower heads containing the major portion. Usually, the grass is cut at a height of 5–8 cm from the ground level, and the whole plant is used for distillation. The maximum yield of oil is obtained when the entire plant is at the full-flowering stage. The flowering tops of palmarosa consist of spikelets, and each spikelet is further composed of racemes and a leaf-like structure called a *spathe*. Both racemes and spathe contain essential oil. Changes in the essential oil content and composition during inflorescence development were studied by Dubey et al. (2000) and Dubey and Luthra (2001). Changes in fresh weight, dry weight, chlorophyll and essential oil content, and its major constituents, that is, geraniol and geranyl acetate, were examined for both racemes and spathe at various stages of spikelet development. The essential oil content was maximum at the unopened spikelets stage and decreased significantly thereafter.

At the unopened spikelets stage, the proportion of geranyl acetate (58.6%) in the raceme oil was relatively greater compared to geraniol (37.2%), whereas the spathe oil contained more geraniol (61.9%) compared to geranyl acetate (33.4%). The relative percentage of geranyl acetate in both the oils, however, decreased significantly with development, and this was accompanied by a corresponding increase in the percentage of geraniol. Analysis of the volatile constituents from racemes, spathes (from mature spikelets), and seeds by capillary gas chromatography (GC) indicated 28 minor constituents besides the major constituent geraniol. Harvesting time, stage, and duration of harvest play a major role in determining the herbage, quality (chemical contents), and productivity of the herb.

3.3.2 HARVESTING

The number of harvests depends on the climatic condition of the place of cultivation and the method of crop management. In the first year, usually one crop is obtained during October–November, whereas two to three crops are obtained in the subsequent years in subtropical areas of the North Indian plains. Four harvests are taken in the tropical areas of South and Northeast India. By about 3½ to 4 months, the plants attain a height of 150–200 cm, and they start producing inflorescence.

TABLE 3.4
Effects of Different Harvesting Intervals On the Herbage, Oil Content, and Oil Yield of Palmarosa

Cutting	Fresh Herb Yield (q/ha)			Oil Content (%)			Oil Yield (L/ha)		
	75[a]	100	125	75[a]	100	125	75[a]	100	125
First	60.3	151.8	254.5	0.78	0.58	0.44	46.6	87.0	100.6
Second	291.1	231.2	166.4	0.51	0.25	0.27	148.4	57.8	45.1
Third	149.5	101.7	—	0.27	0.31	—	88.8	40.5	32.3
CD 5%	67.8	36.1	NS	NS	0.18	NS	32.1	20.8	18.9

Note: NS = Not significant.
[a] Harvesting intervals (in days).
Source: Gill B S et al. 2007. *India Perfumer* 51: 23–27.

The grass is cut 1 week after flowering. Generally, two cuttings are made during the first year of planting. From the second year onward, three to five cuttings are possible. It is recommended to harvest the crop 7–10 days after the opening of flowers. Usually, the grass is cut at a height of 5–8 cm from the ground level, and the whole plant is used for distillation. The maximum yield of oil is obtained when the entire plant is at the full-flowering stage. The harvested herbage is spread in the field for 4–6 h to reduce its moisture by 50%, and such semidry produce can be stacked in shady, cool spaces for a few days without much loss of its oil.

Palmarosa crop should be harvested at the full-flowering stage to seeding stage to obtain a high oil yield of good quality. During this period, the aerial parts, that is, the stem, leaf, inflorescence, and the whole herb, yielded 0.05%, 0.6%, 1.0%, and 0.5% oil with 78.5% to 88.50% geraniol. It was observed that the essential oil, rich in geraniol, had the least geranyl acetate content (Akhila et al. 1984). The delay or postponement of harvesting affected the yield and quality of oil. The delay led to an increase in leaf:stem ratio of the crop.

An experiment conducted in Punjab, India, revealed that each delay in harvesting, from 75 days harvesting interval to 125 days harvesting interval, increased the oil yield in the first cutting (Table 3.4). An oil yield of 100.6 L/ha was produced when the crop was harvested at 125 days' harvesting interval as compared to 87.0 and 46.6 L/ha of 100 and 75 days' harvesting intervals. The oil content in fresh herb decreased with delay in harvesting from 75 days to 125 days (Table 3.4). The oil content in the herb also decreased in the second cutting in all harvesting intervals as compared to the first cutting. Shahidullah et al. (1996) also reported that in palmarosa, essential oil content was highest when harvested in May compared to November. Kuriakose (1989) reported that maximum oil content was obtained when the crop was harvested at 90 days' interval as compared to 50, 60, 70, and 80 days' interval in palmarosa.

3.3.3   STORAGE

Hay storage of *C. martinii* (during summer) either in the shade or in the open increased the essential oil content. A slight difference in geraniol and geranyl acetate contents of the essential oils of *C. martinii* leaves was observed.

3.3.4   YIELD

Palmarosa plantation remains productive for about 8 years. However, the yield of grass and oil starts decreasing from the fourth year onward. It is, therefore, recommended that the plantation be kept only for 4 to 5 years. Normally, 200–250 q/ha of fresh herbage is obtained in the first cutting, and between 250–320 q/ha in second and subsequent harvests up to 3 years under irrigated conditions. On average, 200 kg of oil are received during the growing period of 15–16 months. The yield of oil for the first 4 years is as follows:

First year:	60 kg/ha
Second year:	80 kg/ha
Third year:	80 kg/ha
Fourth year:	80 kg/ha

3.3.5 Distillation

Oil of palmarosa is generally obtained by steam or hydrodistillation, similar to the process for lemongrass mentioned earlier. It takes 2 h to complete one distillation. The average recovery of oil is 0.40%–0.45%. Allowing the cut grass to wilt in the shade for 24 h during premonsoon months and 48 h during the postmonsoon months increases oil recovery. Small cultivators use direct-fire stills, but properly designed stills should be used. From the quality point of view, the grass should be distilled as fresh as possible. Oil obtained from dry or fermented grass is of inferior quality. For economic production of the oil, it is advisable that the harvested material be allowed to dry for a short period. The distillation unit should be clean, rust free, and free of any other odor.

During steam distillation, a part of the essential oil becomes dissolved in the condensate or distillation water and is lost when this water is discarded. A method was developed to recover the dissolved essential oil from condensate water. The distillation water of palmarosa mixed with hexane in 10:1 proportion was thoroughly shaken for 30 min to trap the dissolved essential oil. Hexane was then distilled to yield "secondary" or "recovered" oil. In palmarosa, the "primary" or decanted oil (obtained directly by distilling the crop biomass) accounted for 92%, and the recovered oil accounted for 8% of the total oil yield. The solvent loss in this process was 4%–7%. Experiments conducted in the laboratory with the essential oil showed that the water solubility of palmarosa oil ranged from 0.12% to 0.15% at 31°C and 0.15% to 0.20% at 80°C. Hexane recovered up to 97% of the dissolved essential oil in water. The recovered essential oil was richer in organoleptically important oxygenated compounds, such as linalool (2.6%–3.8%), geraniol (91.8%–92.8%), and geranial (1.8%–2.0%), compared to the primary oil (Rajeswara Rao et al. 2005).

3.3.6 Oil Storage

The oil should be free of sediments, suspended matter, and moisture before storage. The container should be clean and rust free.

3.3.7 Oil Content and Yield

The content and yield of the oil depend on many factors, such as climatic conditions of the place of cultivation; time of harvesting; maturity of the grass; nature of material being distilled, that is, fresh material or wilted material; method of distillation; etc. All parts of the plant contain the essential oil, the maximum oil being present in flowers and the stalks containing a negligible quantity. On average, the oil content in the various parts of the plant is as follows:

Plant Parts	Essential Oil (%)
Whole plant	0.10–0.40
Stalks	0.01–0.03
Flowering tops	0.45–0.52
Leaves	0.16–0.25

TABLE 3.5
Indian Standard Specifications for Palmarosa Oil

Sl. No.	Characteristics	Requirements No.
1.	Solubility	Soluble in 2 volumes of ethyl alcohol (70% by volume)
2.	Color	Light yellow to yellow
3.	Odor	Rosaceous with a characteristic grassy background
4.	Specific gravity	0.8740 to 0.8860 at 30°C/30°C
5.	Optical rotation	−2° to +3°
6.	Refractive index	1.4690 to 1.4735 at 30°C
7.	Acid value (maximum)	3
8.	Ester value	9 to 36
9.	Ester value after acetylation	266 to 280
10.	Total alcohols, calculated as geraniol percent (minimum)	90.0

3.3.8 STANDARD SPECIFICATIONS

Indian Standard Specifications for the oil of palmarosa (IS: 526-1986) are given in Table 3.5. The characteristic features of oil of palmarosa are the following: first, its sweet odor; and second, its solubility test in 70% alcohol (solubility of oil in 2.2–4.2 volumes of alcohol indicates a higher percentage of free geraniol). Oil of palmarosa chiefly contains 75.0%–95.0% alcohols, calculated as geraniol, and a small but varying amount of esters of the same alcohol, principally acetic and caproic acids. Java oils also have almost the same geraniol content, but their ester content is higher.

3.4 CITRONELLA (*CYMBOPOGON WINTERIANUS* JOWITT)

Citronella oil is one of the industrially important essential oils obtained from different species of *Cymbopogon*. The oil is widely used in perfumery, soaps, detergents, industrial polishes, cleaning compounds, and other industrial products. It is classified in the trade into two types: Ceylon citronella oil, which is extracted from *C. nardus;* and Java citronella oil, which is obtained from *C. winterianus.* The major difference between these oils is the proportion of geraniol and citronellal. The higher proportion of geraniol and citronellal in Java-type oil makes it an important source of various derivatives such as citronellol and hydroxycitronellal, which are extensively used in compounding high-grade perfumes. The Ceylon-type citronella oil, which contains a relatively low proportion of geraniol and citronellal, is mainly used in cheaper products rather than for the extraction of derivatives (Pino and Corrla 1996). The citronella oil has export potential, and its production can utilize rural-sector participation (Coronel et al. 1984).

Java citronella (Bordoloi 1982, *C. winterianus*) is a stoloniferous perennial that may grow up to 1 m high. Young shoots, growing from the axillary leaves of the mother plant, develop into large clumps with leaves bending outward. Cultivation is normally undertaken in tropical areas up to an altitude of 600 m, and a well-distributed rainfall of 1500 to 2500 mm is preferable. Citronella adapts to a wide range of soils, but will not withstand waterlogging. Sandy soils that are fairly fertile are ideal. Very fertile soils provide a high biomass yield, but a low oil yield. Java citronella generally flourishes in areas where lemongrass is cultivated and, also, it is less susceptible to humidity-induced rust. Citronella showed positive correlation with irrigation and nitrogen application (Singh et al. 1996a). The beneficial response is obtained with moderate nitrogen application, while a heavy dosage enhances biomass but not oil product yields. The essential oil from *C. winterianus* of Cuban origin contains citronellal (25.04%), citronellol (15.69%), and geraniol (16.85%)

as major constituents. Brazil is classified as one of the major world producers of essential oils and citronella essential oil.

The oil is used mostly in perfumery, both directly and indirectly. Soaps, soap flakes, detergents, household cleansers, technical products, insecticides, etc., are often perfumed exclusively with this oil. It is also a valuable constituent in perfumery for soaps and detergents (Hussain et al. 1988). Citronellal is occasionally used in traces in flower compositions of citrus, cherry, ginger, etc. However, the greatest importance of citronellal lies in its role as a starting material for further derivatives. Hydroxycitronellal can be prepared from citronellal, and it is a key ingredient in compounding. Hydroxycitronellal is one of the most frequently used floralizing perfume materials. It finds its way into almost every type of floral fragrance and a great many nonfloral ones. For soap perfumes, a slightly rougher grade is used. High grade is used in flavor compositions. Java citronella is rich in geraniol (36.0%) and citronellal (42.7%) and shows repellent, antimycotic, and acaricide activities. It is reported to be an air freshener (Guenther 1992; Lawrence 1996).

Cultivation of Java citronella was first introduced in the northeastern region of India by the Regional Research Laboratory (RRL), Jorhat, in 1965. RRL, Jammu, introduced a *C. winterianus* clone that yielded 60–70 t herbage per hectare, 0.6%–0.8% oil, and 60%–70% citronellal, which is extensively cultivated in the northeastern region (Sobti et al. 1982; Lal et al. 1999).

The biosynthesis of secondary metabolites, although controlled genetically, is strongly affected by environmental, harvest, and postharvest factors. Agricultural factors have a critical effect on the quantitative and qualitative characteristics of Java citronella, which affect plant growth and yield. Precipitation, temperature, light, and humidity influence volatile oil yield and the main constituents' content (citronellol, citronellal, and geraniol) (Sarma 2002). Sarma et al. (2001) cultivated Java citronella between 1998 and 2000 in Northeast India. They found a large variation in volatile oil yield and content, depending on seasonal changes and harvesting time. Highest yield and citronellal content were obtained during September and October. Under drought conditions, the geraniol and citronellal contents reduced, whereas citronellol content increased (Fatima et al. 2002). Saikia et al. (2006) studied the variation of essential oil content in leaf and inflorescence of Java citronella. Maximum essential oil accumulation was recorded at a crop age of 69 days during the summer (April–June), when the crop experienced the highest air temperature and the lowest relative humidity. However, the lowest air temperature and moderate relative humidity experienced by the crop during winter (October–January) provided good conditions for citronella oil quality (accumulation of citronellal, citronellol, and geraniol), which was best at a crop age of 63 days (Singh et al. 1996a).

Java citronella cultivated at an altitude of 60 m showed elevated biomass production (24.5 t/ha) and volatile oil content (1.3%). An increase in citronellal content and a decrease in citronellol and geraniol content were observed at this altitude (Sarma et al. 2001). Misra and Srivastava (1994) studied the influence of iron nutrition on chlorophyll contents, photosynthesis, and essential monoterpene oils in Java citronella. Significant positive correlations were observed between herbage, total essential oil yield, and citronellol content. Yellowing and crinkling disease influenced the essential oil yield and composition of citronella oil from sea level. The disease decreased biomass yield in the first and second years of harvesting by 62.80 and 82.70, respectively. The corresponding decreases in essential oil yield per plant were 62.80 and 79.00 (Rajeswara et al. 2004). However, only few studies have been conducted on the effects of harvest time and drying throughout the different harvest seasons in the northeast region of Brazil (Arrigoni-Blank et al. 2005; Carvalho-Filho et al. 2006).

3.4.1 HARVESTING

The time of harvesting affects the yield and quality of the oil. The first harvest is generally obtained after 4–6 months of transplanting. Subsequent harvests take place at intervals of 50–60 days, depending on the fertility of the soil and seasonal factors. Under normal conditions, two to three harvests are possible during the first year and three to four in subsequent years, depending on the

TABLE 3.6

Effects of Different Harvesting Intervals On Fresh Herb Yield, Oil Content, and Oil Yield of Citronella

Cutting	Fresh Herb Yield (q/ha)			Oil Content (%)			Oil Yield (1/ha)		
	75[a]	100	125	75[a]	100	125	75[a]	100	125
First	26.0	29.7	33.0	1.84	1.54	0.84	46.0	44.6	28.5
Second	63.7	89.0	90.1	1.45	1.12	1.05	92.3	99.3	94.8
Third	50.5	10.8	—	1.19	1.18	—	61.4	12.7	—
CD 5%	22.9	16.8	8.6	NS	NS	NS	NS	15.5	NS

Note: NS = Not significant.
[a] Harvesting intervals (in days).
Source: Gill B S et al. 2007. *India Perfumer* 51: 23–27.

management practices followed. Harvesting is done with the help of sickles, the plants being cut close to their bases about 10 cm above ground level. At harvest time, the grass is usually around 1 m high. The cut should be made above the first node in order to avoid the risk of dieback. Harvesting should be undertaken before flowering of the crop as it reduces the oil yields. Some researchers advocate harvesting when the stem bears six adult leaves and the seventh is rolled up. Others recommend cutting when the leaf tips begin to dry. The maximum yield is obtained in the third year, and after the fifth year the yields diminish rapidly. Mechanical harvesting is possible, but more commonly the grass is cut by hand. A dry day is preferable for the operation. After cutting, the grass should be left to wilt for a day, but care must be exercised to prevent fermentation. The oil obtained from this partially dry grass is more fragrant and of better quality than that obtained from grass distilled immediately after cutting.

Studies were conducted at Punjab, India, to standardize the harvest intervals for citronella grass (Gill et al. 2007). The first cutting yielded 46.0, 44.6, and 28.5 L/ha of citronella oil at 75, 100, and 125 days' harvesting intervals (Table 3.6). Citronella should be harvested at 75-day interval for optimum yield. Shahidullah et al. (1996) reported that in citronella, essential oil content was higher when harvested in May compared to November. Weiss (1997) also reported that oil content of leaves differs significantly, and young leaves synthesize and accumulate most of the essential oil in citronella. Virmani et al. (1979) reported that in Java citronella, delay in harvesting causes the leaves to dry up, decreasing the oil yield.

The grass is cut at all times of the year, but the yield varies with season; it is highest during the hot period, and low during the wet season and flowering period. The highest fresh biomass yield was obtained during summer (9326 kg/ha), fall (8174 kg/ha), and spring (8352 kg/ha) (Table 3.7), and the lowest yield during winter (3788 kg/ha). Seasons had significant effects on dry herbage biomass. Higher dry biomass production was achieved during fall and summer (5363 and 4897 kg/ha, respectively). A lower dry biomass production was obtained during winter and spring (1625 and 3189 kg/ha, respectively); however, proportionally higher moisture was observed (61.82% and 57.11%, respectively; Blank et al. 2007). Higher concentrations of fresh and dry biomass also were observed during dry seasons with irrigation.

Another important factor influencing Java citronella volatile oil production is harvesting time. The results obtained are presented in Table 3.8. Most volatile oils were produced at 0900h (2.71%, 2.22%, 2.36%, and 4.24% for summer, fall, winter, and spring seasons, respectively), because volatile oil content was generally found to peak at that time, except during fall, when there were no significant differences between harvest times. The percentage content of volatile oil in dried herbage was not significantly influenced by harvesting time. However, higher contents of volatile oil were obtained during spring (4.24%, fresh biomass; and 3.40%, dry biomass) and fall (3.17%, dry

TABLE 3.7

Effects of Seasons On Java Citronella (*Cymbopogon winterianus* Jowitt) Fresh and Dry Biomass and Moisture[a]

Season	Fresh Biomass Yield (kg/ha)	Dry Biomass Yield (kg/ha)	Moisture Content (%)
Summer	9326[a]	4897[a]	34.4
Fall	8174[a]	5363[a]	47.5
Winter	3788[b]	1625[c]	61.8
Spring	8352[a]	3189[b]	57.1

[a] Biomass obtained in different seasons is not significantly different ($P < 0.05$).

[b] About the same biomass.

[c] Lowest biomass.

Source: Blank Arie F et al. 2007. *Brazilian Journal of Pharmacognosy* 17(4): 557–564.

TABLE 3.8

Influence of Seasons and Harvest Times On Volatile Oil Content (In Percentage) From Fresh and Dry Leaves of *C. winterianus*[a]

Season	Oil Content (%)			Oil Yield (L/ha)		
	0900 h	1200 h	1500 h	0900 h	1200 h	1500 h
			Fresh Leaves			
Summer	2.71 b A	2.35 b AB	1.98 a B	162.50 a	114.82 a	97.17 a
Fall	2.22 b A	1.84 b A	1.71 a A	119.10 a	91.47 a	98.46 a
Winter	2.36 b A	1.98 b AB	1.67 a B	38.39 b	32.13 b	27.13 b
Spring	4.24 a A	3.36 a B	2.43 a C	135.29 a	107.25 a	77.42 a
			Dry Leaves			
Summer	2.60 b A	2.47 a A	3.10 a A	127.32 b	120.79 b	151.80 a
Fall	3.17 ab A	3.00 a A	3.20 a A	169.82 a	160.88 a	171.61 a
Winter	2.57 b A	2.67 a A	2.33 b A	41.70 c	43.32 c	37.91 c
Spring	3.40 a A	3.20 a A	3.50 a A	108.43 b	102.05 b	111.62 b

[a] Means of oil content in same season followed by the same lowercase letter in each column and by the same uppercase letter in each line are not significantly different.

Source: Blank Arie F et al. 2007. *Brazilian Journal of Pharmacognosy* 17(4): 557–564.

biomass). Harvest at 1200h resulted in quite similar volatile oil content throughout all seasons. In general, 1500h harvest significantly reduced volatile oil content, showing the lowest value during winter season (2.33%, dry biomass), which is characterized by higher precipitation volumes (Table 3.8). In accordance with studies on *C. citratus* conducted by Nascimento et al. (2003), harvests performed at 0900h and 1100h provided higher volatile oil contents (5.59% and 5.31%, respectively). The chemotype citral/limonene of *Lippia alba* (Mill) N.E.Br. produces lower volatile oil content during the rainy season, which accounts for a potential metabolic decrease caused by reduction of solar radiation (Nagao et al. 2004). In general, winter harvest from fresh or dry leaves significantly reduced yield independently of harvest time (Table 3.8).

Drying significantly increased volatile oil content and yield; this is probably due to the drying process affecting cell membrane resistance, assisting volatile oil release during hydrodistillation.

However, other species, such as *Ocimum basilicum* L., did not show significant differences between fresh and dry biomass (Carvalho-Filho et al. 2006), and *Lippia alba* showed peak volatile oil content at 1500h in both rainy and dry seasons. Specifically during the rainy season, volatile oil content increased throughout midday, decreasing by 1700h. Daily variations in temperature and luminosity may account for these results. The major components of *C. winterianus* volatile oil were identified as limonene, citronellal, citronellol, neral, geraniol, geranial, and farnesol. The GC-MS results showed that geraniol is an important volatile constituent of the volatile oil of *C. winterianus*, as well as citronellal. The content of these two compounds in fresh biomass varied with the season. However, no significant variation was observed for dry biomass.

In the rainy period, the citronellal in fresh biomass reached the lowest values at 0900h and 1200h, while the geraniol content increased. However, in the dry period (summer), the content of citronellal increased to 30.54% at 1500h, while the content of geraniol decreased to its lowest value (21.78%). Different results were observed by Rocha et al. (2002): maximum production of citronellal was between 0900h and 1100h. The least limonene content (1.17%) was observed in fresh leaves, during fall at 1200h. However, the highest limonene yield and content was achieved in plants harvested during winter. The content of citronellol in fresh biomass significantly decreased at 1500h during winter. The content of citronellol in dry biomass exhibited no significant variation.

Drying provided an increase in citronellol during summer at 1200h, while in the rainy period a reduction was observed at 0900h and 1200h. In general, no significant differences in the content of neral were found in fresh and dry biomass throughout the harvest time and seasons, except for winter at 0900h when neral was absent. Higher content of neral was obtained by drying biomass harvested at 0900h during winter. Similarly, the content of neral increased after drying *Melissa officinalis* L. for 5 days at 40°C (Blank et al. 2007). Amounts of geranial in fresh and dry biomass were relatively low and not significantly different, except for dry herbage harvested at 1200h during the fall. Peak farnesol content was achieved only on fresh herbage during winter at 0900h, which decreased after drying. After drying, herbage harvested at 0900h and 1200h during summer showed the highest farnesol content (5.69% and 6.13%, respectively).

According to Sarma (2002), Java citronella showed better results under high precipitation (100–200 mm), temperatures between 20°C and 30°C, and high moisture. In India, citronellal is obtained in a higher concentration during September and October, periods with favorable conditions locally. Chemical composition is affected in different seasons and length of drying. Time of harvest had little influence on composition. Java citronella showed lower volatile oil yield during winter. Morning harvests provide higher yield.

Variations in citronella oil and its major constituents due to seasonal changes and stages of the crop were studied under the climatic conditions of Jorhat in Assam, India. Rainfall, temperature, sunshine, and relative humidity have a cumulative effect on oil yield and major constituents of the oil, namely, citronellal, citronellol, and geraniol. Postmonsoon months seemed to be favorable, contributing higher oil yield. Citronellal content was higher during September (44.3%) and October (45.7%). It was observed that light rainfall (100–200 mm), moderate temperature (20°C–30°C), sunshine for 5 to 6 h, and high humidity (90%–95%) were the favorable meteorological parameters for higher oil yields and citronellal content in citronella oil. The growing period or crop growth stage also had a profound effect on oil yield and citronella content. Older crops with highly matured leaves were found to yield higher oil and less citronellal. Alcohol content in citronella oil was not affected by seasonal variation. However, the total alcohol percentage was found to decrease, while aldehyde percentage was found to increase (Sarma 2002).

A study was conducted in Brazil to optimize essential oil extraction processes from twigs and leaves by steam distillation. The process variables evaluated in this study were extraction time and raw material state (dry or natural). The essential oil compositions under different conditions were obtained with the objective of analyzing the influence of the variable value on the composition of citronella oil. The maximum yield, 0.942%, was obtained under the following conditions: extraction time—0400h and state—natural plant, and the results obtained from the factorial experimental planning indicate that the variable that influences the essential oil yield more is the state. The principal compounds identified in the citronella essential oil were citronellal, citronellol, geraniol, geranyl acetate, and α-cadinol.

3.4.2 STORAGE OF *C. WINTERIANUS* HAY

Hay storage of *C. winterianus* (during rainy season), either in the shade or in the open, increased the essential oil content of the leaves. Slight differences in the citronellal and citronellol content of the essential oils of *C. winterianus* leaves were observed.

3.4.3 YIELD

Depending on soil and climatic conditions, a citronella plantation lasts, on average, for 6 years. The yield of oil is less during the first year. It increases in the second year and reaches a maximum in the third and fourth years, after which it declines. For economy, the plantation is maintained only for 6 years. Up to 50 t/ha of fresh material is possible annually, but on average, 25–30 t of fresh herbage is harvested per hectare per annum from four to five cuttings, which yield about 80 kg of oil. The oil content of the freshly cut material ranges from about 0.5%–1.0%, and it is enhanced as the grass dries out through natural wilting. The annual oil yield under very favorable conditions can exceed 200 kg/ha, but 150 kg/ha is regarded as good in most situations.

3.4.4 DISTILLATION PROCEDURE

The same distillation units may be used for extraction of citronella essential oil as well as lemongrass oil. The grass is steam-distilled for better recovery of oil and for economical purposes. The harvested grass sometimes contains dead leaves, which should be removed. The remaining leaves are cut into shorter lengths. This reduces the volume of the grass, and facilities firm and even packing within the still. Further, chopping the grass gives a higher yield of oil than with uncut grass. Generally, distillation is complete within 2½ to 3 h under normal pressure starting from the initial condensation of the oil. About 80% of the total oil yield is recovered in the first hour, 19% in the second hour, and about 1% in the third hour of distillation. Larger percentages of the major components of the total oil, such as citronellal, geraniol, citronellol, and geranyl acetate, are recovered in the first hour of distillation. Steam distillation of citronella gave a maximum yield of 0.85% essential oil at 20 psig saturated steam pressure, within 30–120 min distilling time, and at 42% moisture content. The moisture content in citronella has some influence on the amount of essential oil and steam utilized, while the quality of the oil is determined by the distilling time.

Growers cultivating smaller areas can make use of properly designed direct-fired stills, in case they are not able to purchase a boiler. In such cases, the lower portion of the distillation tub is filled with water, and this functions as a boiler. The water in the boiler is separated from the remaining part of the still by means of a false perforated bottom on which the grass rests. In the still, the water does not come in contact with the grass. The tub is heated from below either by wood or coal, and the steam thus produced passes through the grass placed above in the tub, carrying oil vapors with it. However, distillation in such a direct-fired still takes a little more time, and the quality of the oil is also inferior.

3.4.5 STANDARD SPECIFICATIONS

In the EU, buyers use the ISO standard for Java citronella oil (ISO 3848-1976), and its main physicochemical requirements are summarized as follows:

Relative density at 20°C/20°C	0.880–0.895
Refractive index	1.466–1.473
Optical rotation	−5° to 0°
Carbonyl value	Minimum 127, equivalent to 35% as citronellal
Solubility in 80% ethanol at 20°C	1 in 2
Ester value after acetylation	Minimum 250, equivalent to 85% expressed as geraniol

Source: Bureau of Product Standards, Department of Trade and Industry.

In the United States, the FMA has published a standard (FMA #2308) with requirements very similar to those of ISO. Both the ISO and FMA standards include gas chromatography analysis fingerprints for Java citronella oil, and this analytical technique is the first one performed on a sample received by a buyer. The older physicochemical analyzes are used when adulteration or other quality deficiencies are suspected. It is important to recognize that the published standard specifications are the minimum requirements of buyers and users. More demanding in-house quality criteria may be set by end users, and these will include subjectively assessed odor characteristics.

3.5   JAMROSA (*CYMBOPOGON NARDUS* RENDLE)

Jamrosa, a hybrid between palmarosa and citronella, provides an essential oil that is used to impart a rosy-grassy note to natural perfumes. It is a drought-resistant hardy grass that attains a height of 1.5–2.5 m and has a hairy and fibrous shallow root system with long linear lanceolate leaves. It produces large, fawn-colored inflorescence with white, hairy, star-like spiked flowers. It is cultivated in the Indian states of Uttar Pradesh, Chhattisgarh, Madhya Pradesh, Rajasthan, Karnataka, Maharashtra, and Tamil Nadu (Shahi and Tava 1993). A well-drained sandy loam soil, free from waterlogging, with a soil pH of 7.5–8.5 is ideal for cultivation. A warm tropical climate and up to 300 m elevations in the foothills are suitable for cultivation of jamrosa. Temperatures ranging from 10°C–36°C with annual rainfall around 1000–1500 mm and ample sunshine are congenial for its growth. A moist and warm climate throughout the year accelerates its growth. Areas that are affected by severe frost are not suitable, as the frost kills the grass and reduces the oil content.

3.5.1   USES

Oil of jamrosa is used in perfumery, particularly for flavoring tobacco and for blending soaps, because of the lasting rose note it imparts to the blend. It also serves as a source of very high-grade geraniol. Geraniol is highly valued as a perfume and as a starting material for large chemicals, such as geranyl esters, that have a permanent rose-like odor.

3.5.2   HARVESTING

The essential oil is distributed in all parts of the grass, that is, flower heads, leaves, and stems, with the flower heads having the major content. It is recommended to harvest the crop 7–10 days after the opening of flowers. The number of harvests depends on the climatic condition of the place of cultivation and method of crop management. During the first year, usually one crop is obtained during October–November, whereas two to three crops are obtained in the subsequent years in the subtropical areas of the North Indian Plains. Four harvests are taken in the tropical areas of South and Northeast India. Usually, the grass is cut at a height of 5–8 cm from the ground level, and the whole plant is used for distillation. The maximum yield of oil is obtained when the entire plant is at the full-flowering stage. The harvested herbage is spread in the field for 4–6 h to reduce its moisture by 50%, and this semidry produce can be stacked in shady, cool spaces for a few days without much loss of its oil.

The effect of seasonal changes on oil content of a new clone of jamrosa, RL-931 (*C. nardus* var. *confertiflorus* × *C. jwarancusa*), was investigated by Bhan et al. (2003). The time of harvesting was found to be very crucial for the productivity of oil per unit area. Biosynthesis of oil coupled with its main constituents is directly associated with the timing of harvest. RL-931 showed a good deal of variation in oil content from the first harvest to the fourth harvest of the season. The premonsoon and onset of the monsoon period (May–July) was characterized by higher oil content (1.0%), whereas the postmonsoon and winter period (August–December) showed comparatively lower oil content (0.7%). These findings were in accordance with the results obtained for lemongrass (Handique et al. 1984b). The quality of the oil is best in May–July. High temperature and low humidity favor the accumulation of geraniol, geranyl acetate, and a low citral content in the oil. Maximum oil content was recorded when both maximum and minimum temperatures were high.

Central India is more suitable for growing aromatic grasses because of the large area under degraded forest lands. Many commercial houses in Chhattisgarh have entered into agreements to buy back aromatic plant produce, and the state government has announced Chhattisgarh as an herbal state because it has the climatic conditions most suited to aromatic plant production. To promote aromatic plant production, the chief emphasis is to be given to the establishment of processing units for oil extraction. Leaf collection will generate about 1 million man-days, and this will provide enormous employment opportunities. Patchouli and jamrosa keep yielding for 2 years with minimal expenditure in the second year, which will further enhance the income of farmers.

3.5.3 DISTILLATION

Jamrosa grass oil can be obtained by the method mentioned previously for lemongrass.

3.5.4 YIELD

Jamrosa plantations remain productive for about 8 years. However, the yield of grass and oil starts decreasing from the fourth year onward. It is, therefore, recommended that the plantation be kept only for 4 years. Normally, 200–250 q/ha of fresh herbage is obtained in the first cutting and between 250 and 320 q/ha in the second and subsequent harvests for up to 3 years under irrigated conditions. On average, 200 kg of oil is received during the growing period of 15–16 months.

The yield of oil (in kg/ha) for the first 4 years is as follows:

First year	60
Second year	80
Third year	80
Fourth year	80

3.6 CONCLUSION

India is considered to be one of the leading producers of essential oils. However, the oil distilled by traditional raw methods does not fetch a good value in the international market. Improvements in extraction technology and the use of oil as a raw material can change this scene. To increase India's participation in the international market, it is necessary to improve extraction technology in order to obtain products having international quality standards. Research on improving distillation technology is required to increase oil quality and distillation efficiency. Intermediate solutions to save energy using existing equipment, such as methods to insulate the stainless steel drums, should be explored.

REFERENCES

Adams R P. 1970. Seasonal variation of terpenoid constituents in natural population of *Juniperus pinchotii* Sudw. *Phytochemistry* 9: 397–402.

Akhila A, Tyagi B R, Naqvi A. 1984. Variation of essential oil constituents in *Cymbopogon martinii* Wats. var. *motia* at different stages of plant growth. *Indian Perfumer* 28: 126–128.

Anonymous. *Handbook of Medicinal and Aromatic Plants*, Citronella (*Cymbopogon winterianus* Jowitt). NEDFI, pp. 13–18.

Arrigoni-Blank M F, Silva-Mann R, Campos D A, Silva P A, Antoniolli A R, Caetano L C, Sant'Ana A E G, Blank A F. 2005. Morphological, agronomical and pharmacological characterization of *Hyptis pectinata* (L.) Poit germplasm. *Revista Brasileira Farmacognosia* 15: 298–303.

Beauchamp P S, Dev V, Docter D R, Ehsani R, Vita G, Melkani A B, Mathela C S, Bottini A T. 1996. Comparative investigation of the sesquiterpenoids present in the leaf oil of *Cymbopogon distans* (Steud.) Wats. var. *loharkhet* and the root oil of *Cymbopogon jwarancusa* (Jones) Schult. *Journal of Essential Oil Research* 8(2): 117–121.

Beech D F. 1997. Growth and oil production of lemongrass (*Cymbopogon citratus*) in the Ord Irrigation Area, Western Australia. *Australian Journal of Experimental Agriculture and Animal Husbandry* 17(85): 301–307.

Beech D F. 1990. The effect of carrier and rate of nitrogen application on the growth and oil production of lemongrass (*Cymbopogon citratus*) in the Ord Irrigation Area, Western Australia. *Australian Journal of Experimental Agriculture* 30(2): 243–250.

Bhan M K, Rekha Kanti, Kak S N, Agarwal S G, Dhar P L, Thappa R K. 2003. Variation of oil content in a new clone of Jamrosa 'RL-931' (*Cymbopogon nardus* var. *confertiflorus* × *C. jwarancusa*) during one year of crop growth. *New Zealand Journal of Crop and Horticultural Science* 31: 187–191.

Blank Arie F, Costa Andressa G, Arrigoni-Blank Maria de Fátima, Cavalcanti Sócrates C H, Alves Péricles B, Innecco Renato, Ehlert Polyana A D, de Sousa Inajá Francisco. 2007. Influence of season, harvest time and drying on Java citronella (*Cymbopogon winterianus* Jowitt) volatile oil. *Brazilian Journal of Pharmacognosy* 17(4): 557–564.

Bordoloi D N. 1982. Citronella oil industry in North East India. In *Cultivation and Utilization of Aromatic Plants*, Regional Research Laboratory, Jammu, Tawi, India, pp. 318–324.

Boruah P, Misra B P, Pathak M G, Ghosh A C. 1995. Dynamics of essential oil of *Cymbopogon citratus* (DC) Stapf. under rust disease indices. *Journal of Essential Oil Research* 7(3): 337, 338.

Buggle V, Ming L C, Furtado E L, Rocha S F R, Marques M O M. 1997. Influence of different drying temperatures on the amount of essential oils and citral content in *Cymbopogon citratus* (DC) Stapf., Proceedings of the Second World Congress on Medicinal and Aromatic Plants, WOCMAP-2. Biological resources, sustainable use, conservation and ethnobotany, Mendoza, Argentina, November 10–15, 1997 [edited by Caffini N, Bernath J, Craker L, Jatisatienr A, Giberti G]. *Acta Horticulturae*, 1999, No. 571–574.

Carlson Luiz H C, Machado R A F, Spricigo C B, Pereira L K, Bolzan A. 2001. Extraction of lemongrass essential oil with dense carbon dioxide. *Journal of Supercritical Fluids* 21(1): 33–39.

Carvalho-Filho J L S, Blank A F, Alves P B, Ehlert P A D, Melo A S, Cavalcanti S C H, Arrigoni-Blank M F, Silva-Mann R. 2006. Influence of the harvesting time, temperature and drying period on basil (*Ocimum basilicum* L.) essential oil. *Revista Brasileira Farmacognosia* 16: 24–30.

Cassel E, Vergas Rubem M F. 2006. Experiments and modelling of the *Cymbopogon winterianus* essential oil extraction by steam distillation. *Journal of the Mexican Chemical Society* 50(3): 126–129.

Chisowa, E H. 1997. Chemical composition of flower and leaf oils of *Cymbopogon densiflorus* Stapf. from Zambia. *Journal of Essential Oil Research* 9(4): 469, 470.

Chisowa E H, Hall D R, Farman D I. 1998. *Flavour and Fragrance Journal* 13: 29, 30.

Chowdhury J U, Mohammed Yusuf, Jaripa Begum, Mondello L, Previti P, Dugo G. 1998. Studies on the essential oil bearing plants of Bangladesh. Part IV. Composition of the leaf oils of three Cymbopogon species: *C. flexuosus* (Nees ex Steud.) Wats., *C. nardus* (L.) Rendle var. *confertiflorus* (Steud.) N. L. Bor and *C. martinii* (Roxb.) Wats. var. *martinii*. *Journal of Essential Oil Research* 10(3): 301–306.

Choudhury S N, Leclercq P A. 1995. Essential oil of *Cymbopogon khasianus* (Munro ex Hack.) Bor from northeastern India. *Journal of Essential Oil Research* 7(5): 555, 556.

Clark R J, Menary R C. 1980. Environmental effects on peppermint (*Mentha piperita* L.). I. Effect of day length, photon flux density, night temperature and day temperature on the yield and composition of peppermint oil. *Australian Journal of Plant Physiology* 7: 685–692.

Coronel V Q, Anzaldo F E, Recaña M P. 1984. Effect of moisture content on the essential oil yield of lemon-grass and citronella. *NSTA Technology Journal*, Vol. IX No. 3.

Costa L C do B, Correa R M, Cardoso J C W, Pinto J E B P, Bertolucci S K V, Ferri P H. 2005. Yield and composition of essential oil of lemongrass under different drying and fragmentation conditions. *Horticultura Brasileira* 23(4): 956–959.

Dhar A K, Dhar R S, Rekha K, Koul S. 1996. Effect of spacings and nitrogen levels on herb and oil yield, oil concentration and composition in three selections of *Cymbopogon jwarancusa* (Jones) Schultz. *Journal of Spices and Aromatic Crops* 5(2): 120–126.

Dikshit A, Husain A. 1984. Antifungal action of some essential oils against animal pathogens, *Fitoterapia* 55(3): 171–176.

Dubey V S, Mallavarapu G R, Luthra R. 2000. Changes in the essential oil content and its composition during palmarosa (*Cymbopogon martinii* (Roxb.) Wats. var. *motia*) inflorescence development. *Flavour and Fragrance Journal* 15(5): 309–314.

Dubey V S, Luthra R. 2001. Biotransformation of geranyl acetate to geraniol during palmarosa (*Cymbopogon martinii*, Roxb. Wats. var. *motia*) inflorescence development, *Phytochemistry* 57(5): 675–680.

Dutta S C. 1982. Cultivation of *Cymbopogon winterianus* Jowitt for production of citronella (Java) oil. In *Cultivation and Utilization of Aromatic Plants*, Regional Research Laboratory, Jammu, Tawi, India, pp. 325–329.

Fatima S, Farooqi A H A, Sharma S. 2002. Physiological and metabolic responses of different genotypes of *Cymbopogon martinii* and *C. winterianus* to water stress. *Plant Growth Regulation* 37: 143–149.

Figueiredo A C, Barroso J G, Pedro L G, Scheffer J J C. 2008. Factors affecting secondary metabolite production in plants: volatile components and essential oils. *Flavour and Fragrance Journal* 23(4): 213–226.

Gill B S, Salaria A, Kaur Satvinder, Gill G S. 2007. Harvesting management studies in lemongrass, palmarosa and citronella, *India Perfumer* 51: 23–27.

Guenther E. 1950. Oils of citronella, essential of citronella, aceite essential citronella, citronellol, oleum citronellac. In *The Essential Oils*, Vol. IV. pp. 65–82. Van Nostrand, London.

Guenther E. 1961. *The Essential Oils*, Vol. IV. D. Van Nostrand, New York.

Guenther E. 1992. *The Essential Oils*. D. Van Nostrand, Princeton, New Jersey.

Handique A K, Gupta R K, Bordoloi D N. 1984a. Variation of oil content in lemongrass and its genetic aspects. *Indian Perfumer* 24: 13–16.

Handique A K, Gupta R K, Bordoloi D N. 1984b. Variation of oil content in lemongrass as influenced by seasonal changes and its genetic aspects. *Indian Perfumer* 28: 54–63.

Hussain A, Virmoni O P, Sharma A, Mishra L N. 1988. Citronella oil (Java) *Cymbopogon winterianus* Jowitt. In *Major Essential Oil Bearing Plants of India*, pp. 52–57.

Hussain A. 1994. Essential Oil Plants and Their Cultivation. Central Institute of Medicinal and Aromatic Plants, Lucknow, India.

Jha B K, Kumar D, Joshi P P. 2004. Harvesting frequency in lemongrass var. OD-19 under semi-arid ecosystem of Gujarat. *Indian Perfumer* 48(3): 317–321.

Kanjilal P B, Pathak M G, Singh R S, Ghosh A C. 1995. Volatile constituents of the essential oil of *Cymbopogon caesius* (Nees ex Hook. et Arn.) Stapf. *Journal of Essential Oil Research* 7(4): 437–439.

Kiran G, Babu, D, Kaul V K. 2004. Variation in essential oil composition of rose-scented geranium (*Pelargonium* sp.) distilled by different distillation techniques. *Flavour and Fragrance Journal* 20(2): 222–231.

Kuriakose K P. 1989. Stage of harvest trial in palmarosa. *Indian Perfumer* 33(1): 21, 22.

Lal R K, Sharma J R, Misra H O. 1999. Development of new variety Manjari of citronella Java (*Cymbopogon winterianus*). *Journal of Medicinal and Aromatic Plant Sciences* 21(3): 727–729.

Lawrence B M. 1996. Progress in essential oils. *Perfumer and Flavorist* 21(1): 37–45.

Leal T C A B, Freitas S P, Silva J F, Carvalho A J C. 2001. Evaluation of the effect of season variation and harvest time on the foliar content of essential oil from lemongrass (*Cymbopogon citratus* (DC) Stapf.). *Crop Husbandry* 48(278): 445–453.

Leal T C A B, Freitas S P, Silva J F, Carvalho A J C. 2003. Production of biomass and essential oil in plants of lemongrass (*Cymbopogon citratus* (DC) Stapf) of various ages. *Revista Brasileira de Plantas Medicinais* 5(2): 61–64.

Lincoln D E, Langenheim J H. 1978. Effect of light and temperature on monoterpenoid yield and composition in *Satureja douglasii*. *Biochemical Systematics and Ecology* 24: 627–635.

Mahanta J J, Chutia M, Bordoloi M, Pathak M G, Adhikary R K, Sarma T C. 2007. *Cymbopogon citratus* L. essential oil as a potential antifungal agent against key weed moulds of *Pleurotus* spp. spawns. *Flavour and Fragrance Journal* 22(6): 525–530.

Maheshwari S K, Sharma R K, Gangrade S K. 1995. Effect of spatial arrangement on performance of palmarosa (*Cymbopogon martini* var. *motia*)-pigeonpea (*Cajanus cajan*) intercropping on a black cotton soil (vertisol). *Indian Journal of Agronomy* 40(2): 181–185.

Marongiu B, Piras A, Porcedda S, Tuveri E. 2006. Comparative analysis of the oil and supercritical CO(2) extract of *Cymbopogon citratus* Stapf. *Natural Product Research* 20(5): 455–459.

Ming L C, Figueiredo R O, Machado S R, Andrade R M C. 1995. Yield of essential oil of and citral content in different parts of lemongrass leaves (*Cymbopogon citratus* (D.C.) Stapf.) Poaceae. International symposium on medicinal and aromatic plants, Amherst, Massachusetts, August 27–30, 1995 [edited by Craker L E, Nolan L, Shetty K] *Acta Horticulturae* (1996) (No. 426): 555–559.

Misra A, Srivastava N K. 1994. Influence of iron nutrition on chlorophyll contents, photosynthesis and essential monoterpene oil(s) in Java citronella (*Cymbopogon winterianus* Jowitt). *Photosynthetica* 30(3): 425–434.

Moody J O, Adeleye S A, Gundidza M G, Wyllie G, Ajayi-Obe O O. 1995. Analysis of the essential oil of *Cymbopogon nardus* (L.) Rendle growing in Zimbabwe. *Pharmazie* 50(1): 74, 75.

Nagao E O, Innecco R, Mattos S H, Filho S M, Marco C A. 2004. Effect of the harvest time on content and major constituent of essential oil of *Lippia alba* (Mill) N.E.Br., chemotype citral-limonene. *Cienc Agron* 35: 355–360.

Nair E C G, Chinnamma N P, Puspha Kumari R. 1979. A quarter century of research on lemongrass at lemongrass research station. *Indian Perfumer* 23(3): 218, 219.

Nascimento I B do, Innecco R, Marco C A, Mattos S H, Nagao E O. 2003. Effects of cutting time on the essential oil of lemon grass. *Revista Ciencia Agronomica* 34(2): 169–172.

Oyen L P A, Dung N X. 1999. *Plant Resources of South-East Asia No. 19: Essential Oil Plants*. Backhuys Publishers, Leiden.

Pal S, Chandra S, Singh S, Balyan S, Kaul B L. 1990. Harvest management studies on lemongrass (CKP–25)—a new hybrid strain. *Indian Perfumer* 34(3): 213–216.

Panda H. 2005. *Cultivation and Utilization of Aromatic Plants*. Asia Pacific Business Press Delhi, 600 pp.

Pandey A K, Chowdhury A R. 2000. Chemical composition of essential oil of *Cymbopogon flexuosus* from Central India. Proceedings of Centennial conference on Spices and Aromatic Plants, September 20–23, 2000, Calicut, Kerala, India.

Pandey A K, Pandey B K, Banerjee S K, Shukla P K. 2001. Oushdhiya Paudhon Ki Kheti-Lemongrass (*Cymbopogon flexuosus*), 73 ICFRE-BR-15,CFRHRD-BR-4, Chhindwara, Madhya Pradesh, India.

Pandey M C, Sharma J R, Dikshit A. 1996. Antifungal evaluation of the essential oil of *Cymbopogon pendulus* (Nees ex Steud.) Wats. cv. Praman. *Flavour and Fragrance Journal* 11(4): 257–260.

Paranagama P A, Abeysekera K H T, Abeywickrama K, Nugaliyadde L. 2003. Fungicidal and anti-aflatoxigenic effects of the essential oil of *Cymbopogon citratus* (DC.) Stapf. (lemongrass) against *Aspergillus flavus* Link. isolated from stored rice. *Letters in Applied Microbiology* 37(1): 86–90.

Patra D D, Singh R, Singh D V. 1990. Agronomy of *Cymbopogon* species. *Current Research on Medicinal and Aromatic Plants* 12: 72–84.

Peisíno Alan Luppi, Alberto Diogo Laranja, Mendes Marisa, Calçada Luís Américo. 2005. Study of the frying effects in the composition of the essential oil of Lemon grass (*Cymbopogan citratus*). 2nd Mercosur Congress on Chemical Engineering, ENPROMER 2005. Rio de Janeiro, Brazil.

Pino J A, Rosado A, Correa M T. 1996. Chemical composition of the essential oil of *Cymbopogon winterianus* Jowitt from Cuba. *Journal of Essential Oil Research* 8(6): 693, 694.

Rana V, Prakash Om, Joshi C S. 1996. Effect of nitrogen application and cutting frequencies on productivity of lemongrass (*Cymbopogon flexuosus*) under Poplar (*Populus deltoids*) plantation. *Indian Perfumer* 40(3): 67–72.

Raina V K, Srivastava S K, Tandon S, Mandal S, Aggarwal K K, Kahol A P, Sushil Kumar. 1998. Effect of container materials and long storage periods on the quality of aromatic grass oils. *Journal of Medicinal and Aromatic Plant Sciences* 20(4): 1013–1017.

Raina V K, Srivastava S K, Aggarwal K K, Syamasundar K V, Khanuja S P S. 2003. Essential oil composition of *Cymbopogon martinii* from different places in India. *Flavour and Fragrance Journal* 18(4): 312–315.

Rajeswara Rao B R, Chand S, Bhattacharya A K, Kaul P N, Singh C P, Singh K. 1998. Response to lemongrass (*Cymbopogon flexuosus*) cultivars to spacing and NPK fertilizers under irrigated conditions in semi-arid tropics. *Journal of Medicinal and Aromatic Plant Sciences* 20: 407–412.

Rajeswara Rao B. R. 2001. Biomass and essential oil yields of rainfed palmarosa (*Cymbopogon martinii* (Roxb.) Wats. var. *motia* Burk.) supplied with different levels of organic manure and fertilizer nitrogen in semi-arid tropical climate. *Industrial Crops and Products* 14(3): 171–178.

Rajeswara Rao B R, Bhattacharya A K, Mallavarapu G R, Ramesh S. 2004. Yellowing and crinkling disease and its impact on the yield and composition of the essential oil of citronella (*Cymbopogon winterianus* Jowitt). *Flavour and Fragrance Journal*, 19(4): 344–350.

Rajeswara Rao B R, Kaul P N, Syamasundar K V, Ramesh S. 2005. Chemical profiles of primary and secondary essential oils of palmarosa (*Cymbopogon martinii* (Roxb.) Wats var. *motia* Burk.). *Industrial Crops and Products* 21(1): 121–127.

Rao B L, Verma V, Sobti S N. 1980. Genetic variability in citral containing *Cymbopogon khasianus*. *Indian Perfumer* 24: 13–16.

Rao B L, Lala S. 1992. Optimization of oil yield in CKP-25. *Indian Perfumer* 36(4): 246, 247.

Rao E V S P, Rao R S G, Puttanna K, Ramesh S. 2005. Significance of harvest intervals on oil content and citral accumulation in variety Krishna of lemongrass (*Cymbopogon flexuosus*). *Journal of Medicinal and Aromatic Plant Sciences* 27(1): 1–3.

Rao P G. 2006. *Citronella*. Regional Research Laboratory, Jorhat, India.

Reverchon E. 1997. Supercritical fluids extraction and fractionation of essential oils and related products. *Journal of Supercritical Fluids* 10: 1–37.

Rocha M F A, Borges N S S, Inneco R, Mattos S H, Nagao E O. 2002. Influencia do Larario de Corte Sobre o Citronellol do oleo essencial (*Cymbopogon winterisnus*). *Hortio Bras* 20(Suppl.): 1–4.

Rozzi N L, Phippen W, Simon J E, Singh R K. 2002. Supercritical fluid extraction of essential oil components from lemon-scented botanicals. *Lebensmittel-Wissenschaft und–Technologie*, 35(4): 319–324.

Rudloff V E, Underhill W E. 1965. Seasonal variation in the volatile oil from *Tanacetum vulgare* L. *Phytochemistry* 4: 11–17.

Sahoo S. 1994. Variability, selection and mutation for palmarosa oil production in India. *Journal of Herbs, Spices & Medicinal Plants* 2(3): 49–57.

Saikia R C, Sarma A, Sarma T C, Baruah P K. 2006. Comparative study of essential oils from leaf and inflorescence of Java citronella (*Cymbopogon winterianus* Jowitt). *Journal of Essential Oil Bearing Plants* 9(1): 85–87.

Sandra P, Bicchi C. 1987. *Capillary Gas Chromatography in Essential Oil*, Heidelberg, Basel, New York.

Sargenti S R, Lancas F M. 1997. Supercritical fluid extraction of *Cymbopogon citratus* (DC.) Stapf. *Chromatographia* 465: 6285.

Sarma A, Sarma T C, Handique A, Baruah A K S. 2003. Variation in major chemical constituents in the oil of lemon grass (*Cymbopogon flexuous* [Stud. Wats]), association in different seasons under Brahmaputra valley agro-climatic conditions, *FAFAI Journal* 5(2): 43–49.

Sarma T C, Sharma R K, Adhikari R K, Saha B N. 2001. Effect of altitude and age on herb yield, oil and its major constituents of Java citronella (*Cymbopogon winterianus* Jowitt.). *Journal of Essential Oil Bearing Plants* 4: 77–81.

Sarma T C. 2002. Variation in oil and its major constituents due to season and stage of the crop in Java citronella (*Cymbopogon winterianus* Jowitt). *Journal of Spices and Aromatic Crops* 11(2): 97–100.

Shahi A K, Tava A. 1993. Essential oil composition of three *Cymbopogon* species of Indian Thar Desert. *Journal of Essential Oil Research* 5: 639–643.

Shahi A K, Sharma S N, Tava A. 1997. Composition of *Cymbopogon pendulus* (Nees ex Steud) Wats, an elemicin-rich oil grass grown in Jammu region of India. *Journal of Essential Oil Research* 9(5): 561–563.

Shahidullah M, Huq M F, Islam U, Karim M A, Begum F. 1996. Effect of number of harvests on herbage yield and oil content of citronella and palmarosa grasses. *Bangladesh Journal of Scientific and Industrial Research* 31(1): 127–135.

Singh A, Balyan S S, Shahi A K. 1978. Harvest management studies and yield potentiality of Jammu lemongrass. *Indian Perfumer* 22(3): 189–191.

Singh A K, Naqvi A A, Ram G, Singh K. 1994. Effect of hay storage on oil yield and quality in three *Cymbopogon* species (*C. winterianus*, *C. martinii* and *C. flexuosus*) during different harvesting seasons. *Journal of Essential Oil Research* 6(3): 289–294.

Singh A K, Ram G, Sharma S. 1996. Accumulation pattern of important monoterpenes in the essential oil of citronella Java (*Cymbopogon winterianus*) during one year of crop growth. *Journal of Medicinal and Aromatic Plant Sciences* 18(4): 803–807.

Singh K, Kothari S K, Singh D V, Singh V P, Singh P P. 2000. Agronomic studies in cymbopogons—a review. *Journal of Spices and Aromatic Crops* 9(1): 13–22.

Singh M, Chandrasekhara G D, Rao E V S P. 1996b. Oil and herb yields of Java citronella (*Cymbopogon winterianus* Jowitt) in relation to nitrogen and irrigation regimes. *Journal of Essential Oil Research* 8(5): 531–534.

Singh M. 1997. Growth, herbage, oil yield, nitrogen uptake and nitrogen utilization efficiency of different cultivars of lemongrass (*Cymbopogon flexuosus*) as affected by water regimes. *Journal of Medicinal and Aromatic Plant Sciences* 19(3): 695–699.

Singh M, Shivaraj B, Sridhara S. 1996c. Effect of plant spacing and nitrogen levels on growth, herb and oil yields of lemongrass (*Cymbopogon flexuosus* (Steud.) Wats. var. *Cauvery*). *Journal of Agronomy and Crop Science* 177(2): 101–105.

Singh M. 2001. Long-term studies on yield, quality and soil fertility of lemongrass (*Cymbopogon flexuosus*) in relation to nitrogen application. *Journal of Horticultural Science and Biotechnology* 76(2): 180–182.

Singh N, Luthra R, Sangwan R S. 1989. Effect of leaf position and age on the essential oil quantity and quality in lemongrass (*Cymbopogon flexuosus*) *Planta Medica* 55(3): 254–256.

Singh N, Luthra R. 1988. Sucrose metabolism and essential oil accumulation during lemongrass (*Cymbopogon flexuosus* Stapf.) leaf development. *Plant Science* 57: 127–133.

Sobti S N, Verma V, Rao B L. 1982. Scope for development of new cultivars of *Cymbopogon* as a source of terpene chemicals. In *Cultivation and Utilization of Aromatic Plants*, Regional Research Laboratory, Jammu, Tawi, India, pp. 302–307.

Sushil K, Samresh D, Kukreja A K, Sharma J R, Bagchi G D. 2000. *Cymbopogon: The Aromatic Grass Monograph*. Central Institute of Medicinal and Aromatic Plants, Lucknow, India, 380 pp.

Taskinen J, Mathela D K, Mathela C S. 1983. Composition of the essential oil of *Cymbopogon flexuosus*. *Journal of Chromatography A* 262: 364–366.

Thomas J, Geetha K, Joy P P. 1980. Comparative performance of lemon grass species. *Indian Perfumer* 34: 171, 172.

Virmani O P, Singh K K, Singh P. 1979. Java Citronella and its Cultivation in India. Central Institute of Medicinal and Aromatic Plants, Farm Bulletin No. 1, Lucknow, India.

Voirin B, Brun N, Bayet C. 1990. Effect of day length on the monoterpene composition of leaves of *Mentha piperita*. *Phytochemistry* 29: 749–755.

Wannissorn B, Jarikasem S, Soontorntanasart T. 1996. Antifungal activity of lemon grass oil and lemon grass oil cream. *Phytotherapy Research* 10(7): 551–554.

Weurman C. 1969. *Journal of Agricultural and Food Chemistry* 17: 370.

Weiss E A. 1997. *Gramineae*. In *Essential Oil Crops*. CAB International, Wallingford, U.K., pp. 59–137.

4 Biotechnological Studies in Cymbopogons

Current Status and Future Options

Ajay K. Mathur

CONTENTS

4.1 INTRODUCTION

Human dependence on plants for food, shelter, flavors, fragrance, colors, and health is prehistoric. Recent years have witnessed an unprecedented preference for natural herbals in health, nutraceuticals, and cosmetic sectors, and the gap between demand and supply is fast widening. Since the majority of medicinal and aromatic plants are still collected from the wild, several natural populations are being threatened or extinction. This changing scenario demands two immediate measures: (1) bringing more and more medicinal and aromatic plants under organized cultivation/domestication and (2) upgrading conventional plant improvement techniques with modern tools of plant biotechnology (Galili 2002; Gomez-Galera et al. 2007; Kole 2007). The human quest to improve the yield and quality of bioresources dates back to more than 10,000 years, with simple selection of traits quantifiable at the naked-eye level (Jauhar 2006). Such empirical selective breeding approaches then moved toward channelization of useful genes through conscious hybridization to induction of novel characters through deliberate mutation and polyploid breeding. With the advent of cell and tissue culture approaches coupled with tools of genetic engineering, the state of the art has reached to a level where crops are being designed by borrowing genes of interest from across the taxonomic boundaries, spread over plant, animal and microbial kingdoms (Vasil 2005; Datta 2007; Jullien 2007). Flava-savor tomato, golden rice and Bt-cotton are just a few examples of the tremendous advantages that these tools of plant biotechnology can offer if properly amalgamated with traditional breeding approaches.

Cymbopogons, belonging to the aromatic genus *Cymbopogon* of the angiosperms, are an attractive plant resource for the aroma industry because of their ability to grow under diverse environments, possession of significant chemical polymorphism in their essential oils, and the ease of low-input cultivation in even the most neglected rural setting (Croteau and Gershenzon 1994; Khanuja et al. 2005). The cymbopogons, despite their perennial nature coupled with erratic flowering and seed-setting behavior, an extremely narrow genetic base, and high susceptibility to environmental stress, constitute a storehouse of useful terpenes of high premium that value make these aromatic grasses an ideal candidate for biotechnological prospecting and scrutiny. Biotechnological approaches that are being vigorously pursued in *Cymbopogon* species today include the following: (1) somaclonal breeding for varietal improvement, (2) in vitro cloning of elites, (3) molecular taxonomy, (4) enzymatic/genetic dissection of the biogenetic pathways associated with terpenoid synthesis, (5) isolation, cloning, and expression of pathway related genes, (6) production of designer genotypes through pathway engineering and transgenic route, and (7) high-throughput screening of various constituents of the essential oils for novel bioactivities and their diversified usage (Mathur et al. 1988a, 1988b; Sreenath and Jagadishchandra 1991; Patnaik et al. 1999; Fiehn 2002; Hall et al. 2002; Khanuja et al. 2005; Oksman-Caldentey and Saito 2005; Shasany et al. 2007; Kumar et al. 2007a, 2007b; Tyo et al. 2007; Edris 2007; Gomez-Galera et al. 2007).

This chapter provides an overview of the developments in biotechnological research carried out in *Cymbopogon* species over the last two decades. An attempt has also been made here to highlight the unaddressed or unfinished challenges in the study of this important group of aromatic grasses that plant biotechnologists should take up first.

4.2   *CYMBOPOGON*: THE AROMATIC GENUS

Stapf (1906) gave *Cymbopogon* the status of genus under the tribe Andropogoneae of the family Poaceae. Taxonomically, 140 species of *Cymbopogon* are known today. They have been classified into three series: Schoenanthi, Rusae, and Citrati (Soenarko 1977). Members of these three series are distinct in possessing thin, subcordate, or lanceolate type of leaves, respectively. Nearly 60 species of *Cymbopogon* are odoriferous grasses that are naturally adapted to hot-moist regions of the oriental tropics, subtropical Africa, and Southeast Asia (Jagadishchandra 1975; Soenarko 1977; Robbins 1983; Sreenath and Jagadishchandra 1991; Kuriakose 1995; Shasany et al. 2000). Most of the *Cymbopogon* species are recognized by the characteristic odor note of their essential oils. Lemongrass oil, citronella oil, palmarosa oil, gingergrass oil, and karnkusa oil obtained from *C. flexuosus* and *C. winterianus*, *C. martinii* var. *motia/C.martinii* var. *sofia*, and *C. jwarancusa*, respectively, are the five most widely traded essential oils in the aroma sector.

Endowed with a differential blend of more than 50 terpenoidal constituents, these oils are in high demand in the aroma industry as a perfumery agent per se or as a source of lead molecules to derive more useful value-added products required for high-grade cosmetics or drugs. For example, lemongrass oil, which is used in a range of industrial products where a lemon flavor is required, can also provide citral that can be modified into β-ionone and methyl-ionone and serves as a starting precursor for vitamins A and E synthesis (Robbins 1983). Similarly, citronella oil is a basic ingredient in soaps and mosquito-repellent preparations, and is also a good source of citronellal, which can impart a lily aroma upon hydroxylation or can be converted into l-menthol for the toothpaste industry. Likewise, palmarosa oil rich in geraniol often fetches a high price as a permissible adulterant of costly rose oil. Tremendous value addition can be made in geraniol by converting it into nerol and laevo-citronellol (Thapa et al. 1971; Maheshwari and Mohan 1985). The essential oil of gingergrass is a good source of perillyl alcohols, which can be chemically converted into perillyl acetate to get a spearmint aroma (Zutchi 1982). Karnkusa oil from *C. jwarancusa* is used for its antipyretic and asthmolytic activity. Piperitone obtained from this oil can now be synthetically converted into thymol to fetch very high returns (Shasany et al. 2007).

Among the six commercially cultivated species of *Cymbopogon*, namely, *C. flexuosus* (East Indian lemongrass), *C. winterianus* (Java citronella), *C. martinii* (palmarosa), *C. nardus* (Ceylon citronella), *C. citratus* (West Indian lemongrass), *and C. pendulus* (North Indian lemongrass), the first three are most widely used as a primary source of citral, geraniol and citronellol, citronellal, linalool, 1,8-cineole, limonene, beta-caryophyllene, geranyl acetate, and geranyl formate in the perfumery world (Sobti et al. 1982; Sharma and Ram 2000; McGarvey and Croteau 1995; Aharoni et al.

2005). The majority of these cultivated species are predominantly propagated through vegetative slips, except *C. martinii,* which is multiplied by transplanted seedlings. However, seed progeny thus raised in case of *C. martinii* show a lot of heterogeneity because of its outcrossing nature (Soenarko 1977) and, hence, vegetative propagation is also resorted to in this species when clonal multiplication is desired.

Essential oil content in *Cymbopogon* species has been shown to be a quantitative trait with a high degree of narrow sense heritability and distinct coinheritance and correlation patterns with other plant traits such as height, tiller number, biomass yield, etc. (Kulkarni 1994; Kole 1985; Kulkarni and Rajgopal 1986; Sharma and Ram 2000; Tripathy et al. 2007). As a result of these tight linkages, coupled with problems associated with erratic flowering and seed setting, traditional breeding methods are often found inadequate to fix genes through selfing to develop a superior genotype in *Cymbopogon* species. Genetic improvement in these grasses is therefore confined to recurrent selection and introduction of superior clones out of the vegetative populations. Though a few workers have resorted to induced mutation breeding approach, success has been limited and the genetic advantage was meager (Sharma and Ram 2000).

4.3 BIOTECHNOLOGICAL STUDIES IN CYMBOPOGONS

4.3.1 TISSUE CULTURE STUDIES

On account of the problems associated with their difficult reproductive behavior in applying traditional breeding approaches, the genetic base of cymbopogons is shrinking at a fast pace. Complementation of these limited breeding options with modern tools of plant biology was, therefore, considered necessary in the early 1980s and the first biotechnological tool pressed into the service was plant tissue culture (Jagadishchandra 1982; Jagadishchandra and Sreenath 1986; Mathur et al. 1988a, 1988b, 1989a; Yadav et al. 2000; Zheng et al. 2007). In the initial phase, tissue culture techniques were applied to standardized methods for in vitro cloning of elite selections via axillary shoot culturing in *C. martinii, C. nardus, C. citratus,* and *C. jwarancusa* (Jagadishchandra 1982). Both rhizome and nodal explants readily responded on a hormone-free Murashige and Skoog (1962) medium. Additional supplementation of 6-benzyladenine in the medium at the 0.1–0.5 mg/L level was found beneficial for the growth of 5–8 new leaves from the preformed meristems present in the explants in 3–4 weeks' time.

The focus of the plant tissue culturist was then shifted toward the callus cultures, with two primary goals: (1) to develop somatic embryogenesis-based rapid clonal propagation system and (2) to enhance the spectrum of variability through in vitro mutagenesis, followed by calli cloning to produce somaclonal variations for desired traits. A variety of explants such as zygotic embryos, mesocotyl pieces of seedling, stem, rhizome, and leaf sheath base were tested in different *Cymbopogon* species for the purpose of callus induction (see Sreenath and Jagadishchandra 1991). MS, B5 (Gamborg et al. 1968), and LS (Linsmaier and Skoog 1965) basal media were found suitable for inducing a callus response from these explants. Young unemerged inflorescences with early stages of flower ontogeny were also found to be responsive for callus initiation in *C. martinii* (Baruah and Bordoloi 1989). Most of the workers faced initial problems in initiating an axenic culture in *Cymbopogon* species. Leaching of phenolics from the injured ends of the explants and high degree of contamination due to endophytic microbes present deep inside the explants were the two major hurdles encountered. Addition of antioxidants such as ascorbic acid (40–80 mg/L) in the callus initiation medium and frequent shifting of explants onto the fresh medium during the initial 2 weeks of culturing were found to be very effective in circumventing the problem of browning of explants (Mathur et al. 1988a, 1989a). These workers also showed that soaking of explants in aqueous solution of 50 mg/L chloramphenicol for 8–12 h prior to their plating on the culture medium could also bring about a reduction in contamination frequency. Among the various growth regulators tested, 2,4-dichlorophenoxyacetic acid (2,4-D) at 1.0–5.0 mg/L was found to be most effective in eliciting a callus response in cymbopogons. BAP and α-naphthaleneacetic acid (NAA) at low doses (0.1–1.0 mg/L) in 2, 4-D containing medium were often found helpful in making the induced callus organogenetic in subsequent passages. Once induced, the callus of all *Cymbopogon* species was found to be very fast proliferating with high regenerative potential (Mathur et al. 1988a, 1988b; Nayak et al. 1996; Patnaik et al. 1999). Considering the reports published on callus induction in *Cymbopogon* species so far, the leaf sheath base and seedling explants were found most totipotent for this response. Mature leaves, young rolled leaves, and internodal stem explants were found least responsive. Callus initiated from former set of explants normally sustained their regeneration potential for a longer time (2–4 years). Callus originating from rhizome pieces, on the other hand, depicted a very low organogenetic potential.

Callus cultures of cymbopogons normally showed two types of morphology: (1) organogenic and (2) nonorganogenic. The organogenic callus is usually nodular, fragile, yellowish, and opaque in external appearance, whereas their nonorganogenic counterpart is nonnodular, pale white, slimy, and translucent (Mathur et al. 1989a; Jagadishchandra 1982; Sreenath and Jagadishchandra 1989, 1991). Nonorganogenic calli are further characterized by the presence of loosely arranged isodiametric parenchyma. In comparison, the organogenic callus has a central core of parenchyma surrounded by a peripheral nodular zone consisting of meristematic cells. Friable callus also goes fast into a suspension mode (Sreenath and Jagadishchandra 1980; Patnaik et al. 1995, 1996, 1997, 1999). A 1:5 dilution every 2–3 weeks was found suitable for maintaining these suspension cultures in vitro.

High-frequency regeneration of somatic embryos or adventive shoot buds is a common occurrence in *Cymbopogon* calli. While *C. martinii* and *C. citratus* calli predominantly regenerate via the somatic embryogenic route, *C. winterianus* callus showed a regeneration mode involving de novo formation of shoot bud primordia (Nayak et al. 1996; Patnaik et al. 1999; Mathur et al. 1988a). *Cymbopogon jwarncusa* calli showed both routes of plant regeneration. Though regeneration can often occur on the callus multiplication medium itself, removal of 2,4-D from the media was found essential in some *Cymbopogon* species (Mathur et al. 1989a, 1989b). These workers observed that *C. winterianus* callus cultures did not enter the shoot regeneration stage unless they were shifted from 2,4-D containing medium to a medium supplemented with 3-indole acetic acid (IAA) and BAP or kinetin (Kn). The regenerated shoots showed root initiation on the shoot multiplication medium, but a short exposure of individually excised shoots on medium with 0.1–0.5 mg/L NAA or IAA alone was found to enhance the root quality and subsequent establishment of plantlets in soil (Mathur et al. 1988a, 1988b). Following these protocols, efficient plant regeneration has been reported in palmarosa grass (Baruah and Bordoloi 1989; Patnaik et al. 1995, 1997, 1999). Plants derived through somatic embryos, in general, showed better genetic uniformity in comparison to those obtained through shoot bud organogenesis. Though a few workers have also reported a wide range of chromosomal variations in palmarosa callus (Patnaik et al. 1996; Sreenath and Jagadishchandra 1991) but embryogenesis and subsequent plantlet regeneration were found to occur from cells with normal diploid constitution only. These protocols have been employed for large-scale clonal propagation of this aromatic herb and have important implications for in vitro maintenance of inbreds and male sterile lines of palmarosa for harnessing the hybrid vigor inbreeding program involving sexual hybrids, synthetics, and composites (Ahuja et al. 2000). Nayak et al. (1996) reported a method of rapid in vitro propagation in *C. flexuosus* through somatic embryogenesis. Somatic embryos were formed on callus cultures grown on MS medium with 5.0 mg/L 2,4-D, 0.1 mg/L NAA, and 0.5 mg/L Kn. Embryo-to-plantlet conversion occurred on medium supplemented with 3.0 mg/L BAP, 1.0 mg/L GA$_3$, and 0.1 mg/L NAA. The regeneration potential of these lemongrass cultures was stably sustained for >2 years. These somatic embryo-derived plants, when examined under field conditions, did not show any significant variation over the base material.

The scope and potential of tissue culture technology for widening the narrow genetic base in cymbopogons was first demonstrated from the author's laboratory (Mathur et al. 1988a). Realizing the constraints of applying conventional breeding systems (sexual mating and recombination through meiosis) to create variability in *C. winterianus*, our group at CIMAP initiated an attempt to exploit the tools of somaclonal breeding in this grass. Conceptualized by Larkin and Scowcroft (1981) in wheat and sugarcane, somaclonal variations are defined as variations that arise in vitro during the culturing of a somatic cell or tissue. Such phenotypic variations generally result either from the preexisting genetic variability in the explants or through meiotically stable epigenetic DNA modifications induced by the culture environment (Philips et al. 1994; Bajaj 1990; Duncan 1997). Cryptic chromosomal changes and change in degree of DNA methylation during culturing have been found to be the main factors behind such variations. In our study in citronella, more than 700 callus-derived plantlets (regenerated through adventive shoot buds) were screened under field conditions, out of which 230 calli clones were advanced to plant to row stability assessment for several productivity-linked traits such as herb yield, tiller number, diameter of the bush, dry-matter production, and essential oil content. Variations were also recorded for six important constituents of the oil, namely, citronellal, citronellol, geraniol, citronellylacetate, geranylacetate, and elemol (the negative component in the oil). Correlation analysis between these agronomic characters indicated a strong negative linkage between herb yield and oil content.

Interestingly, we could recover five somaclones with high herb and high oil content, showing thereby that even tight genetic linkages can be broken through somaclonal variation. Recurrent testing of five high herb and oil content bearing type clones at different locations ultimately led to the release of a somaclone (designated CIMAP/Bio-13) as India's first plant variety improved through the somaclonal breeding approach. This variety has shown wide climatic adaptability in rainfed to subtropical to hot and moist conditions. Beside other agronomic traits of commercial importance, the variety registered a remarkable improvement in the initial establishment frequency of its vegetative slips (>85% in comparison to <60% in case of other cultivars in use) and was readily accepted by citronella farmers in the country. The improved strain today occupies more than 60% of land under citronella cultivation in the country and is still the first choice of farmers after almost two decades of its introduction. Following the protocol developed by us, Patnaik et al. (1999) reported the selection of useful somaclones from among the 120 clones, regenerated from cell suspensions in palmarosa (*C. martinii*). Eight clones were advanced to detailed field evaluation for herb and oil productivity. Three such clones with high geraniol content exhibited genetic stability for five clonal cycles. No new cultivar was, however, reported to be released as a new variety from this group.

Recently, two reports (Quiala et al. 2006; Tapia et al. 2007) have described a very effective temporary immersion system (TIS) for biomass production in 1.0 L capacity fermenters for shoot cultures of *Cymbopogon citratus*. Plantlets were regenerated from shoot apices explants and were subjected to immersion in culture medium for 2, 4, or 6 times per day. The highest multiplication coefficient of 12.3 was recorded when 6 times immersion/day treatment was employed. Though maximum fresh weight of 62.2 and 66.2 g was observed when 4 and 6 immersions per day were employed mean dry weight value of 6.4 g and shoot height of 8.97 cm were maximum under the 4 immersion/day treatment. This study has added a new dimension to the in vitro clonal propagation technology for cymbopogons.

4.3.2 PHYLOGENETIC STUDIES

Conventional taxonomic tools for estimation of genetic diversity and phylogenetic scaling of plants primarily rely on morphological, chemotypic, and reproductive characteristics. Molecular prospecting based on genetic maps, on the other hand, can also depict cryptic differences at genetic and functional levels. Molecular markers such as RFLP, AFLP, SSRs, etc., that can show extensive polymorphism at the DNA level are now being used to taxonomically demarcate the closely related taxa (Pecchioni et al. 1996; Croteau et al. 2000; Galili 2002; Jain et al. 2003; Jauhar 2006; Datta

2007; Jullien 2007; Shasany et al. 2007; Tyo et al. 2007; Xu and Crouch 2008). Such molecular markers today find extensive usage in settling taxonomic controversies and phylogenetic analysis of closely similar taxa. Besides, use of these DNA markers can also help plant breeders in their varietal development programs through marker-assisted selection of desired genotypes with a designed combination of genes that control the yield and quality traits (Khanuja et al. 1999; Sangwan and Sangwan 2000; Aharoni et al. 2006; Kumar et al. 2007a, 2007b; Xu and Crouch 2008). These DNA markers can, therefore, enhance the speed of any hybridization or transgenomic program aimed at producing designer specialty crops.

Due to their large number and wider occurrence coupled with a prevalent introgressive hybridization mechanism, the precise phylogenetic identification of *Cymbopogon* species has always been a difficulty (Jagadishchandra 1975; Khanuja et al. 1999, 2005; Sangwan and Sangwan 2000; Sangwan et al. 2000; Shasany et al. 2000, 2007; Sharma and Ram 2000; Kumar et al. 2007a, 2007b). Primary taxonomic delimitation is basically based on chemotaxonomic differences between the species. Isozyme patterns have also been employed to demarcate *Cymbopogon* species (Ganjewala and Luthra 2007). According to Shasany et al. (2007), the genus *Cymbopogon* is today known to have 140 species. Out of these, 52 are described from Africa, 45 from India, 6 each from Australia and South America, 4 from Europe, and the remaining species are distributed in different parts of Southeast Asia. A literature survey of taxonomy of the *Cymbopogon* species indicates that many species were classified on the basis of their morphological and oil quality profile alone. Information on their phylogenesis has been obscure. Distinct chemical features of otherwise morphologically similar *C. martinii* var. *sofia* and *C. martinii* var. *motia* on the one hand and morphologically distinct but chemically identical *C. citratus* and *C. flexuosus* on the other are typical examples of the complexity of such phenotypic/chemotaxonomic classification. Such an approach may lead to erroneous identity and evolutionary relationships between *Cymbopogon* species. Addressing this critical issue, Shasany et al. (2000, 2007) and Khanuja et al. (2005) conducted a detailed study to establish the ancestry of several *Cymbopogon* members. Phylogeny was traced by construction of RAPD/AFLP maps. These workers could successfully estimate the extent of molecular diversity among 19 *Cymbopogon* taxa belonging to 11 taxonomically classified species, 2 released varieties, 1 hybrid taxon, and 4 unidentified wild collections (Khanuja et al., 2005). Citral content was employed as a base marker for chemotypic clustering. Based on their results, the workers have proposed the elevation of *C. flexuosus* var. *microstachy* to a separate species level. They also concluded that *C. travancorensis*, which was earlier merged with *C. flexuosus*, also deserves a separate species status. Their data also substantiated the independent species status for *sofia* and *motia* varieties of *C. martinii*. Several unidentified wild collections included in this molecular tagging work also required a separate taxonomic identity as intermediate forms in the evolution of new taxa.

Advancing the scope of molecular markers for assigning phylogenetic relationships between different species of *Cymbopogon*, Kumar et al. (2007a, 2007b) have recently reported the development of a set of simple sequence repeat (SSR) markers. They used a genomic library of *C. jwarancusa* to develop these markers for the precise identification of *Cymbopogon* species up to the accession level. The SSRs containing genomic DNA clones of *C. jwarancusa* contained a total of 32 SSRs with a range of 1–3 SSRs per clone. About 68.8% of the 32 SSRs had dinucleotide repeat motifs, followed by 21.8% SSRs with trinucleotide and other higher-order repeat motifs. Eighteen of the 32 designed primers for the SSRs, amplified the products to anticipated sizes when tried with genomic DNA of source species *C. jwarancusa*. Thirteen of these 18 functional primers detected polymorphism among *C. flexuosus, C. pendulus,* and *C. jwarancusa* and amplified a total of 95 alleles with a PIC value of 0.44 to 0.96 per SSR. The workers have postulated that higher allelic range and high level of polymorphism depicted by these SSRs can be put to use in a variety of applications in genetic improvement of *Cymbopogon* species through genotype/species authentication and mapping or tagging the genes controlling the productivity-linked traits in marker-assisted breeding endeavors.

4.3.3 METABOLIC ENGINEERING STUDIES

Metabolic engineering is a relatively new area of plant biotechnology research; the term was coined for the first time by Bailey (1991). The concepts in metabolic engineering normally revolved around the identification of the major limiting blocks in the synthesis and accumulation of a desired plant metabolite, followed by systematically removing these blocks with the help of gene manipulation options available today (Verpoorte and Alfermann 2000; Endt et al. 2002; Martin et al. 2003; Horn et al. 2004; Liao 2004; Koffas and Cardayre 2005). Three conditions must be met in a plant (cell) before a significant increment in the biogenesis of a metabolite can be expected to occur. They are (1) induction of genes/enzymes involved in the targeted metabolic pathway at right time and place; (2) availability and supply of the precursors has to be assured; and (3) there should be sufficient sink capacity for storage of the desired metabolite (Mathur et al. 2006). Since the primary goal in a metabolic engineering effort is to divert or channelize the pathway flux toward the synthesis and accumulation of a required phytomolecule, the experimental approaches that are the focus of such a program include the following:

1. Downregulation of the flux in undesired route at branch points of a pathway by antisense or RNAi approach
2. Overexpression of rate-limiting enzyme
3. Decrease catabolism of end products
4. Increase in metabolite-producing or metabolite-storing cell types
5. Manipulation of regulatory elements acting as transcriptional activator/repressor of several genes associated with the target pathway
6. Generation of heterologous expression systems capable of operating minipathways of a broad complex, multistep pathway to obtain intermediates that can be converted into a desired molecule using tools of semisynthetic chemistry or cell-free biotransformations.

Metabolic engineering, therefore, is a sum of all the optimization efforts, culminating in the upgradation of a required biochemical/genetic expression of a phenotype in a defined genotypic background (Tyo et al. 2007). Evidence from the protein engineering and system biology approaches are further advancing the precision of metabolic engineering experiments in plants. Clearly, the pathway manipulation strategies are now evolving from single-gene perturbation to global transcription machinery engineering (gTME). The coming years will see more and more focus on those phytomolecules that are exceptionally costly, are exclusively of plant origin, and that otherwise defy the rules of synthetic chemistry due to complex chirality (Hall et al. 2002; Fiehn 2002; Brent 2004; Oksman-Caldentey and Saito 2005; Yonekura-Sakakibara and Saito 2006; Hall 2006; Gomez-Galera et al. 2007). It is, therefore, no surprise that plant terpenoids are attracting most attention from metabolic engineers (Verpoorte et al. 2000; Verpoorte and Memelink 2002; Trapp and Croteau 2001; Wallaart et al. 2001; Mahmoud and Croteau 2002; Schijlen et al. 2004; Akhila 2007; Chang et al. 2007; Petersen 2007). Initial efforts have indicated that multiple-step terpene pathway engineering across several cellular compartments is feasible (Aharoni et al. 2005).

An enormous wealth of information on chemistry, biological activity, synthesis, and regulation of plant terpenoid metabolism exists in the literature (Akhila 1985, 1986; Croteau and Gershenzon 1994; Penuelas et al. 1995; McGarvey and Croteau 1995; Dixon et al. 1996; Lichtenthaler 1999; Verpoorte et al. 2000; Hallahan 2000; Bourgaud et al. 2001; Eisenreich et al. 2001; Trapp and Croteau 2001; Pichersky and Gershenzon 2002; Holopainen 2004; Cheng et al. 2007). Terpenoids, in general, are a structurally varied class of natural products that are commercially in demand as flavoring, perfumery, pharmaceuticals, insecticidal, and antimicrobial agents (Martin et al. 2003). In plants, they play important roles in plant–plant, plant–environment, and plant–microbe or plant–insect interactions, and impart an ecological fitness to a plant (Aharoni et al. 2005, 2006).

A generalized scheme of biosynthesis of terpenoidal molecules starts with the condensation of iso-pentyl diphosphate (IDP) and its allylic isomer dimethyl-allyl diphosphate (DMADP). The sequential head-to-tail addition of IDP units with DMADP yields geranyl diphosphate (GDP), farnesyl diphosphate (FDP), and geranylgeranyl diphosphate (GGDP). Under the influence of synthases or cyclases, these three metabolites then serve as starting precursors for mono-, sesqui-, and diterpenoids, respectively. The terminal decoration of the basic terpenoidal skeleton is brought about by an array of substitution reactions such as hydroxylation, dehydrogenation, reduction, glucosylation, and methyl or acyl transfers to generate the wide spectrum of terpenoid diversity that we encounter in plants. Terpenoid biosynthesis occurs both in the cytosol and the plastids via mevalonate and MEP pathways, respectively (Lichtenthaler 1999; Eisenreich et al. 2001; Wallaart et al. 2001; Akhila 2007; Cheng et al. 2007).

A literature survey of metabolic engineering efforts in *Cymbopogon* species reflects a vacuum. This is partly because of the difficulty associated with the DNA isolation and genetic transformation protocols in these grasses (Khanuja et al. 1999). Since upstream steps of terpenoid biogenesis in *Cymbopogon* species are the same as those in other aromatic plants such as mints, and sufficient knowledge of their in vitro handling is available today, it may be anticipated that the void will be filled soon. According to Sangwan and Sangwan (2000), application of metabolic tools in cymbopogons would find increasing attention in the following three areas:

1. Bringing alterations in magnitude of flux of pyruvate and glyceraldehydes-3 phosphate toward isopentenyl pyrophosphate (IPP) generation.
2. Modulation of geranyl pyrophosphate synthase or farnesyl pyrophosphate synthase, or their overexpression, to divert the flux toward monoterpene or sesquiterpene accumulation.
3. Secondary modification of parental terpene molecules via individual or a combination of isomerization, reduction, oxidation, lactonization, or epoxidation reaction to derive value-added phytomolecules with enhanced organoleptic or pharmaceutical properties.

Coining a new term *terpenomics* to refer to the ongoing metabolic engineering approaches in aroma crops, Khanuja (2003) also emphasized the need to focus on modulation of terpene synthases' class of enzymes in *Cymbopogon* and *Mentha* species to facilitate unique regiochemical/stereochemical configuration in intermediates of terpene metabolism to derive newer activities. It is hoped that these new bioactives will equip our plants with improved disease resistance, weed/pest control, and more powerful pollination mechanism.

4.4 FUTURE OPTIONS IN BIOTECHNOLOGY OF CYMBOPOGONS

Avenues for future research in the biotechnology of *Cymbopogon* species are enormous. Some of the areas that demand priority efforts are discussed below to ignite the excitement of, and the challenges for, future researchers:

- Somaclonal breeding work in *Cymbopogon* species deserves more scrutiny and attention. With the documented success story of CIMAP/Bio-13 variety of citronella Java in the kitty, future efforts should be specifically focused on enhancing the volumetric yield of important essential oils and their industrially important pure constituents. Cultivars resistant to water stress or flooding stress are required to utilize *Cymbopogon* species in marginal lands or wasteland. Somaclones with better resistance to nutritional and microbial diseases are required through directed in vitro mutagenesis and cell line selections.
- A sound understanding of handling *Cymbopogon* species in tissue cultures can also be put to use in understanding the influence of microbial associations on expression of the biogenetic pathways in these grasses. In vitro tissues maintained at different morphogenetic levels (callus, cell suspensions, shoots, roots, or whole plantlets) would not only

permit a better understanding of the role of tissue differentiation in advancing the pathway but would also provide the ideal aseptic scenario to assess the influence of individual microorganisms (isolated from plants' rhizosphere) on essential oil synthesis following their deliberate incorporation in the cultured tissue. One such effort has recently made in *Vetiveria zizanioides* (vetiver). This study by DelGuidice et al. (2008) has opened up an entirely new avenue for research in aromatic grasses. Using a tissue culture approach, these workers have reported that essential oil biosynthesis in vetiver is confined to the first cortical layer outside the root endodermis that also harbors a community of 10 endophytic bacteria. When these bacteria were cultured on medium containing vetiver oil as sole carbon source, they were able to metabolize the oil, and each endophyte was able to release into the medium a large number of derivatives that were absent in the raw oil fed to them. Besides opening a new line of investigation on in vitro bioconversions, this approach will be useful in understanding the microbial factors responsible for the final expression of essential oil quality in *Cymbopogon* plants.

- Another potentially fruitful area of research in the biotechnology of cymbopogons would be the time course analysis of volatile terpene emission in the headspace of the culture vessels (Tholl et al. 2006; Predieri and Rapparini 2007). This can prove to be a powerful nondestructive method of monitoring the physiological status of the plants in relation to the synthesis of the oil constituents under a controlled environment (Alonzo et al. 2001). The biogenic emission of terpenes in field-grown plants has been found to be associated with oxidative chemistry of the surrounding environment and the plant itself. It will be interesting to trace these volatile emission changes following various stress and/or elicitation treatments given to a controlled culture in vitro (Maes and Debergh 2003; Loreto et al. 2006; Sharkey and Yeh 2001).

- Most of the *Cymbopogon* species, similar to other aromatic crops, synthesize their essential oils in the epidermal cells of the leaves and store the synthesized oil in glandular trichomes (Shankar et al. 1999; Hallahan 2000). It is probably this tight linkage between oil synthesis and tissue differentiation that has deterred researchers from exploiting cell culture approaches for in vitro metabolite production in these species. However, with recent advancements made in the molecular understanding of terpenoidal pathways in plant and refinements in cell culture technology, it is desirable to revisit this research area (Tisserat and Vaughan 2008). Since microshoots of cymbopogons reared in vitro also depict this tissue specialization, a modest beginning can be made by upscaling them in bioreactors for harvesting of desired metabolite. Such an in vitro production system will be particularly relevant for producing oil constituents that are otherwise synthesized in very low amounts in plants. Alternately, they can be used for the production of a single predominant terpene, as has been done for (−) carvone in *Mentha spicata* (Jullien 2007). Until this stage is reached with cymbopogons, the cell cultures of aromatic grasses must be vigorously screened as a source of common enzymes and intermediates of terpene metabolism to carry out useful biotransformation for value addition of aroma phytoceuticals.

- In spite of major advances made in tissue culturing of *Cymbopogon* species, no effort has so far been made to standardize transgenic protocols in these grasses. It seems to be more of a mental block rather than a technical hurdle. Development of an efficient somatic embryogenic-based transformation method in cymbopogons must be attempted to realize the full potential of "-omic" research.

- Anther and pollen culture is another research area that can complement conventional breeding work in cymbopogons. Haploid plants thus produced will be of immense utility for producing inbred lines and for inducing desired mutations in seed-propagated *Cymbopogon* species.

- Chemomolecular prospecting of various essential oil constituents of *Cymbopogon* species to devise new therapeutics has enormous potential (Delespaul et al. 2000; Ojo et al. 2006; Adeneye and Agbaje 2007; Agbafor and Akubugnov 2007; Shen et al. 2007; Masuda et al.

2008). Essential oils of these grasses are already in use in aroma therapies (see Edris 2007). The bioactivities found to be associated with these oils ranged from antibacterial to antiviral, antioxidant, and antidiabetic to antineoplastic. Their capacity to enhance the permeability of stratum corneum of human skin makes them a good adjuvant in transdermal drug delivery formulations (Williams and Barry 2004). Citral, a major constituent of lemongrass oil, has been found to be a potent inducer of glutathione-*s*-transferase class of enzymes, which provide protection to healthy hepatocytes against apoptosis during chemotherapy of liver cancers (Nakamura et al. 2003). This protective capacity of citral was, however, confined to its *trans* configuration (geraniol) only. Its *cis* configuration (neral) had no such effect. Selection of plant cells capable of converting *cis* configuration to *trans* form should be an exciting goal for plant/microbial biotechnologists. Another constituent of lemongrass oil, namely, perillyl alcohol (POH), which is a hydroxylated analog of d-limonene, is also gaining a lot of attention as a powerful radio-sensitizer of tumor cells. It can drastically reduce the radiation doses to control breast, colon, and prostrate cancers (Duetz et al. 2003; Rajesh et al. 2003). Geraniol, the primary constituent of palmarosa grass, has been found to make tumor cells vulnerable to 5-flurouracil (5-FU) by reducing the activity of thymidylate synthase and thymidine kinase, which are responsible for 5-FU toxicity in mice. Bioprospection of essential oils of *Cymbopogon* species would yield agents against several plant and human pathogens such as *Salmonella, Staphylococcus, Shigella, Aspergillus, Fusarium, Penicillium,* etc. (Burt 2004; Wannissorn et al. 2005; Cimanga et al. 2002; Pandey et al. 2003). Bankole and Joda (2004) have reported that a very effective method of controlling storage deterioration of melon seeds due to *Aspergillus* species is mixing the seeds with powdered dry leaves of lemongrass (*C. citratus*). Complete inhibition of *A. flavus* and *A. niger* was realized, and the efficacy was found to be at par with fungicide (iprodione) treatment. Similarly, Duamkhanmanee (2008) has shown the total control of postharvest anthracnose disease of mango fruit caused by *Colletotrichum gloeosporioides* by *C. citratus* oil (4000 ppm in hot water). The anthracnose disease of cowpea crop in the field was demonstrated to be controlled by cold water leaf extract of lemongrass (Amadioha and Obi 1999). Lemongrass oil has also been shown to inhibit growth of *Candida albicans* at 2.0 μL/mL level in the culture medium (El-Khair 2007). Oil treatment kills the microbial cells due to enhanced K^+ depletion from the cells following a fall in membrane lipid content. A similar inhibitory mechanism was also observed in *Saccharomyces cerevisiae* challenged with geraniol of palmarosa oil.

The mechanisms by which essential oils of *Cymbopogon* species inhibit microbial growth are far from completely understood. Their hydrophobicity has often been cited as a possible mode of action that helps them to get partitioned into the lipid bilayer of the microbial cell membrane, making it permeable enough to allow leakage of vital cellular content (Burt 2004; Pattnaik et al. 1997; Delaquis et al. 2002; Prashar et al. 2003). There is tremendous scope to advance this research at enzymatic and molecular levels to develop more ecofriendly and safe bioprotectants for plant and animal use.

4.5 CONCLUSION

Biotechnological studies so far carried out in *Cymbopogon* species should be integrated now for synergism, coordination, and upscaling to derive commercial gains. Availability of efficient tissue culture protocols for plant regeneration on the one hand and fairly explicit understanding of the biosynthetic pathways of their essential oils at the level of genes and enzymes on the other have set the stage to move toward the redesigning of these aroma crops with transgenic approaches. It is hoped that the vast variety of high-value terpenes that these grasses can harbor and the exclusivity of the chemical reactions that their enzymatic and genetic machinery is capable of carrying out, both in planta and in culture, would motivate chemists and biologists alike to convert them into efficient green bioreactors for aroma and pharmaceutical molecules in the near future. In vitro and in vivo bioprospecting of *Cymbopogon* species for deriving novel bioactives is another area that will keep the scientists intensely engaged with these aromatic grasses.

REFERENCES

Adeneye AA, Agbaje EO. 2007. Hypoglycemic and hypolipidemic effects of fresh leaf aqueous extract of *Cymbopogon citratus* Stapf. in rats. *J Ethanopharmacol.* **112:** 440–444.

Agbafor KN, Akubugwo EL. 2007. Hypocholesterelemic effect of ethanolic extract of fresh leaves of *Cymbopogon citratus* (lemongrass). *African J. Biotechnol.* **6:** 596–598.

Aharoni A, Jongsma MA, Bouwmeester HJ. 2005. Volatile science? Metabolic engineering of terpenoids in plants. *Trends Plant Sci.* **10:** 594–602.

Aharoni A, Jongsma MA, Kim T, Ri M, Giri AP, Verstappen FWA, Schwab W, Bouwmeester HJ. 2006. Metabolic engineering of terpenoid biosynthesis in plants. *Phytochem. Review* **5:** 49–58.

Ahuja PS, Mathur AK, Kukreja AK, Pandey B. 2000. Potential of plant tissue culture for the improvement of *Cymbopogon* species. In: Kumar S, Dwivedi S, Kukreja AK, Sharma JR, Bagchi, GD (Eds.), *Cymbopogon: The Aromatic Grass: A Monograph*, pp. 185–198. CIMAP (CSIR) Publication, Lucknow, India.

Akhila A. 1985. Biosynthetic relationship of citral *trans* and citral *cis* in *Cymbopogon flexuosus* (lemongrass). *Phytochemistry* **24:** 2585–2587.

Akhila A. 1986. Biosynthesis of monoterpenes in *Cymbopogon winterianus*. *Phytochemistry* **25:** 421–424.

Akhila A. 2007. Metabolic engineering of biosynthetic pathways leading to isoprenoids: Mono- and sesquiterpenes in plastids and cytosol. *J. Plant Interact.* **2:** 195–204.

Alonzo G, Saiano F, Tusa N, Del Bosco SF. 2001. Analysis of volatile compounds released from embryogeneic cultures and somatic embryos of sweet oranges by head space SPME. *Pl. Cell Tiss. Org. Cult.* **66:** 31–34.

Amadioha AC, Obi VI. 1999. Control of anthracnose disease of cowpea by *Cymbopogon citrates* and *Ocimum gratissimum*. *Acta Phytopathologica et Entomologica Hungarica* **34:** 685–690.

Bailey JE. 1991. Towards a science of metabolic engineering. *Science* **252:** 1668–1675.

Bajaj YPS. 1990. Somaclonal variation—Origin, induction, cryopreservation and implication in plant breeding. In: Bajaj YPS (Ed.) *Biotechnology in Agriculture and Forestry, Vol 11: Somaclonal variation in crop improvement*, pp. 3–48. Springer-Verlag, Berlin.

Bankole SA, Joda AO. 2004. Effect of lemon grass (*Cymbopogon citrates* Stapf.) powder and essential oil on mould deterioration and aflatoxin contamination of melon seeds (*Colocynthis citrullus* L.). *African J. Biotechnol.* **3:** 52–59.

Baruah A, Bordoloi DN. 1989. High frequency plant regeneration of *Cymbopogon martini* (Roxb.) Wats. by somatic embryogenesis and organogenesis. *Pl. Cell Rept.* **8:** 483–485.

Bourgaud F, Gravot A, Milesi S, Gontier E. 2001. Production of plant secondary metabolites: A historical perspective. *Pl. Sci.* **161:** 839–851.

Brent R. 2004. A partnership between biology and engineering. *Nature Biotechnol.* **22:** 1211–1214.

Burt S. 2004. Essential oils: their antibacterial properties and potential applications in food—A review. *Intl. J. Food Microbiol.* **94:** 223–253.

Chang MCY, Eachus RA, Trieu W, Ro D, Keasling JD. 2007. Engineering *Escherichia coli* for production of functionalized terpenoids using plant P-450s. *Nature Chem. Biol.* **3:** 274–277.

Cheng A, Lou Y, Mao Y, Lu S, Wang L, Chen X. 2007. Plant terpenoids: Biosynthesis and ecological functions. *J. Integrative Pl. Biol.* **49:** 179–186.

Cimanga K, Kambu K, Tona L, Apers S, DeBruyne T, Hermans N, Totte J, Pieters L, Vlietinck AJ. 2002. Correlation between chemical composition and antibacterial activity of essential oils of some aromatic plants growing in Democratic Republic of Congo. *J. Ethnopharmacol.* **79:** 213–222.

Croteau R, Gershenzon J. 1994. Genetic control of monoterpene biosynthesis in mints (*Mentha*: Lamiaceae): Genetic engineering of plant secondary metabolism. *Recent Adv. Phytochem.* **28:** 193–229.

Croteau R, Kutchan TM, Lewis NG. 2000. Natural products (Secondary metabolites). In: Buchana B, Gruissem W, Jones R (Eds.), *Biochemistry and Molecular Biology of Plants*, pp. 1250–1318. *Amer. Soc. Pl. Pathologists*, Rockville, Maryland.

Datta SK. 2007. Impact of plant biotechnology in agriculture. In: Pua EC, Davey MR (Eds.), *Biotechnology in Agriculture and Forestry*, Vol. 59. *Transgenic Crops IV*, pp. 3–34. Springer-Verlag, Berlin, Heidelberg.

Delaquis PJ, Stanich K, Girard B, Mazza G. 2002. Antimicrobial activity of individual and mixed fraction of dill, cilantro, coriander and eucalyptus essential oils. *Intl. J. Food Microbiol.* **74:** 101–109.

Delespaul Q, de Billerbeck VG, Roques CG, Michel O. 2000. The antifungal activity of essential oils as determined by different screening methods. *J. Essen. Oil Res.* **12**: 256–266.

DelGiudice L, Massardo DR, Pontieri P, Bertea CM, Carata E, Tredici SM, Tala A, Destefamo M, Maffei M, Alifano P. 2008. The microbial community of vetiver root is necessary for essential oil biosynthesis. *Proc. 52nd Italian Soc. Agric. Genet. Annual Cong.*, poster No. C.01, September 14–17, 2008, Italy.

Dixon RA, Lamb CJ, Masoud S, Sewalt VJH, Paiva NL. 1996. Metabolic engineering: Prospects for crop improvement through the genetic manipulation of phenylpropanoid biosynthesis and defense responses—A review. *Gene* **179**: 61–71.

Duamkhanmanee R. 2008. Natural essential oils from lemongrass (*Cymbopogon citrates*) to control postharvest anthracnose of mango fruit. *Intl. J. Biotechnol.* **10**: 104–108.

Duetz W, Bouwmeester H, VanBeilent J, Witholt B. 2003. Biotransformation of limonene by bacteria, fungi, yeast and plants. *Appl. Microbiol. Biotechnol.* **61**: 269–277.

Duncan RR. 1997. Tissue culture induced variation and crop improvement. In: Sparks DL (Ed.), *Advances in Agronomy*, Vol. 58, pp. 201–210. Academic Press, London.

Edris AE. 2007. Pharmaceutical and therapeutic potential of essential oils and their individual volatile constituents: A review. *Phytother. Res.* **21**: 308–323.

Eisenreich W, Rohdich F, Bacher A. 2001. Deoxyxylulose phosphate pathway to terpenoids. *Trends Plant Sci.* **6**: 78–84.

El-Khair EK. 2007. Effect of *Cymbopogon citrates* L. essential oil on growth and morphogenesis of *Candida albicans*. *Egyptian J. Biotechnol.* **25**: 145–158.

Endt DV, Kijne JW, Memelink J. 2002. Transcription factors controlling plant secondary metabolism: What regulates the regulators?, *Phytochemistry* **61**: 107–114.

Fiehn O. 2002. Metabolomics—The link between genotypes and phenotypes. *Pl. Mol. Biol.* **48**: 155–171.

Galili G. 2002. Genetic, molecular and genomic approaches to improve the value of plant foods and feeds. *Crit. Rev. Pl. Sci.* **21**: 167–204.

Gamborg OL, Miller RA, Ojima K. 1968. Nutrient requirements of suspension cultures of soybean root cells. *Exp. Cell Res.* **50**: 151.

Ganjewala D, Luthra R. 2007. Identification of *Cymbopogon* species and *C. flexuosus* (Nees Ex. Steud.) Wats. cultivars based on polymorphism in esterases isoenzymes. *J. Pl. Sci.* **2**: 552–557.

Gomez-Galera S, Pelacho AM, Gene A, Capell T, Christon P. 2007. The genetic manipulation of medicinal and aromatic plants. *Pl. Cell Rept.* **26**: 1689–1715.

Hall R, Beale M, Fiehn O, Hardy N, Summer L, Bino R. 2002. Plant metabolomics: The missing link in functional genomics strategies. *Plant Cell* **14**: 1437–1440.

Hall RD. 2006. Plant metabolomics: From holistic hope to hype to hot topic. *New Phytol.* **169**: 453–468.

Hallahan DL. 2000. Monoterpenoid biosynthesis in glandular trichomes of Labiateae plants. *Adv. Bot. Res.* **31**: 77–120.

Holopainen JK. 2004. Multiple functions of inducible plant volatiles. *Trends Plant Sci.* **9**: 529–533.

Horn ME, Woodard SL, Howard JA. 2004. Plant molecular farming: Systems and products. *Pl. Cell Rept.* **22**: 711–720.

Jagadishchandra KS. 1975. Recent studies on *Cymbopogon* Spreng. (aromatic grasses) with special reference to Indian taxa: Taxonomy, cytogenetics, chemistry and scope. *J. Plantation Crops* **2**: 43–57.

Jagadishchandra KS. 1982. In vitro culture and morphogenetic studies in some species of *Cymbopogon* Spreng. (aromatic grasses). In: Fujiwara A (Ed.) *Plant Tissue Culture*, pp. 703, 704. Maruzen, Tokyo.

Jagadishchandra KS, Sreenath HL. 1986. In vitro culture of rhizome in Cymbopogon Spreng and Vetiveria Bory. In: Reddy GM (Ed.) *Proceedings of National Symposium on Recent Advances in Plant Cell and Tissue Culture of Economically Important Plants*, pp. 199–208. Osmania University, Hyderabad, India.

Jain N, Shasany AK, Sundaresan V, Rajkumar S, Darokar MP, Bagchi GD, Gupta AK, Khanuja SPS. 2003. Molecular diversity in *Phyllanthus amarus* assessed through RAPD analysis. *Curr Sci.* **85**: 1454–1458.

Jauhar PP. 2006. Modern biotechnology as an integral supplement to conventional plant breeding: The prospects and challenges. *Crop Sci.* **46**: 1841–1859.

Jullien F. 2007. Mints. In: Pua EC, Davey MR (Eds.), *Biotechnology in Agriculture and Forestry* vol 59. *Transgenic Crops IV*, pp. 435–466. Springer-Verlag, Berlin, Heidelberg.

Khanuja SPS. 2003. Revisiting the fragrances—the terpenomics way. *J. Med. Arom. Pl Sci.* **25**: 335.

Khanuja SPS, Shasany AK, Darokar MP, Kumar S. 1999. Rapid isolation of DNA from dry and fresh samples of plants producing large amounts of secondary metabolites and essential oils. *Pl. Mol. Biol. Rept.* **17**: 1–7.

Khanuja SPS, Shasany AK, Pawar A, Lal RK, Darokar MP, Naqvi AA, Rajkumar S, Sundaresan V, Lal N, Kumar S. 2005. Essential oil constituents and RAPD markers to establish species relationship in *Cymbopogon* Spreng. (Poaceae). *Biochem Syst. Ecol.* **33**: 171–186.

Koffas M, Cardayre SD. 2005. Evolutionary metabolic engineering. *Metabolic Eng.* **7:** 1–3.

Kole C. 1985. Improvement of *Cymbopogon winterianus* Jowitt. through mutagenesis. *Ind. Perfumer* **29:** 129–138.

Kole C. 2007. *Genome Mapping and Molecular Breeding in Plants*, Vol. 6—*Technical Crops* (Ed.). Springer-Verlag, Berlin.

Kulkarni RN, Rajagopal K. 1986. Broad and narrow sense heritability estimates of leaf yield, leaf width, tiller number and oil content in East Indian lemongrass. *J. Pl. Breed.* **96:** 135–139.

Kulkarni RN. 1994. Phenotypic recurrent selection for oil content in East Indian lemongrass. *Euphytica* **78:** 103–107.

Kumar J, Verma V, Qazi GN, Gupta PK. 2007a. Genetic diversity in *Cymbopogon* species using PCR based functional markers. *J. Pl. Biochem. Biotechnol.* **16:** 119–122.

Kumar J, Verma V, Sahai AK, Qazi GN, Balyan HS. 2007b. Development of simple sequence repeat markers in *Cymbopogon* species. *Planta Med.* **73:** 262–266.

Kuriakose KP. 1995. Genetic variability in East Indian lemongrass (*Cymbopogon flexuosus* Stapf.). *Ind. Perfumer* **39:** 76–83.

Larkin PJ, Scowcroft WR. 1981. Somaclonal variation—a novel source of variability from cell cultures for plant improvement. *Theor. Appl. Genet.* **60:** 197–214.

Liao JC. 2004. Custom design of metabolism. *Nature Biotechnol.* **22:** 29–36.

Lichtenthaler HK. 1999. The 1-deoxy-d-xylulose-5-phosphate pathway of isoprenoid biosynthesis in plants. *Annu. Rev. Pl. Physiol. Pl. Mol. Biol.* **50:** 47–66.

Linsmaier EM, Skoog, F. 1965. Organic growth factor requirement of tobacco tissue cultures. *Physiol. Plant.* **18:** 100–127.

Loreto F, Batra C, Brilli F, Nogues I. 2006. On the induction of volatile organic compound emissions by plants as a consequence of wounding or fluctuation of light and temperature. *Pl. Cell Environ.* **29:** 1820–1828.

Maes K, Debergh PC. 2003. Volatiles emitted from in vitro grown tomato shoots during abiotic and biotic stress. *Pl. Cell Tiss. Org. Cult.* **75:** 73–78.

Maheshwari ML, Mohan J. 1985. Geranyl formate and other esters in palmarosa oil. *Pafai J.* **7:** 21–26.

Mahmoud SS, Croteau R. 2002. Startegies for transgenic manipulation of monoterpene biosynthesis in plants. *Trends Plant Sci.* **7:** 366–373.

Martin VJJ, Pitera DJ, Withers ST, Newman JD, Keasling JD. 2003. Engineering the mevalonate pathway in *Escherichia coli* for production of terpenoids. *Nat. Biotechnol.* **21:** 1–7.

Masuda T, Osaka Y, Ogawa N, Nakamoto K, Kuminaga H. 2008. Identification of geranic acid—a tyrosinase inhibitor in lemongrass. *J. Agric. Food Chem.* **56:** 597–601.

Mathur AK, Ahuja PS, Pandey B, Kukreja AK, Mandal S. 1988a. Screening and evaluation of somaclonal variations for quantitative and qualitative traits in an aromatic grass. *Cymbopogon winterianus* Jowitt. *Pl. Breed.* **101:** 321–334.

Mathur AK, Ahuja PS, Pandey B, Kukreja AK, Mathur A. 1988b. Development of superior strains of an aromatic grass *Cymbopogon winterianus* Jowitt, through somaclonal variation. In: *Proc. International Conference on Research in Plant Sciences and its Relevance to the Future*, p. 160. Delhi University Press, Delhi, India.

Mathur AK, Ahuja PS, Pandey B, Kukreja A. 1989a. Potential of somaclonal variation in the genetic improvement of aromatic grasses. In: Kukreja AK, Mathur AK, Ahuja PS, Thakur RS (Eds.), *Tissue Culture and Biotechnology of Medicinal and Aromatic Plants*, pp. 79–90. Paramount Publishing House, New Delhi, India.

Mathur AK, Ahuja PS, Kukreja AK, Pandey B, Mathur A. 1989b. Progress in the application of tissue culture for the improvement of aromatic grasses. In: *Proc. XIII Annual Meeting of Plant Tissue Culture Association of India*, pp. 31. North-Eastern Hill University, Shillong, India.

Mathur AK, Mathur A, Seth R, Verma P, Vyas D. 2006. Biotechnological interventions in designing speciality medicinal herbs for 21st century: Some emerging trends in pathway modulation through metabolic engineering. In: Sharma RK, Arora R (Eds.), *Herbal Drugs: A Twenty First Century Perspective*, pp. 83–94. Jaypee Brothers Medical Publishers, New Delhi, India.

McGarvey DJ, Croteau R. 1995. Terpenoid metabolism. *Pl. Cell* **7:** 1015–1026.

Murashige T, Skoog, F. 1962. A revised medium for rapid growth and bioassays with tobacco tissue cultures. *Physiol. Plant.* **15:** 473–497.

Nakamura Y, Miyamoto M, Murakami A, Ohigashi H, Osawa T, Uchida R. 2003. A phase II detoxification enzyme inducer from lemongrass: identification of citral and involvement of electrophilic reaction in the enzyme induction. *Biochem. Biophys. Res. Commun.* **302:** 593–600.

Nayak S, Debata, BK, Sahoo S. 1996. Rapid propagation of lemongrass [*Cymbopogon flexuosus* (Nees) Wats.] through somatic embryogenesis in vitro. *Pl. Cell Rept.* **15:** 367–370.

Ojo OO, Kabutu FR, Bello M, Babayo U. 2006. Inhibition of paracetamol-induced oxidative stress in rats by extracts of lemongrass (*Cymbopogon citrates*) and green tea (*Camellia sinensis*) in rats. *African J. Biotechnol.* **5:** 1227–1232.

Oksman-Caldentey KM, Saito S. 2005. Integrating genomics and metabolomics for engineering metabolic pathways. *Curr. Opin. Biotechnol.* **16:** 174–179.

Pandey AK, Rai MK, Acharya D. 2003. Chemical composition and antimycotic activity of the essential oils of corn mint (*Mentha arvensis*) and lemongrass (*Cymbopogon flexuosus*) against human pathogenic fungi. *Pharmaceut. Biol.* **41:** 421–425.

Patnaik J, Sahoo S, Debata BK. 1995. Somatic embryogenesis and cytological variations in long-term callus cultures of *Cymbopogon martinii* (Roxb.) Wats. *An assessment. Pl. Sci. Res.* **17:** 1–5.

Patnaik J, Sahoo S, Debata BK. 1996. Cytology of callus, somatic embryos and regenerated plants of palmarosa grass [*Cymbopogon martinii* (Roxb.) Wats.]. *Cytobios.* **87:** 79–88.

Patnaik J, Sahoo S, Debata BK. 1997. Somatic embryogenesis and plantlet regeneration from cell suspension cultures of Palmarosa grass (*Cymbopogon martinii*). *Pl. Cell. Rept.* **16:** 430–434.

Patnaik J, Sahoo S, Debata BK. 1999. Somaclonal variation in cell suspension culture-derived regenerants of *Cymbopogon martinii* (Roxb.) Wats. var. *motia. Pl. Breed.* **118:** 351–354.

Pattnaik S, Subramanyam VR, Bajpai M, Kole CR. 1997. Antibacterial and antifungal activity of aromatic constituents of essential oils. *Microbios* **89:** 39–46.

Pecchioni N, Faccioli P, Monetti A, Stanca AM, Terzi V. 1996. Molecular markers for genotype identification in small grain cereals. *J. Genet. Breed.* **50:** 203–219.

Penuelas J, Llusia J, Estiarte M. 1995. Terpenoids: A plant language. *Trends Ecol. Evol.* **10:** 289.

Petersen M. 2007. Current status of metabolic phytochemistry. *Phytochemistry* **68:** 2847–2860.

Philips RL, Kaepler SM, Olhoft P. 1994. Genetic instability of plant tissue cultures: breakdown of normal control. *Proc. Natl. Acad. Sci. USA* **91:** 5222–5226.

Pichersky E, Gershenzon J. 2002. The formation and function of plant volatiles: perfumes for pollinator attraction and defense. *Curr. Opin. Pl. Biol.* **5:** 237–242.

Prashar A, Hili P, Veness RG, Evans CE. 2003. Antimicrobial action of palmarosa oil (*Cymbopogon martinii*) on *Saccharomyces cerevisiae. Phytochemistry* **63:** 569–575.

Predieri S, Rapparini F. 2007. Terpene emission in tissue culture. *Pl. Cell Tiss. Org. Cult.* **91:** 87–95.

Quiala E, Barbon R, Jimenez E, DeFeria M, Chavez M, Capote A, Perez N. 2006. Biomass production of *Cymbopogon citrates* (D.C.) Stapf. A medicinal plant, in temporary immersion system. *In Vitro Cell. Dev. Biol. Plant* **42:** 298–300.

Rajesh D, Stenzel R, Howard S. 2003. Perillyl alcohol as a radio/chemo-sensitizer in malignant glioma. *J. Biol. Chem.* **278:** 35968–35978.

Robbins SRJ. 1983. *Selected Markets for the Essential Oils of Lemongrass, Citronella and Eucalyptus. Tropical Products.* Institute Publication, London.

Sangwan NS, Naqvi AA, Sangwan RS. 2000. A novel chemotype of aromatic grass *Cymbopogon*. In: Kumar S, Kukreja AK, Dwivedi S, Singh AK (Eds.), *Proc. Of the National Seminar on Research and Development in Aromatic Plants: Current Trends in Biology, Uses, Production and Marketing of Essential Oils*, pp. 483, 484. *J. Med. Pl. Sci.* Vol. 22 (Suppl. 1B), CIMAP Publication, Lucknow, India.

Sangwan RS, Sangwan NS. 2000. Metabolic and molecular analysis of chemotypic diversity in *Cymbopogon*. In: Kumar S, Dwivedi S, Kukreja AK, Sharma JR, Bagchi, GD (Eds.), *Cymbopogon: The Aromatic Grass: A Monograph*, pp. 223–247. CIMAP (CSIR) Publication, Lucknow, India.

Schijlen EG, Ric de Vos CH, Van Tunen AJ, Bovy AG. 2004. Modification of flavonoid biosynthesis in crop plants. *Phytochemistry* **65:** 2631–2648.

Shanker S, Ajaykumar PV, Sangwan NS, Kumar S, Sangwan RS. 1999. Essential oil gland number and ultrastructure during *Mentha arvensis* leaf ontogeny. *Biol. Plant.* **42:** 379–387.

Sharkey T, Yeh S. 2001. Isoprene emission from plants. *Annual Rev. Pl. Physiol. Pl. Mol. Biol.* **52:** 407–436.

Sharma JR, Ram RS. 2000. Genetics and genotype improvement of Cymbopogon species. In: Kumar S, Dwivedi S, Kukreja AK, Sharma JR, Bagchi, GD (Eds.), *Cymbopogon: The Aromatic Grass: A Monograph*, pp. 85–128. CIMAP (CSIR) Publication, Lucknow, India.

Shasany AK, Lal RK, Darokar MP, Patra NK, Garg A, Kumar S, Khanuja SPS. 2000. Phenotypic and RAPD diversity among *Cymbopogon winterianus* Jowitt. Accessions in relation to *Cymbopogon nardus* Rendle. *Genet. Resour. Crop. Evol.* **47:** 553–559.

Shasany AK, Shukla AK, Khanuja SPS. 2007. Medicinal and aromatic plants. In: Kole C (Ed.), *Genome Mapping and Molecular Breeding in Plants*, Vol. 6—*Technical Crops*, pp. 175–196. Springer-Verlag, Berlin.

Shen T, Wan W, Yuan H. 2007. Secondary metabolites from *Cymbopogon opobalsamum* and their anti-proliferative effect on human prostrate cancer cells. *Phytochemistry* **68**: 1331–1337.

Sobti SN, Verma V, Rao BL. 1982. Scope for development of new cultivars of *Cymbopogon* as a source of terpene chemicals. In: Atal CK, Kapur BM (Eds.), *Cultivation and Utilization of Aromatic Plants*, pp. 302–307. Regional Research Laboratory (CSIR) Jammu, India.

Soenarko S. 1977. The genus *Cymbopogon* Sprengel (Gramineae). *Reinwardtia* **9**: 225–375.

Sreenath HL, Jagadishchandra KS. 1980. Callus induction and growth in two varieties of *Cymbopogon nardus* (L) Rendle. *Curr. Sci.* **49**: 437, 438.

Sreenath HL, Jagadishchandra KS. 1989. Somatic embryogenesis and plant regeneration from inflorescence culture of Java citronella (*Cymbopogon winterianus* Jowitt). *Ann Bot.* **64**: 211–215.

Sreenath HL, Jagadishchandra KS. 1991. Cymbopogon Spreng (Aromatic Grasses): In vitro culture, regeneration and the production of essential oils. In: Bajaj YPS (Ed.), *Biotechnology in Agriculture and Forestry*, Vol. 15 *Medicinal and aromatic plants III*, pp. 211–236. Springer-Verlag, Berlin, Heidelberg.

Stapf O. 1906. The oil grasses of India and Ceylon. *Kew Bull.* **8**: 297–463.

Tapia A, Georgirev M, Bley T. 2007. Free radical scavengers from *Cymbopogon citrates* (D.C.) Stapf. Plants cultivated in bioreactor of the temporary immersion principle. *Z. Naturforsch. Sect. C J Biosci.* **62**: 447–457.

Thapa RK, Bradu BL, Vashit VN, Atal CK. 1971. Screening of *Cymbopogon* species for useful constituents. *Flavour India* **2**: 49–51.

Tholl D, Boland W, Hansel A, Loreto F, Rose USR. 2006. Practical approaches to plant volatile analysis. *The Plant J.* **45**: 540–560.

Tisserat B, Vaughn SF. 2008. Growth, morphogenesis and essential oil production in *Mentha spicata* L. plantlets in vitro. In Vitro *Cell. Dev. Biol.-Plant* **44**: 40–50.

Trapp SC, Croteau R. 2001. Genomic organization of plant terpene synthases and molecular evolutionary implications. *Genetics* **158**: 811–832.

Tripathy SK, Kole C, Lenke PC. 2007. Isolation of some useful mutants in citronella. *Orisa J. Hort.* **35**: 100–102.

Tyo ET, Alper HS, Stephanopoulos G. 2007. Expanding the metabolic engineering toolbox: more option to engineer cells. *Trends Biotechnol.* **25**: 132–137.

Vasil IK. 2005. The story of transgenic cereals: The challenge, the debate and the solution—A historical perspective. In Vitro *Cell. Dev. Biol.-* Plant **41**: 577–583.

Verpoorte R, Alfermann AW. 2000. *Metabolic Engineering of Plant Secondary Metabolism*. Kluwer, Dordrecht, Netherlands.

Verpoorte R, Memelink J. 2002. Engineering secondary metabolite production in plants. *Curr. Opin. Biotechnol.* **13**: 181–187.

Verpoorte R, van der Heijden R, Memelink J. 2000. Engineering the plant cell factory for secondary metabolite production. *Transgenic Res.* **9**: 323–343.

Wallaart TE, Bouwmeester HJ, Hille J, Poppinga L, Maijers NCA. 2001. Amorpha-4,11-diene synthase: Cloning and functional expression of a key enzyme in the biosynthetic pathway of the novel antimalarial drug artemisinin. *Planta* **212**: 460–465.

Wannissorn B, Jarikasem S, Siriwangchai T, Thubthimthed S. 2005. Antibacterial properties of essential oils from Thai medicinal plants. *Fitoterapia* **76**: 233–236.

Williams A, Barry B. 2004. Penetration enhancers. *Adv. Drug Delivery Res.* **56**: 603–618.

Xu Y, Crouch JH. 2008. Marker-assisted selection in plant breeding: From publication to practice. *Crop Sci.* **48**: 391–407.

Yadav U, Mathur AK, Sangwan NS. 2000. In vitro calli-clone regeneration to enhance qualitative variability in essential oil of a wild *Cymbopogon* species. In: Kumar S, Kukreja AK, Dwivedi S, Singh AK (Eds.), *Proc. of the National Seminar on Research and Development in Aromatic Plants: Current Trends in Biology, Uses, Production and Marketing of Essential Oils*, pp. 485–487. *J. Med. Pl. Sci.* Vol. 22 (Suppl. 1B), CIMAP Publication, Lucknow, India.

Yonekura-Sakakibara K, Saito K. 2006. Genetically modified plants for the promotion of human health. *Biotechnol. Lett.* **28**: 1983–1991.

Zheng Z, Zhou S, Zhang J, Qing GJ, Xu GZ, Guilin X. 2007. Study on the in vitro culture of African lemongrass. *Acta Agric. Shanghai* **23**: 61–64.

Zutchi NL. 1982. Essential oils—isolates and semi-synthetics. In: Atal CK, Kapur BM (Eds.), *Cultivation and Utilization of Aromatic Plants*, pp. 38, 39. Regional Research Laboratory (CSIR), Jammu Publication.

5 The Trade in Commercially Important *Cymbopogon* Oils

Rakesh Tiwari

CONTENTS

5.1 INTRODUCTION

Among the various species of cymbopogons numbering approximately 120, only four are of commercial importance, namely, *Cymbopogon winterianus*, *C. nardus* (citronella), *C. flexuosus* (lemongrass), and *C. martinii* var. *motia* (palmarosa). Essential oils extracted from mainly leaves and aboveground parts of these plants are valued commercially and traded globally (Anonymous 1986). Accurate production, trade, and export and import figures from third world countries, and even some Western countries, are obscure (Lawrence 1985, 1986, 1993). It has been mentioned in some old references that reliable trade data are available for oils produced in the United States only (Simon 1990). However, export and import data from India are available up to country level. The statistics for export and import data for India are collected from available entry points, which include seaports, airports, and land routes. The export data, thus, is inclusive of both exports and re-exports.

5.2 CITRONELLA OIL

Citronella oil is an industrially important essential oil obtained from leaves and stems of *C. winterianus* and *C. nardus*. The oil is regarded as one of the 20 most important essential oils (Lawrence 1993) found in the world trade. Citronella oil is an important source of perfumery chemicals such as citronellol, citronellal, and geraniol, which are widely used in perfumery, soaps, detergents, industrial polishes, cleaning compounds, and other industrial products (Anonymous 1986; Lawless 1995). In the trade, citronella oil is classified into two chemotypes: Ceylon-type citronella oil and Java-type citronella oils. Ceylon-type citronella oil is extracted from *C. nardus* Rendle, and Java-type citronella oil is obtained from *C. winterianus* Jowitt (Torres and Tio [2003]). The name *C. winterianus* was given to this selected variety to commemorate Winter, an important oil distiller of Ceylon (now Sri Lanka), who first cultivated and distilled the Maha Pangeri type of citronella in Ceylon (Chang 2007).

The main difference between the Ceylon and Java types of oil is the relative proportion of geraniol and citronellal. Java-type citronella oil is characterized by a high proportion of geraniol (11%–13%) and citronellal (32%–45%), making it an important source of derivatives such as citronellol and hydroxycitronellal, which are extensively used in compounding high-grade perfumes. Other major constituents of the oil are geranyl acetate (3%–8%) and limonene (1%–4%). The chemistry of the oil is discussed in another chapter 2 of this book.

Ceylon-type citronella oil contains a relatively low proportion of geraniol (18%–20%) and citronellal (5%–15%), and is mainly used as such in lower-grade products. Unlike Java-type oil, it is rarely used for the extraction of derivatives. The other major constituents of Ceylon-type oil are limonene (9%–11%), methyl isoeugenol (7%–11%), and citronellol (6%–8%).

Citronella oil (both Ceylon and Java types) is also a renowned plant-based insect repellent, and has been registered for use in the United States since 1948 as "McKesson's Oil of Citronella" for human application to repel gnats and mosquitoes. The U.S. Environment Protection Agency (EPA) considers oil of citronella as a biopesticide with a nontoxic mode of action (Anonymous 1997, 2001, 2004, 2006, 2007; Trongtokit et al. 2005). It is registered as an insect repellent, feeding depressant, and animal repellent. Research also shows that citronella oil has strong antifungal properties and is effective in calming barking dogs.

5.2.1 World Market, Demand, and Production

The data on world market size, market price, world production, global area under cultivation, number of cultivators engaged, etc., are not easily available due to lack of reliable documentation for such items. The problem is further aggravated because of the lack of scientific documentation in many of the major producer countries. The current world production of citronella oil is estimated at 5000 t, valued at about 20 million USD. The majority of the oil is Java type, with Indonesia as the major supplier in commercial quantities (Robbins 1983). Production of Ceylon-type oil is restricted to Sri Lanka, where it peaks out to approximately 200 t (Oyen and Nguyen 1999). Major producers of citronella oil are China and Indonesia, which account for more than 40% of the world production. The oil is also produced in Taiwan, Guatemala, Honduras, Brazil, Sri Lanka, India, Argentina, Ecuador, Jamaica, Madagascar, Mexico, and South Africa (Lawrence 1985). Citronella oil is used in a variety of products in India. India was a net importing country 60 years ago, but currently produces approximately 600 t of the oil (Singh et al. 2000).

The leading exporters of citronella oil are China, Indonesia, Taiwan, Guatemala, Sri Lanka, Argentina, and Brazil. Officially, Indonesia is the leading exporter of oil, probably because statistics on China's export volume are not available. European countries such as France and United Kingdom play an important role in the world's export of citronella oil; presumably a significant proportion of their import is being re-exported. The same probably is true of Singapore, as this country has a considerable role as an entry port in the citronella oil trade.

Citronella oil has been witnessing demand and price fluctuations due to proliferation of inexpensive synthetic isolates in the market (Robbins 1983). Earlier, Java citronella oil was the most important source of geraniol and citronellal, but the advent of pinene chemistry and production of these isolates in reasonable quantity reduced the demand and reliance for Java-type citronella oil. Further, due to its rich citronellal content, oil from *Eucalyptus citriodora* has become another major competitor of citronella oil in the world market (Anonymous 1986). Thus, the overall demand for citronella oil has been affected by synthetic isolates produced from turpentine oil and *E. citriodora* oil. These isolates and substitutes are generally cheaper than citronella oil, making them the preferred

TABLE 5.1

Consolidated Export and Import Figures of Citronella Oil for India

	Citronella Oil			
	Exports		Imports	
Year	Quantity (t)	Value (USD)	Quantity (t)	Value (USD)
2007	617.502	7,490,027	270.787	1,669,257
2006	440.065	5,322,190	189.462	1,013,885
2005	308.043	2,325,885	115.888	863,690
2004	100.179	1,418,954	122.765	724,823
2003	72.200	616,819	42.558	502,509
2002	33.477	174,241	76.087	574,743

Source: Monthly Statistics of Foreign Trade of India, Vol. 1 and 2.

choice when price is the main criterion. However, natural citronella oil and its derivatives are still the preferred choice of the perfumery industry, mainly because of its unique olfactory and stable properties, which are vital in blending perfumes and compounding industrially important essences.

The state of Assam leads in the production of citronella oil in India owing to climatic conditions suited for the cultivation of the crop. The area under cultivation is around 10,000 hectares (ha). Export and import of citronella oil in India are well documented by Ministry of Commerce and Industry publications (Anonymous 2003–2007a, 2003–2007b).

5.2.2 WORLD DEMAND

Earlier, there were no official data available on the current world demand for citronella oil. International Trade Centre has published data on citronella-oil-importing countries in 1981, which was used in predicting possible trends for the market of citronella oil. However, the situation has improved significantly since then.

The United States of America is the world's largest importer of citronella oil, followed by European countries, namely, France, United Kingdom, Germany, and the Netherlands. The European countries are the major trading hub for Java-type oil because of the presence and proximity of the world-famous perfumery industry, notably in France and Germany. In the Asian region, the largest importers of citronella oil are Japan and Hong Kong. Hong Kong does a lot of re-export of citronella oil to the Philippines.

The consolidated export–import data of citronella oil with respect to India is given under Table 5.1, and countrywise breakup of exports and imports is reproduced in Tables 5.2 and 5.3.

5.3 LEMONGRASS OIL

Lemongrass oil is obtained by steam distillation of leaves and flowering tops of *C. flexuosus* and *C. citratus*. The oil is a commercially important commodity with antifungal properties. The name *lemongrass* is attributed to the lemon-like odor of the essential oil, which is due to high citral content. Two chemotypes of lemongrass oil are well recognized in the trade, namely, East Indian and West Indian lemongrass oil. East Indian lemongrass, obtained from *C. flexuosus*, is also called *Cochin grass* or *Malabar grass*. It is native to Cambodia, India, Sri Lanka, Burma, and Thailand.

TABLE 5.2
Countrywise Import of Citronella Oil by India

Country	2002 Qty (kg)	2002 Value (USD)	2003 Qty (kg)	2003 Value (USD)	2004 Qty (kg)	2004 Value (USD)	2005 Qty (kg)	2005 Value (USD)	2006 Qty (kg)	2006 Value (USD)	2007 Qty (kg)	2007 Value (USD)
Australia	—	—	—	—	20	395	660	9,759	110	4,556	252	52,209
Austria	125	3,191	—	—	—	—	—	—	—	—	115	4,416
Brazil	—	—	8,335	50,303	13	218	6,780	29,218	2	53	8,482	62,888
Canada	1,000	3,704	—	—	30	341	—	—	—	—	—	—
China	7,302	38,388	2,340	18,740	79,118	335,885	45,206	231,737	117,649	445,273	193,650	881,181
France	1,473	27,250	6,523	64,726	1,315	63,728	4,482	97,189	3,646	55,110	678	27,290
Germany	2,122	17,041	2,650	29,072	820	27,298	843	29,634	2,633	48,534	6,599	71,771
Hong Kong	—	—	—	—	—	—	—	—	520	5,606	—	—
Hungary	—	—	25	1,205	—	—	—	—	—	—	—	—
Indonesia	2,090	67,914	105	2,636	1,650	6,120	20,540	83,200	27,520	103,900	1,650	20,409
Ireland	—	—	—	—	—	—	—	646	30	180	—	—
Israel	165	1,691	20	258	5,352	22,455	2,118	21,682	—	—	120	3,887
Italy	870	6,374	13	579	205	8,530	380	13,612	1,027	30,818	1,151	33,867
Japan	596	16,888	111	507	—	—	—	—	—	—	2	12
Malaysia	—	—	—	—	10	75	—	—	4	252	—	—
Nepal	24,498	33,624	4,000	20,592	17,200	47,655	7,730	87,136	2,400	12,121	1,445	15,493

Netherlands	40	1,192	—	—	1	48	25	504	—	—	39	1,297
Paraguay	—	—	—	—	2,090	21,280	—	—	570	5,504	2,000	15,891
Poland	—	—	815	1,189	138	1,678	713	4,561	40	2,204	2	374
Singapore	2,291	10,510	8,865	120,551	2,300	14,072	175	1,343	4,307	35,610	795	24,131
Slovenia	—	—	—	—	75	5087	—	—	—	—	—	—
Spain	545	1,690	1,506	25,302	1,100	31,386	4,944	21,130	11,401	46,841	21,189	98,421
Sri Lanka	—	—	—	—	280	4,164	—	—	—	—	—	—
Swaziland	—	7,840	—	—	—	—	25	—	—	—	241	3,458
Switzerland	840	—	—	—	1,280	3,044	—	1,512	480	7,226	386	8,950
Tanzania	—	—	—	—	—	—	—	—	200	13,087	1,249	77,109
Thailand	42	—	1137	—	—	—	—	—	—	—	95	3,171
UAE	—	—	—	—	202	1,532	52	529	2	23	—	—
U.K.	2,776	39,089	1,247	32,498	1,066	35,243	1,789	93,782	1,496	36,255	3,931	52,479
U.S.	25,712	275,506	9,500	130,291	7,201	84,528	15,826	123,247	13,625	153,526	24,716	196,726
Uruguay	—	—	—	—	1,199	2,835	—	—	—	—	—	—
Unspecified	—	—	—	—	100	7,226	—	—	—	—	—	—
Vietnam	3,600	21,714	5,040	4,060	—	—	3,600	13,269	1,800	7,206	2,000	13,827
Total	**76,087**	**574,743**	**42,558**	**502,509**	**122,765**	**724,823**	**115,888**	**863,690**	**189,462**	**1,013,885**	**270,787**	**1,669,257**

Source: Monthly Statistics of Foreign Trade of India, Vol. 1 and 2.

TABLE 5.3
Countrywise Export of Citronella Oil by India

	2002		2003		2004		2005		2006		2007	
Country	Qty (kg)	Value (USD)	Qty (kg)	Value (USD)	Qty (kg)	Value (USD)	Qty (kg)	Value (USD)	Qty (kg)	Value (USD)	Qty (kg)	Value (USD)
Afghanistan	—	—	—	—	—	—	—	—	—	—	100	456
Angola	—	—	—	—	—	—	—	—	—	—	2,000	42,912
Australia	67	255	—	—	300	1,507	951	9,193	630	11,054	1,840	24,376
Austria	16	5,033	16	905	—	—	700	1,740	—	—	300	2,066
Bahrain	—	—	—	—	—	—	—	—	178	15,844	1,450	55,539
Bangladesh	2,500	5,574	—	—	—	—	1,800	4,108	4,500	7,777	5,410	10,029
Belgium	—	—	50	2,605	80	551	180	1,328	—	—	—	—
Bulgaria	—	—	—	—	20	1,065	1,530	15,913	—	—	—	—
Cambodia	—	—	—	—	—	—	—	—	2,080	12,737	25	5,978
Canada	40	1,904	9,021	24,539	3,261	26,597	189	3,656	231	2,261	230	10,001
Chile	—	—	—	—	700	5,753	—	—	—	—	—	—
China	—	—	14,450	95,189	—	—	11	147	45,200	430,406	76,874	1,088,379
Chinese Taipei	—	—	107	11,569	—	—	—	—	—	—	—	—
Colombia	—	—	—	—	—	—	—	—	360	5,819	117	3,352
Congo P REP	—	—	—	—	5,600	26,342	—	—	—	—	26,773	79,226
Costa Rica	—	—	—	—	—	—	—	—	40	867	—	—
Czech Republic	—	—	—	—	—	—	11	1,153	—	—	1,005	17,847
Denmark	720	3,841	—	—	30	827	50	1,409	—	—	—	—
Djibouti	—	—	—	—	—	—	—	—	400	12,101	—	—
Egypt	—	—	—	—	45	998	75	1,458	70	1,342	—	—
Ethiopia	—	—	—	—	3,060	7,268	1,620	8,533	—	—	—	—
Fiji	—	—	—	—	—	—	—	—	—	—	17	746
Finland	—	—	—	—	—	—	—	—	—	—	5,860	100,293
France	1,407	22,365	1,257	22,476	1,180	13,597	8,225	152,677	5,376	117,235	2,908	39,223
Gambia	—	—	—	—	—	—	56	3068	—	—	—	—
Germany	35	2,395	3,280	46,450	13,330	84,528	500	36,404	4,560	59,299	11,235	316,242
Ghana	—	—	—	—	—	—	16,850	39,311	—	—	11,600	108,785
Greece	—	—	—	—	600	3,523	—	—	—	—	—	—

Country												
Guyana	—	—	—	—	—	—	—	—	—	—	150	2,023
Hong Kong	—	—	—	—	145	4,066	7,500	251,457	40	1,499	28,895	347,037
Hungary	—	—	100	795	—	—	—	—	26	1,963	565	12,063
Iceland	—	—	—	—	—	—	—	—	—	—	31	2,230
Indonesia	—	—	1,000	10,937	10	358	—	—	190	2,960	—	—
Iran	—	—	—	—	50	1,163	50	1,317	—	—	—	—
Ireland	—	—	1,000	—	—	—	140	2,160	30	254	180	1,343
Israel	—	—	1,854	5,863	20	213	—	—	—	—	685	12,648
Italy	—	—	—	26,273	2	177	14,543	98,830	30	1,587	24	772
Jamaica	—	—	—	—	—	—	—	—	—	—	150	2,159
Japan	—	—	—	—	5,102	438,600	4,320	36,859	4,430	64,159	4,071	124,168
Jordan	—	—	—	—	—	—	—	—	—	—	50	652
Kenya	—	—	—	—	15,40	76,777	34,783	199,318	31,819	263,200	40,750	337,873
Korea DP RP	—	—	—	—	—	—	1,000	6,952	—	—	228	4,907
Korea RP	—	—	—	—	1,500	7,599	—	—	1,566	23,138	215	6,015
Kuwait	—	—	20	1,429	—	—	298	9,933	299	10,420	118	11,164
Latvia	—	—	—	—	—	—	—	—	333	5,169	101	4,836
Malaysia	1,496	17,917	100	245	1,324	7,629	4,082	13,539	265	4,834	4,039	24,910
Maldives	—	—	40	1,192	430	4,285	100	511	—	—	—	—
Mauritius	501	—	85	1,642	65	1,107	852	6,081	3,540	15,440	110	3,701
Mexico	—	—	—	—	—	—	—	—	50	375	—	—
Mozambique	8,000	6,149	—	—	—	—	—	—	720	1686	170	3503
Myanmar	—	20,123	1,100	4,126	5,000	54,727	2,000	30,853	—	—	—	—
Nepal	804	—	359	5,612	10	61	1,061	6,672	4,421	30,561	5,474	36,963
Netherlands	—	7,843	4,000	33,633	865	21,158	169	13,812	1,100	14,873	10,141	171,511
New Caledonia	—	—	—	—	—	—	—	—	—	—	100	143
New Zealand	—	—	9	395	—	—	2	96	670	971	150	5,751
Nigeria	2,000	—	24	570	4,182	52,693	—	—	105,891	288,324	90,118	347,096
Oman	—	2,953	—	—	65	1,058	—	—	200	1,489	55	1,247
Pakistan	—	—	—	—	25	1,955	—	—	—	—	145	7,607
Panama Republic	—	—	—	—	42	3,034	171	9,072	—	—	—	—
Paraguay	—	—	—	—	—	—	—	—	—	—	4,000	86,736
Peru	—	—	—	—	—	—	1,000	26,056	—	—	—	—
Philippines	—	—	—	—	—	—	—	—	—	—	—	—

(continued on next page)

TABLE 5.3 (continued)

Countrywise Export of Citronella Oil by India

Country	2002 Qty (kg)	2002 Value (USD)	2003 Qty (kg)	2003 Value (USD)	2004 Qty (kg)	2004 Value (USD)	2005 Qty (kg)	2005 Value (USD)	2006 Qty (kg)	2006 Value (USD)	2007 Qty (kg)	2007 Value (USD)
Poland	—	—	—	—	—	—	—	—	445	7,269	70	207
Reunion	—	—	—	—	—	—	—	—	—	—	—	—
Romania	—	—	—	—	—	—	—	—	—	—	—	—
Russia	—	—	—	—	—	—	110	7,843	1,100	5,714	1,400	13,883
Rwanda	—	—	—	—	—	—	500	1,381	—	—	—	—
Saudi Arabia	6,250	13,724	—	—	1,925	6,036	1,170	24,321	2,605	66,550	1,287	45,409
Singapore	12	171	1,408	24,902	3,358	30,928	13,111	88,392	710	21,360	30,294	292,650
South Africa	—	—	600	3,761	460	10,357	1,870	30,024	2,500	50,539	2,462	50,139
Spain	600	17,784	16,346	112,975	193	2,443	1,654	118,038	4,570	60,387	9,400	128,492
Sri Lanka	450	5,575	6,540	39,485	65	2,514	2,835	14,553	2,370	8,777	86	4,048
Sudan	—	—	—	—	—	—	11,500	15,716	300	3,708	10	602
Switzerland	—	—	—	—	190	980	1,252	15,471	750	63,819	2,000	49,887
Syria	—	—	—	—	—	—	—	—	—	—	100	2,482
Taiwan	—	—	—	—	120	3,597	151	4,028	11	381	6	778
Tanzania	1,500	2,783	—	—	34,290	186,623	84,913	348,669	59,200	366,188	35,530	171,104
Thailand	—	—	500	1,132	—	—	—	—	2,700	37,477	43	5,749
Tunisia	—	—	—	—	—	—	—	—	—	—	934	107,816
Turkey	526	1,498	—	—	—	—	—	—	25	245	325	8,644
Uganda	4,960	13,923	—	—	460	2,500	600	5,215	20	419	1,080	12,586
UAE	—	—	4,591	20,912	6,125	39,496	4,003	44,813	9,716	996,468	6,206	137,707
U.K.	—	—	458	25,668	500	23,242	26,104	225,236	26,562	444,360	64,912	756,278
U.S.	1,593	22,431	3,885	91,539	3,790	257,425	46,140	368,914	98,430	1,735,440	121,872	2,231,351
Yemen	—	—	—	—	—	—	—	—	8,795	43,064	1,020	4,929
Zambia	—	—	—	—	—	—	3,160	7,339	—	—	—	—
Unspecified	—	—	—	—	—	—	4,000	8,288	—	—	—	—
Total	**33,477**	**174,241**	**72,200**	**616,819**	**100,179**	**1,418,954**	**308,043**	**2,325,885**	**440,065**	**5,322,190**	**617,502**	**7,490,027**

Source: Monthly Statistics of Foreign Trade of India, Vol. 1 and 2.

West Indian lemongrass, obtained from *C. citratus*, is assumed to be native to Malaysia. While both can be used interchangeably, *C. citratus* is more suited for cooking. In India, *C. citratus* is used both as a medicinal herb and in perfumes. The main difference between these two types of oil is the relative percentage of citral. East Indian lemongrass oil is higher in citral content, which ranges up to 90%. West Indian lemongrass oil has lower citral content and lower solubility in alcohol. The lower solubility is attributed to the presence of myrcene, which polymerizes on exposure to air and light. In the trade, West Indian oil is considered inferior to East Indian oil and has meager trade. The name *West Indian lemongrass oil* is a misnomer in the sense that this grass is not indigenous to West Indies and, currently, the production of lemongrass in West Indies is very low (Husain 1993).

The lemongrass oil finds widespread use in soap, perfumery, cosmetics, and the beverages industry. Additionally, it is an important natural source of citral, which is an important starting material for the synthesis of beta ionone. Beta ionone is further used for synthesis of a number of aroma chemicals widely used in perfumery and cosmetics. Beta ionone is also used to produce vitamin A. Due to its distinct lemony flavor, the herb itself finds use in imparting citrus flavor in fresh, chopped and sliced, or dried and powdered forms. Lemongrass is commonly used in teas, soups, and curries. In India, the record of medicinal use of lemongrass oil dates back to more than 2000 years, though its distillation started only in 1890.

5.3.1 WORLD PRODUCTION AND DEMAND

During the early 1950s, India produced over 1800 t/annum of lemongrass oil and held monopoly both in production and world trade. This has changed, as Guatemala, China, Mexico, Bangladesh, etc., have developed large-scale cultivation of lemongrass.

Currently, the world production of oil of lemongrass ranges between 800 and 1300 t/annum (Singh et al. 2000). However, another 600 t of a substitute oil, that is, *Litsea cubeba* (rich in citral), is exported by China, which limits the scope for any faster growth in export trade of lemongrass oil (Lawrence 1985). Synthetic citral available at a relatively lower rate competes with lemongrass oil and natural citral in the market.

5.3.2 PRODUCTION

India is a major producer of the oil, and about 80% of the produce is exported, mainly to West Europe, United States, and Japan. The situation changed during the Second World War because of problems of production and supply logistics. Consequently, Guatemala, Haiti, Madagascar, Zaire, Cambodia, Vietnam, and Laos started producing lemongrass oil to meet the demand, and continue to do so. Guatemala, China, Mexico, Bangladesh, etc., have developed its cultivation over large areas. India has moved to systematic cultivation of lemongrass; earlier, it was mainly collected from wild forests. East Indian lemongrass oil is preferred due to its high citral content. Though exact information on production and trade from the producing countries is not available, the current world production of the oil is estimated at 1300 t/annum, with India contributing to the tune of 350 t. The consolidated export–import data of lemongrass oil with respect to India is given in Table 5.4, and the countrywise breakup of the export and import figures is presented in Tables 5.5 and 5.6.

In India, the crop grows in an area of about 3000 ha, largely in the states of Kerala, Karnataka, Tamil Nadu, Maharashtra Uttar Pradesh and Assam. Supply of cheaper citral by China produced from alternative sources results in periodic fluctuation in demand for lemongrass oil. However, in general, the global demand for the oil is robust.

Currently, other major lemongrass-oil-producing countries are Brazil, China, Guatemala, Haiti, Nepal, Russia, and Sri Lanka.

TABLE 5.4
Consolidated Export and Import Figures of Lemongrass Oil for India

	Lemongrass oil			
	Exports		Imports	
Year	Quantity (t)	Value (USD)	Quantity (t)	Value (USD)
2007	71.44	808,522	0.283	5,333
2006	67.493	641,149	0.776	8,472
2005	36.183	319,006	0.1	2,847
2004	27.079	290,073	0.276	4,534
2003	45.387	475,284	0.755	4,389
2002	46.778	457,524	1.68	12,206

Source: Anonymous 2003–2007a and b.

TABLE 5.5
Countrywise Import of Lemongrass by India

	2002		2003		2004		2005		2006		2007	
Country	Qty (kg)	Value (USD)	Qty (kg)	Value (USD)	Qty (kg)	Value (USD)	Qty (kg)	Value (USD)	Qty (kg)	Value (USD)	Qty (kg)	Value (USD)
Bhutan	1,080	9,588	—	—	—	—	—	—	—	—	—	—
France	—	—	—	—	200	3,271	—	—	—	—	—	0
Germany	—	—	—	—	—	—	—	—	35	404	5	108
Indonesia	—	—	25	353	—	—	—	—	—	—	—	—
Italy	—	—	—	—	—	—	—	—	250	3,627	250	3,975
Singapore	—	—	—	—	—	—	—	—	391	2,811	10	744
Spain	600	2,618	—	—	—	—	—	—	—	—	—	—
U.K.	—	—	368	1,941	—	—	75	2,033	—	—	18	506
U.S.	—	—	362	2,095	—	—	25	814	100	1,630	—	—
Unspecified	—	—	—	—	76	1,263	—	—	—	—	—	—
Total	1,680	12,206	755	4,389	276	4,534	100	2,847	776	8,471	283	5,333

Source: Anonymous 2003–2007a and b.

5.4 PALMAROSA OIL

Palmarosa oil of commerce is obtained by steam distillation of ground, freshly harvested or partially dried flowering shoot biomass leaves and stems of *C. martinii* var. *motia*. The oil is valued for its aroma chemical geraniol (75%–90%), which is separated through fractional distillation of the essential oil. Geraniol is widely used in flavor, fragrance, and pharmaceutical industries. The oil is extensively used as perfumery raw material in soaps, floral rose-like perfumes, cosmetics preparations, and in the manufacture of mosquito repellent products. It is used for flavoring tobacco products, foods, and nonalcoholic beverages. In medicine, the volatile oil is used as a remedy for lumbago, stiff joints, skin diseases, and for bilious complaints. Another variety, *C. martinii* var. *sofia*, which has lower geraniol content (<70%), is referred to as *gingergrass oil*. It is distilled from wild growth in Madhya Pradesh, Maharashtra, and Andhra Pradesh (states of India). The plant is native to Southeast Asia (India and Pakistan). The other popular name for the oil is Indian geranium oil. Palmarosa oil is valued for its geraniol content, which finds widespread applications in

TABLE 5.6
Countrywise Export of Lemongrass Oil by India

Country	2002 Qty (kg)	2002 Value (USD)	2003 Qty (kg)	2003 Value (USD)	2004 Qty (kg)	2004 Value (USD)	2005 Qty (kg)	2005 Value (USD)	2006 Qty (kg)	2006 Value (USD)	2007 Qty (kg)	2007 Value (USD)
Australia	5,636	45,685	1,980	21,291	1,760	23,673	1,600	16,055	3,520	36,946	5,040	60,926
Austria	20	336	—	—	—	—	—	—	—	—	—	—
Belgium	—	—	—	—	—	—	—	—	540	5,235	1,080	15,118
Bosnia	—	—	—	—	22	1,162	—	—	—	—	—	—
Brazil	—	—	—	—	—	—	—	—	600	6,850	400	4,531
Canada	400	4,031	415	4,323	400	3,986	35	2,149	800	4,935	600	6,514
China	—	—	10	191	—	—	—	—	—	—	95	1,464
Denmark	—	—	—	—	—	—	420	9,623	100	679	—	—
Ecuador	—	—	—	—	—	—	2	34	10	161	—	—
Egypt	—	—	20	506	—	—	—	—	100	796	100	791
French South and Antarctic Territories	—	—	—	—	—	—	—	—	—	—	1	9
Fiji	—	—	—	—	—	—	30	126	—	—	—	—
France	6,000	63,216	4,709	49,137	540	5,962	—	—	1,982	20,855	5,640	70,816
Germany	4,400	46,962	7,733	91,140	6,360	62,518	3,206	30,524	4,820	50,520	5,250	58,583
Ghana	—	—	25	218	—	—	—	—	—	—	—	—
Greece	750	3,400	—	—	—	—	—	—	—	—	—	—
Guatemala	—	—	—	—	—	—	—	—	1,080	11,620	1,260	14,460
Hong Kong	—	—	10	105	—	—	—	—	—	—	—	—
Indonesia	—	—	25	325	—	—	—	—	—	—	—	—
Israel	20	330	45	655	145	2031	20	185	940	8,220	—	—
Italy	—	—	56	4,243	—	—	—	—	—	—	410	4,667
Japan	1,620	17,380	—	—	740	9,047	540	5,033	540	4,740	390	4,338
Kenya	150	2,299	1,135	7,262	—	—	—	—	—	—	100	1,360
Korea RP	30	471	6	58	—	—	1,000	2,110	—	—	200	3,025
Maldives	—	—	—	—	—	—	—	—	43	493	—	—

(continued on next page)

TABLE 5.6 (continued)
Countrywise Export of Lemongrass Oil by India

Country	2002 Qty (kg)	2002 Value (USD)	2003 Qty (kg)	2003 Value (USD)	2004 Qty (kg)	2004 Value (USD)	2005 Qty (kg)	2005 Value (USD)	2006 Qty (kg)	2006 Value (USD)	2007 Qty (kg)	2007 Value (USD)
Mauritius	—	—	—	—	10	150	—	—	50	375	—	—
Mexico	—	—	—	—	360	3,145	—	—	—	—	—	—
Myanmar	—	—	—	—	1,000	30,143	30	346	—	—	—	—
Nepal	—	—	—	—	—	—	100	663	—	—	—	—
Netherlands	—	—	180	1,942	—	—	5,000	43,671	2,000	16,711	—	—
New Zealand	—	—	—	—	—	—	—	—	—	—	100	1,043
Oman	—	—	—	—	210	1,406	—	—	—	—	—	—
Philippines	—	—	—	—	—	—	—	—	150	347	—	—
Singapore	1,200	13,542	1,880	20,873	367	4,741	1,960	17,871	3,965	35,172	2,760	29,741
Slovenia	80	848	—	—	—	—	—	—	—	—	—	—
South Africa	—	—	—	—	—	—	50	632	—	—	—	—
Spain	900	8,890	—	—	1,080	10,511	—	—	1,190	13,089	—	—
Sri Lanka	1,140	10,828	210	2,726	20	176	1,315	10,929	584	6,411	975	11,325
Surinam	—	—	—	—	420	5,135	—	—	—	—	—	—
Switzerland	—	—	3	114	—	—	—	—	1,080	10,561	6,888	73,801
Taiwan	—	—	—	—	30	522	50	485	—	—	2,180	24,565
Thailand	20	209	2,150	23,067	500	5,491	625	4,891	220	1,971	601	5,232
Tanzania	—	—	—	—	50	466	—	—	—	—	—	—
U.K.	13,575	135,961	12,670	128,514	7,340	70,775	6,800	55,703	18,760	171,513	12,056	135,112
U.S.	10,837	103,136	12,125	118,594	5,025	48,911	13,400	117,976	24,419	232,949	25,148	279,271
Unspecified	—	—	—	—	700	122	—	—	—	—	—	—
Vietnam	—	—	—	—	—	—	—	—	—	—	100	438
Yemen	—	—	—	—	—	—	—	—	—	—	66	1,392
Total	**46,778**	**457,523**	**45,387**	**475,284**	**27,079**	**290,072**	**36,183**	**319,004**	**67,493**	**641,147**	**71,440**	**808,521**

Source: Anonymous 2003–2007a.

traditional medicine and household uses. The essential oil has a scent similar to that of rose oil, and hence the name *palmarosa*.

5.4.1 PRODUCTION

Historically, the first attempt at its cultivation dates back to 1924 in West Punjab, which now is in Pakistan. Later, its cultivation in India was started in the early 1950s in Dehra Dun (Uttarakhand). Currently, the crop is cultivated in several states of India, namely, Uttar Pradesh, Uttarakhand, Assam, Andhra Pradesh, and Karnataka. The grass has been introduced to the Central American countries of Guatemala, Honduras, Indonesia, and Brazil (Husain 1993).

The current Indian production is around 100 t of oil per year, with consumption figures of 40–60 t. India enjoyed a virtual monopoly in palmarosa oil production until the 1990s. Indian palmarosa oil was rated as premium quality oil with 80%–90% geraniol content. Brazil was the second largest producer of the oil (Lawrence 1985). Later, Brazil, Indonesia, Honduras, and Guatemala started producing and supplying better quality oil in the world market. With the appearance of synthetic geraniol, the world market witnessed a weakening in demand for natural palmarosa oil. The consolidated export–import data of palmarosa oil for India is given in Table 5.7, and country-wise breakup of the export and import is presented in Tables 5.8 and 5.9. The United States is the major importer of the oil, followed by Europe (France, Germany, the Netherlands, and the United Kingdom) and Australia.

TABLE 5.7
Consolidated Export and Import Figures of Palmarosa Oil for India

	Palmarosa Oil			
	Exports		Imports	
Year	Quantity (t)	Value (USD)	Quantity (t)	Value (USD)
2007	20.684	365,323	0	0
2006	17.883	212,466	0.56	24,852
2005	7.257	79,313	0.027	226
2004	14.795	183,779	0.11	7,949
2003	22.828	318,365	0.5	24,405
2002	11.620	166,449	—	—

Source: Anonymous 2003–2007a and b.

TABLE 5.8
Country-wise Breakup of the Import Figures of Palmarosa Oil for India

	2003		2004		2005		2006	
Country	Qty (kg)	Value (USD)	Qty (kg)	Value (USD)	Qty (kg)	Value (USD)	Qty (kg)	Value (USD)
Germany	500	24,406	—	—	—	—	200	11,116
Indonesia	—	—	—	—	—	—	360	13,736
UAE	—	—	—	—	3	118	—	—
U.K.	—	—	110	7,950	—	—	—	—
U.S.	—	—	—	—	24	108	—	—
Total	500	24,406	110	7,950	27	226	560	24,852

TABLE 5.9
Countrywise Export of Palmarosa Oil by India

Country	2002 Qty (kg)	Value (USD)	2003 Qty (kg)	Value (USD)	2004 Qty (kg)	Value (USD)	2005 Qty (kg)	Value (USD)	2006 Qty (kg)	Value (USD)	2007 Qty (kg)	Value (USD)
Argentina	—	—	—	—	10	118	—	—	—	—	20	351
Australia	—	—	410	5,387	161	2,981	—	—	—	—	25	386
Bahrain	—	—	—	—	—	—	—	—	—	—	25	1,469
Canada	1,000	14,358	—	—	—	—	—	—	—	—	10	467
France	3,860	55,877	2,338	33,458	2,080	22,070	540	5,838	1,105	16,394	3,740	67,284
Germany	360	4,916	5,220	70,770	—	—	485	5,062	1,880	24,615	2,365	37,042
Hong Kong	—	—	—	—	—	—	100	1,798	300	4,963	145	8,408
Ireland	—	—	360	4,694	—	—	—	—	—	—	360	6,472
Japan	—	—	—	—	—	—	1,080	10,754	20	506	4	107
Mexico	—	—	—	—	50	589	—	—	—	—	—	—
Netherlands	—	—	1,800	25,422	4,680	63,232	3,060	32,600	2,880	44,090	2,700	54,114
New Zealand	130	3,542	—	—	—	—	—	—	—	—	—	—
Singapore	—	—	260	3,372	240	3,419	—	—	—	—	200	2,738
South Africa	—	—	—	—	—	—	10	126	—	—	—	—
Spain	900	11,002	1,260	17,919	1,580	19,526	1,080	12,708	5,287	33,534	680	12,435
Switzerland	540	8,451	720	12,992	—	—	720	9,140	1,410	26,128	—	—
Taiwan	—	—	—	—	20	271	—	—	—	—	—	—
Tanzania	—	—	—	—	—	—	—	—	—	—	30	319
Thailand	—	—	—	—	—	—	2	35	200	4,296	—	—
UAE	—	—	—	—	24	235	—	—	—	—	500	9,191
U.S.	4,830	68,303	10,460	144,351	5,950	71,338	180	1,252	4,801	57,940	9,880	164,540
Total	**11,620**	**166,449**	**22,828**	**318,365**	**14,795**	**183,779**	**7,257**	**79,313**	**17,883**	**212,466**	**20,684**	**365,323**

REFERENCES

Anonymous. 1997. U.S. Environmental Protection Agency Fact Sheet. 1997. Prevention, Pesticides and Toxic Substances (7508W), Re registration Eligibility Decision Sheet EPA-738-F-97-002 (February 1997).

Anonymous. 1986. Essential oils and oleoresins: A study of selected producers and major markets. Int. Trade Centre, UNCTAD/GATT, Geneva.

Anonymous. 2001. WHO International Programme on Chemical Safety: Guidance document for the use of chemical specific adjustment factors (CSAFs) for interspecies differences and human variability in dose concentration response assessment. 76 pp. World Health Organization, Geneva.

Anonymous. 2007. Citronella (Oil of Citronella (021901) Fact Sheet), U.S. Environmental Protection Agency. Issued 11/99; Updated October 22, 2007.

Anonymous. 2003–2007a. Monthly Statistics of Foreign Trade of India. Vol. I, Exports and Re-Exports March, DG CI&S, Ministry of Commerce and Industry, Govt. of India, Kolkata.

Anonymous. 2003–2007b. Monthly Statistics of Foreign Trade of India. Vol. II, Imports, Directorate General of Commercial Intelligence and Statistics, Ministry of Commerce and Industry, Govt. of India, Kolkata.

Anonymous. 2004. Re-evaluation of citronella oil and related active compounds for use as personal insect repellants, Proposed Acceptability for Continuing Registration PACR 3004–36, September 17, 2004, Pest Management Regulatory Agency, Ontario, Canada.

Anonymous. 2006. Report of an independent science panel on citronella oil used as an insect repellent. March 16, 2006, Canada.

Anonymous 2008. Essential oil dictionary, detailed reference guide and herbal encyclopaedia. http://www.deancoleman.com/essentialref.htm.

Chang Yu S. 2007. Eight MAP species from Malaysia for ICS. Forest Research Institute Malaysia, Workshop on NFP, 28029 May 2007, Nanchang, China.

Husain A. 1993. *Essential Oil Plants and Their Cultivation*. Central Institute of Medicinal and Aromatic Plants, Lucknow, India.

Lawless J. 1995. *The Illustrated Encyclopedia of Essential Oils: Complete Guide to the Use of Oils in Aromatherapy and Herbalism,* Thorson, U.K.

Lawrence B. M. 1985. A review of the world production of essential oils (1984). *Perfum Flav.* 10: 1–16.

Lawrence B. M. 1986. Essential oil production: A discussion of influencing factors. In T. H. Parliament and R. Croteau (Eds.). *Biogeneration of Aromas*, pp. 363–369. ACS Symposium Series 317 Amer. Chem. Soc., Washington, DC.

Lawrence B. M. 1993. A planning scheme to evaluate new aromatic plants for the flavour and fragrance industries, pp. 620–627. In J. Janick and J. E. Simon (Eds.). *New Crops*, Wiley, New York.

Oyen L. P. A. and Nguyen X. D. 1999. *Plant Resources of South East Asia No. 19: Essential Oil Plants*. Backhuys Publishers, Leiden.

Robbins S. R. J. 1983. Selected markets for essential oil of lemongrass, citronella and eucalyptus. Tropical Products Research Institute 6171, London.

Rosalinda C. Torres and Barbara D. J. Tio. Citronella oil industry: challenges and breakthroughs.

Simon J. E. 1990. Essential oils and culinary herbs. *Advances in New Crops*, pp. 472–483. In J. Janick and J. E. Simon (Eds.). Timber Press, Portland.

Singh A. K., Gauniyal A. K., and Virmani O. P. 2000. Essential oil of important Cymbopogons: Production and trade. In Kumar S. et al. (Eds.). *Cymbopogon: The Aromatic Grass Monograph*. Central Institute of Medicinal and Aromatic Plants, Lucknow, India.

Smith R. L., Cohen S. M., Doull J., Feron V. Y., Goodman J. I., Marnett L. J., Portoghese P. S., Waddel W. J., Wagner B. M., Hall R. L., Higley N. A., Lucas-Gavin C., and Adams T. B. 2005. A procedure evaluation of natural flavour complexes used as ingredients in food: Essential oils. *Food Chem. Toxicol.* 43: 345–363.

Torres R. C. and Tio B. D. 2001. Citronella (*Cymbopogon winterianus*) oil industry challenge.

Trongtokit Y., Rongsriyam Y., Komalamisra N., and Apiwathnasorn C. 2005. Comparative repellency of 38 essential oils against mosquito bites. *Phytother Res.* 19(4): 303–309.

6 In Vitro Antimicrobial and Antioxidant Activities of Some *Cymbopogon* Species

Watcharee Khunkitti

CONTENTS

6.1 INTRODUCTION

Essential oils have been widely used in antimicrobial, antiviral, antiparasitical, insecticidal, medical, and cosmetic applications (Bakkali et al. 2008). *Cymbopogon* species are well known as a source of commercially valuable compounds, such as geraniol, geranyl acetate, citral (neral and geranial), citronellal, piperitone, eugenol, etc. (Shahi and Tava 1993). Bioactivity of *Cymbopogon* species such as lemongrass (*C. citratus*), Indian lemongrass (*C. flexuosus*), Indian palmarosa (*C. martinii*), Java citronella (*C. winterianus*), and Ceylon citronella (*C. nardus*) has been reported. Essential oils from *Cymbopogon* species and their components are known for their antimicrobial (de Billerbeck et al. 2001; Pattnaik et al. 1995a; Pattnaik et al. 1995b) and antioxidant activities (Hierro et al. 2004; Ruberto and Baratta 2000; Lertsatittanakorn et al. 2006). Bioactivity of the same essential oils may be markedly different when using different strains of the same microorganism and different sources of essential oils. According to Oussalah et al. (2006), the major constituents of lemongrass oil are citral (77%) and limonene (8.5%); Indian lemongrass has citral (77%); palmarosa has geraniol (80%) and geranyl acetate (8.6%) as main constituents; Java citronella oil contains citronellal (34%), geraniol (21.5%), and citronellol (11.5%) as major components; and Ceylon citronella oil contains geraniol (19.1%), limonene (9.9%), and camphene (9.0%). Since the oils are poorly soluble in water, many factors affect the results of their activities, for example, oil solubilizers, vehicles, method of testing, etc. Therefore, this chapter reviews the common methods of testing antimicrobial and antioxidant activities of some *Cymbopogon* species and their major components, as well as factors affecting their activities.

6.2 FACTORS AFFECTING ANTIMICROBIAL ACTIVITY OF ESSENTIAL OILS

6.2.1 SOLUBILIZING AGENTS

The main problem in the study of essential oil bioactivities is that they are poorly soluble in water. In order to overcome this problem, many authors have used various solvents. Examples of solubilizing agents used are provided in Table 6.1.

However, interaction between essential oil components and solubilizer has to be taken into consideration. For example, Tween 20 (polyoxyethylene (2) sorbitan monolaurate) and Tween 80 (polysorbate 80), which are the most commonly used nonionic solublizers, may cause either enhancement or reduction of antimicrobial activity. Interaction of essential oil with nonionic surfactants could be the result of either micellar solubilization or complex formation between the two molecules. The activity of essential oil depends on an adequate concentration of free oil existing in aqueous phase outside the micelle. At concentrations above critical micelle concentration (CMC) of the surfactants, antimicrobials solubilized within the micelles do not contribute to the antimicrobial activity, whereas at low concentrations below CMC, antimicrobial activity increases because of an increase in bacterial cell permeability to the antimicrobial compound (Russell et al. 1992). Reduction of the bioactivity of tea tree oil and thyme oil has been reported. It is possible that Tween 80 might inactivate phenolic compounds in those essential oils (cited in Mann and Markham 1998; Manou et al. 1998). In addition, Hili et al. (1997) reported that the reduction of antimicrobial activity of cinnamon oil in the presence of dimethylsulphoxide (DMSO) might be due to the partitioning of the oil between the aqueous phase and DMSO, distancing the oil from cells. This effect, however, did not

---

TABLE 6.1
Essential Oil-Solubilizing Agents

Oil-Solubilizing Agent	Concentration Used	Organism	Assay Method	References
Tween 20	0.001% v/v	Fungi	Broth microdilution method	Devkatte et al. (2005)
Tween 20	0.1% w/v	Fungi	Agar dilution method	Tampieri et al. (2005)
Tween 20	5% v/v	Bacteria	Agar dilution method	Hammer et al. (1999)
Tween 80	1% w/v	Bacteria	Agar diffusion method	Jirovetz et al. (2007)
		Yeast	(Paper disk 6 mm)	Lertsatithanakorn et al. (2006)
DMSO	0.2% v/v	Bacteria	Broth microdilution method	Ohno et al. (2003)
DMSO	1% v/v	Fungi	Agar dilution method	Inouye et al. (2001)
Ethanol	2%v/v	Fungi	Broth microdilution method	Tullio et al. (2007)
Ethyl acetate	Pure	Fungi	Vapor contact method	Inouye et al. (2001)
DMSO + Tween 80	10% v/v DMSO + 0.5% v/v Tween 80	Bacteria	Agar diffusion method (Paper disk 6 mm)	Prabuseenivasan et al. (2006)
Ethanol + Tween 80	5% Ethanol + 5% Tween 80	Bacteria Yeast	Broth microdilution method	Unpublished data

occur when low concentrations (0.15%–0.2% w/v) of bacteriological agar were used as a stabilizer of the oil–water mixture (Mann and Markham 1998; Remmal et al. 1993).

6.2.2 Type of Organism

The action of biocides depends on the type of microorganisms, which is mainly related to their cell wall structure and the outer membrane arrangement (Kalemba and Kunicka 2003; Russell et al. 1992).

6.2.2.1 Gram-Positive Bacteria

Gram-positive bacteria are more sensitive to biocides, particularly essential oil, than Gram-negative bacteria. Probably the main reason for this difference in sensitivity is the relative composition of the cell envelope. In general, the cell wall of Gram-positive bacteria is composed basically of peptidoglycan, which forms a thick, fibrous layer. Many antimicrobial agents much penetrate the outer and cytoplasm membranes to reach their site of action. The effects of various disinfectants, antiseptics, and preservatives on Gram-positive bacteria have been well documented. The action of essential oils against Gram-positive bacteria and fungi appears to be similar. The oil components destroy the bacterial and fungal cell wall and cytoplasmic membrane, causing a leakage of cytoplasm and coagulation. They also inhibit the synthesis of DNA, RNA, proteins, and polysaccharides in fungal and bacterial cells (Himejima and Kubo 1993; Zani et al. 1991).

6.2.2.2 Gram-Negative Bacteria

Gram-negative bacteria, especially *Escherichia coli, Klebsiella* spp., *Proteus* spp., *Pseudomonas aeruginosa,* and *Seratia macescens*, appear to be increasingly implicated as hospital pathogens. Gram-positive bacteria are more sensitive to essential oils than Gram-negative bacteria. *Pseudomonas aeruginosa*, for example, is resistant to a wide variety of essential oils due to the hydrophilic surface of their outer membrane, which is rich in lipopolysaccharide molecules. Thus, essential oil constituents are unable to penetrate the membrane barrier (Nikaido 1994). Essential oils containing phenolic compounds such as carvacrol and thymol cause the outer cell membrane damage (Helander et al. 1998). However, palmarosa, lemongrass, peppermint, and eucalyptus oils are found to be bactericidal to *Escherichia coli* strain SP-11. Only peppermint and palmarosa oils induced the formation of elongated filamentous forms of *E. coli* (Pattnaik et al. 1995b).

6.2.2.3 Mould and Yeast

Yeast and mould comprise important groups of microorganisms that are responsible for several infections and for causing spoilage of foods, pharmaceutical products, and cosmetic products. Some fungal species are of agricultural and industrial importance and many cause disease in plants. Antifungal activity of some *Cymbopogon* species has been reported. For example, the mycelium growth of *Aspergillus niger* is inhibited by *C. nardus* essential oil. It causes morphological changes such as hyphal diameter and hyphal wall thinning, plasma membrane disruption, and mitochondria structure disorganization (de Billerbek et al. 2001). Lemongrass oil is an effective postharvest fungitoxicant of higher-order plant origin, potentially suitable for protection of foodstuffs against storage fungi (Mishra and Dubey 1994). Palmarosa oil had some inhibitory activity against 12 fungi. The response of this oil is dependent on the species of fungi (Pattnaik et al. 1996). Moreover, palmarosa oil also has antimicrobial against yeast cell, *Saccharomyces cerevisiae,* by passively entering the plasma membrane to initiate membrane disruptions followed by accumulation in the plasma membrane resulting in the inhibition of cell growth (Prashar et al. 2003).

6.2.3 The Correlation between Oil Components and Activity

Biological activity of an essential oil is in strict direct relationship to its chemical composition. The relation between components and activity may be attributed both to their major components and to the minor ones present in the oils. The antibacterial action of essential oils depends on the chemical structure of their components. Phenolic compounds such as eugenol and carvacrol have an aromatic nucleus and phenolic OH group that is known to be reactive and to form hydrogen bonds with active sites of target enzymes (Farag et al. 1989). Mahmoud (1994) reported that geraniol, nerol, and citronellol, which are aliphatic alcohols, are broad-spectrum antifungal agents. The effectiveness of essential oil is higher than the activity of each component. It is possible that major components and minor compounds may act together synergistically to contribute to the activity (Milos et al. 2000).

6.2.4 METHODS COMMONLY USED FOR ANTIMICROBIAL ASSESSMENT OF ESSENTIAL OIL

Although there are the nonconventional methods for determining the minimum inhibitory concentration of essential oils (Kalemba and Kunicka 2003; Mann and Markham 1998), only the basic techniques commonly used for assessment of antibacterial and antifungal activities of essential oil, which are serial broth dilution method and agar dilution method, are stated. In the case of estimation of essential oil activity on vapor contact assay, the agar diffusion method is slightly modified. Antimicrobial activity of some *Cymbopogon* species and their major components is shown in Tables 6.2 and 6.3, respectively. However, Janssen et al. (1987) noted that the antimicrobial activity from different methods is not necessarily comparable.

6.2.4.1 The Serial Broth Dilution Method

Essential oil is dissolved in a solubilizing agent, which is nontoxic to microorganisms, before being serially diluted in an appropriate broth or agar. The effectiveness of essential oil is generally expressed as minimum inhibitory concentration (MIC), which is defined as the lowest concentration of essential oil that inhibits the visible growth, and as minimum bactericidal concentration (MCC) and minimum fungicidal concentration (MFC), which are defined as the lowest concentration of essential oil that causes more than 99.9% reduction of microorganism number or 3 log reduction of microorganisms (Kalemba and Kunicka 2003).

6.2.4.2 The Agar Diffusion Method

The agar diffusion method is generally recommended as a prescreening method for a large number of essential oils since it is easy to perform and requires a small amount of essential oil. An assay can be performed accordingly; an appropriate nutrient agar plate is inoculated with microorganisms, either by adding the organism to the agar before it is poured or by streaking the organisms across the surface of the plate. The amount of essential oil tested can be accurately incorporated either on paper disk or into the well. The effectiveness of essential oil is demonstrated by the size of the microorganism inhibition zone around the disk or well, and it is usually expressed as the diameter of the zone. However, the disadvantages of this method are that essential oils are likely to evaporate with solvent during the incubation period, and they may show limited agar diffusion. Moreover, the inhibition zone of the oils depends on the characteristics of the oil components partitioned through the agar (Kalemba and Kunicka 2003; Southwell et al. 1993).

6.2.4.3 Vapor Contact Assay

This method is used for evaluating the activity of essential oils that are to be employed as atmospheric biocides (Lopez 2005; Tullio et al. 2006). An appropriate nutrient agar plate is inoculated with microorganisms. Essential oil is diluted in organic solvent such as ethyl acetate, ethyl ether, and ethanol. The exact amount of diluted essential oil is added to a paper disk, attached to the lid of a petri dish, and completely sealed; the lid is then inverted and incubated. The results are expressed as minimum inhibitory dose (MID), which is defined as the lowest concentration of essential oil (mg/L in air) that inhibits visible growth. In some studies, the soaked paper disk is placed in the airtight box next to the agar plate (Inouye et al. 2001; Nakahara et al. 2003).

TABLE 6.2
In Vitro Antimicrobial Activity of *Cymbopogon* Essential Oils

Organisms	Cymbopogon citratus (lemongrass)	Cymbopogon flexuosus (Indian lemongrass)	Cymbopogon giganteus (tsauri grass)	Cymbopogon martinii (palmarosa)	Cymbopogon nardus (Ceylon citronella)	Cymbopogon winterianeous (Java citronella)	References
Bacteria							
Acinetobacter baumaii	MIC 0.03% v/v			MIC 0.12% v/v	MIC 0.25% v/v		Hammer et al. (1999)
Aeromonas sobria	MIC 0.12% v/v			MIC 0.12% v/v	nd		Hammer et al. (1999)
Bacillus brevis	MIC 0.16 µL/mL				MIC[a] 1.66 µL/mL		Kalemba and Kunicka (2003)
Campylobacter jejuni	15 µL in 6 mm paper disk: Inhibition zone 90 mm				15 µL in 6 mm paper disk: Inhibition zone 40 mm		Wannissorn et al. (2005)
Clostridium perfringens	15 µL in 6 mm paper disk: Inhibition zone 90 mm				15 µL in 6 mm paper disk: Inhibition zone 39.5 mm		Wannissorn et al. (2005)
Enterococcus faecalis	MIC 0.12% v/v		Leaf and stem MIC 60 and 600 ppm	MIC 0.25% v/v	MIC 1.0% v/v		Hammer et al. (1999)
Escherichia coli	MIC 0.06% v/v MIC > 0.8% v/v MCC 0.12% v/v MCC 1.66 µL/mL MID 100 mg/L, air 15 µL in 6 mm paper disk: Inhibition zone 12 mm	MIC > 0.8% v/v	Leaf and stem MIC 60 ppm	MIC 0.06–0.12% v/v MIC 0.12% v/v MIC 0.2% v/v MCC 1.66 µL/mL	MIC 0.25–0.5% v/v MIC > 0.8% v/v MCC 0.25% v/v 15 µL in 6 mm paper disk: Inhibition zone 10.5 mm	MIC > 0.8% v/v	Hammer et al. (1999) Inouye et al. (2001) Oussalah et al. (2007) Patnaik (1995a) Wannissorn et al. (2005)
Haemophilus influenzae	MID 1.56 mg/L, air						Inouye et al. (2001)
Helicobacter pylori	MIC <1 µg/mL MCC 0.1% v/v						Ohno et al. (2003)

(continued on next page)

TABLE 6.2 (continued)
In Vitro Antimicrobial Activity of *Cymbopogon* Essential Oils

Organisms	Cymbopogon citratus (lemongrass)	Cymbopogon flexuosus (Indian lemongrass)	Cymbopogon giganteus (tsauri grass)	Cymbopogon martinii (palmarosa)	Cymbopogon nardus (Ceylon citronella)	Cymbopogon winterianeous (Java citronella)	References
Klebsiella pneumoniae	MIC 0.25% v/v		Leaf and stem MIC 60 ppm	MIC 0.25% v/v	MIC 1.0% v/v		Hammer et al. (1999)
Listeria monocytogenes	MIC 0.4% v/v	MIC 0.4% v/v		MIC 0.2% v/v	MIC 0.8% v/v	MIC 0.4% v/v	Oussalah et al. (2007)
Propionibacterium acnes	MIC 0.6 μL/mL MCC 0.6 μL/mL					MIC 0.005–0.3 μL/mL MCC 0.625–1.2 μL/mL	Lertsatitthanakorn et al. (2006)
Pseudomonas aeruginosa	MIC 1.0% v/v MIC 1.3 μL/mL		Leaf and stem MIC 60 ppm	MIC > 2.0% v/v	MIC > 2.0% v/v		Hammer et al. (1999) Kalemba and Kunicka (2003)
Pseudomonas putida	MIC 0.8% w/v	MIC 0.8% w/v		MIC 0.2% w/v	MIC 0.4% w/v	MIC > 0.8% w/v	Oussalah et al. (2006)
Salmonella typhimurium	MIC 0.25% v/v MIC 0.8% v/v MIC 1.66 μL/mL 15 μL in 6 mm paper disk: Inhibition zone 24 mm	MIC 0.4% v/v		MIC 0.5% v/v MIC 0.2% v/v MIC 0.80 μL/mL	MIC > 2.0% v/v MIC 0.8% v/v 15 μL in 6 mm paper disk: Inhibition zone 24 mm	MIC 0.4% v/v	Hammer et al. (1999) Kalemba and Kunicka (2003) Oussalah et al. (2007) Wannissorn et al. (2005)
Serratia marcescens	MIC 0.25% v/v			MIC 0.25% v/v	MIC > 2.0% v/v		Hammer et al. (1999)
Staphylococcus aureus	MIC 0.06% v/v MIC 0.1% v/v MIC 0.3 μL/mL MCC 0.06% v/v MID 12.5 mg/L, air	MIC 0.1% v/v	Leaf and stem MIC 60 ppm	MIC 0.12% v/v MIC 0.1% v/v MIC 0.66 μL/mL MCC 0.12% v/v	MIC 0.12–0.25% v/v MIC 0.4% v/v MCC 0.25% v/v	MIC 0.05% v/v	Hammer et al. (1999) Inouye et al. (2001) Kalemba and Kunicka (2003) Oussalah et al. (2007)

Organism						References
Streptococcus pyogenes	MID 6.25 mg/L, air					Inouye et al. (2001)
Streptococcus pneumoniae	MID 6.25 mg/L, air					Inouye et al. (2001)
Streptococcus enteritidis	15 µL in 6 mm paper disk: Inhibition zone 11 mm				15 µL in 6 mm paper disk: Inhibition zone 12.8 mm	Wannissorn et al. (2005)
Vibrio cholerae	MIC 0.3 µL/mL				MIC 0.66 µL/mL	Kalemba and Kunicka (2003)
Yeast and Fungi						
Aspergillus niger	5 µL in 5 mm paper disk: Hyphae inhibition zone 21 mm; Spore inhibition zone 33 mm		5 µL in 5 mm paper disk: Hyphae inhibition zone 8 mm; Spore inhibition zone 12 mm	MID 800 mg/L, air		Pawar and Thaker (2006)
Aspergillus flavus	MIC 0.6 mg/mL; MFC 1.0 mg/mL			MID 250 mg/L, air		Paranagama et al. (2003)
Candida albicans	MIC 0.06% v/v; MIC 322 µg/mL; MFC 0.06–0.12% v/v	MIC 500 ppm; MIC > 0.2 mg/mL	Leaf and stem MIC 60 ppm; MIC 0.06–0.12% v/v; MFC 0.12% v/v	MIC 0.12% v/v; 0.5–1.0% v/v; MFC[a] 0.12% v/v; 2.0% v/v; Mycelium growth inhibited at 800 mg/L; MIC 0.6 mg/mL	MIC 0.6 mg/mL	Nakahara et al. (2003); Hammer et al. (1999); Devkatte et al. (2005); Lertsatithanakorn et al. (2005); Tampieri et al. (2005); Duarte et al. (2005); Dutta et al. (2006); Jirovetz et al. (2007); De Billerbeck et al. (2001)
Eurotium amstelodami				MID 250 mg/L, air		Nakahara et al. (2003)

(continued on next page)

TABLE 6.2 (continued)
In Vitro Antimicrobial Activity of *Cymbopogon* Essential Oils

Organisms	Cymbopogon citratus (lemongrass)	Cymbopogon flexuosus (Indian lemongrass)	Cymbopogon giganteus (tsauri grass)	Cymbopogon martinii (palmarosa)	Cymbopogon nardus (Ceylon citronella)	Cymbopogon winterianeous (Java citronella)	References
Eurotium chevalieri					MID 250 mg/L, air		Nakahara et al. (2003)
Penicillium adametzii					MID 250 mg/L, air		Nakahara et al. (2003)
Penicillium citrinum					MID 250 mg/L, air		Nakahara et al. (2003)
Penicillium griseofulvum					MID 250 mg/L, air		Nakahara et al. (2003)
Penicillium islandicum					MID 250 mg/L, air		Nakahara et al. (2003)
Saccharomyces cerevisiae	MIC 0.2 mg/mL	MFC 0.1% v/v					Prashar et al. (2003) Sacchetti et al. (2005)
Tricophyton mentagrophytes	MID 1 µg/mL, air MFD 5.2 µg/mL, air MFD 1.56 µg/mL, air MIC 50 µg/mL; 0.25 µL/mL MFC 15.2 µg/mL						Inouye et al. (2001) Inouye et al. (2006) Kalemba and Kunicka (2003)
Tricophyton rubrum	MID 1 µg/mL, air MFD 5.2 µg/mL, air MIC 50 µg/mL MFC 15.2 µg/mL						Inouye et al. (2001)

a Does not indicate species.

TABLE 6.3

In Vitro Antimicrobial Activity of the Major Components of *Cymbopogon* Essential Oils

	Alcohols		Aldehydes		Ester	Terpene Hydrocarbons		References
	Geraniol	Citronellol	Citral	Citronellal	Geranyl acetate	Limonene	Mycene	
Bacillus cereus	MIC 0.7 mg/mL MIC 51.9 mM	MIC 0.70 mg/mL		MIC >207.5 ± 45.4 mM	MIC 163.0 ± 29.0 mM	MIC 234.9 ± 0.0 mM		Rosato et al. (2007) Van Zyl et al. (2006)
Bacillus subtilis		MIC 0.35 mg/mL						
Escherichia coli	MIC 0.8 mg/mL MID >25 mg/L air MIC 1.4 mg/mL MBC 283–300 μg/mL MIC 25.9 ± 0.0 mM	MIC 1.4 mg/mL	MID >12.5 mg/L air	MIC 207.5 ± 64.8 mM	MIC >163.0 ± 51.0 mM	MID >800 mg/L air MIC 176.2 ± 40.4 mM		Inouye et al. (2001) Rosato et al. (2007) Si et al. (2006) Van Zyl et al. (2006)
Haemophilus influenzae	MID 6.25 mg/L air		MID 3.13 mg/L air			MID 200 mg/L air		
Listeria monocytogenes	MIC 1000 μg/mL		MIC 500 μg/mL					Kalemba and Kunicka (2003)
Pseudomonas aeruginosa	MIC 0.375 mg/mL							Kalemba and Kunicka (2003)
Streptococcus pyogenes	MID 12.5 mg/L air		MID 3.13 mg/L air			MID 400 mg/L air		
Streptococcus pneumoniae	MID 6.25 mg/L air		MID 6.25 mg/L air			MID 200–400 mg/L air		
Candida albicans	MIC 100 ppm MIC 19.5 ± 0.0 mM		MIC 100 ppm	MIC 77.8 ± 0.0 mM	MIC 163.0 ± 0.0 mM	MIC 1000 ppm MIC 73.4 ± 0.0 mM		Tampieri et al. (2005) Van Zyl et al. (2006)
Salmonella typhimurium	MIC 500 μg/mL MBC 367 μg/mL		MIC 500 μg/mL					Kalemba and Kunicka (2003) Si et al. (2006)

(continued on next page)

TABLE 6.3 (continued)
In Vitro Antimicrobial Activity of the Major Components of *Cymbopogon* Essential Oils

	Alcohols		Aldehydes		Ester	Terpene Hydrocarbons		References
	Geraniol	Citronellol	Citral	Citronellal	Geranyl acetate	Limonene	Mycene	
Staphylococcus aureus	MID > 25 mg/L, air MIC 0.7 mg/mL MIC 38.9 ± 18.2 mM	MIC 0.70 mg/mL	MID 12.5 mg/L, air	MIC 129.7 ± 55.1 mM	MIC > 163.0 ± 35.7 mM	MID 800 mg/L, air MIC 176.2 ± 40.4 mM		Inouye et al. (2001) Rosato et al. (2007) Van Zyl et al. (2006)
Staphylococcus epidermidis	MIC 0.125 mg/mL							Kalemba and Kunicka (2003)
Aspergillus candidus	No activity	No activity	No activity	Potent activity MID 28 mg/L, air	No activity	No activity	No activity	Nakahara et al. (2003)
Aspergillus flavus	MIC 500 ppm	MIC 500 ppm	No activity	Potent activity MID 56 mg/L, air	No activity	No activity	No activity	Kalemba and Kunicka (2003) Nakahara et al. (2003)
Aspergillus versicolor	No activity	No activity	No activity	Potent activity MID 28 mg/L, air	No activity	No activity	No activity	Nakahara et al. (2003)
Aspergillus nidulans	1.0 mM growth inhibition > 20 days	1.0 mM growth inhibition > 20 days	1.0 and 2.0 mM growth inhibition 3 and > 20 days, respectively	2.0 mM growth inhibition 0 days		2.0 mM growth inhibition 0 days	2.0 mM growth inhibition 0 days	Kurita et al. 1981
Eurotium amstelodami	No activity	No activity	No activity	Potent activity MID 28 mg/L, air	No activity	No activity	No activity	Nakahara et al. (2003)

							Reference
Eurotium chevalieri	No activity	Moderate activity	Moderate activity	Potent activity MID 14 mg/L, air	No activity	No activity	No activity	Nakahara et al. (2003)
Penicillium adametzii	No activity	No activity	No activity	Potent activity MID 56 mg/L, air	No activity	No activity	No activity	Nakahara et al. (2003)
Penicillium citrinum	No activity	No activity	No activity	Potent activity MID 28 mg/L, air	No activity	No activity	No activity	Nakahara et al. (2003)
Penicillium griseofulvum	No activity	No activity	No activity	Potent activity MID 56 mg/L, air	No activity	No activity	No activity	Nakahara et al. (2003)
Penicillium islandicum	Moderate activity	No activity	Moderate activity	Potent activity MID 14 mg/L, air	No activity	No activity	No activity	Nakahara et al. (2003)
Tricophyton mentagrophytes			MIC 25 µg/mL MFC 7.7 ± 1.85 µg/mL MID 0.5 µg/L, air MFC 3.9 ± 1.2 µg/L, air					Inouye et al. (2001)
Tricophyton rubrum	1.0 mM growth inhibition > 20 days	1.0 mM growth inhibition > 20 days	1.0 mM growth inhibition > 20 days	1.0 mM growth inhibition 0 days 2.0 mM growth inhibition 2 days		2.0 mM growth inhibition 0 days	2.0 mM growth inhibition 0 days	Kurita et al. 1981

6.3 IN VITRO ANTIOXIDANT ACTIVITY OF SOME *CYMBOPOGON* SPECIES AND THEIR MAJOR COMPONENTS

Essential oils and their components are gaining increasing interest because of their potential for multipurpose functional use. Apart from antimicrobial and antifungal properties, antioxidant and radical-scavenging properties are of great interest to health and food science researchers. An antioxidant may be defined as any substance that delays or inhibits oxidation of that substrate. In vitro methods provide a useful indication of antioxidant activity. Due to the differences between in vitro testing and biological environments, the data obtained from in vitro methods are difficult to apply to biological systems (Antolovich et al. 2002). Antioxidant activity can be divided into two classes: primary (chain-breaking or radical-scavenging antioxidants) and secondary (preventive antioxidants) (Laguerre et al. 2007). Frankel and Meyer (2000) pointed out that multifaceted testing of antioxidant activity is needed because antioxidants often act via mixed mechanisms. Essential oils are a very complex mixture that can contain two or three major components at fairly high concentrations and other components in trace amounts (Bakkali et al. 2008). The antioxidant activity of essential oils, therefore, would be associated with various mechanisms. In this review, only the common in vitro antioxidant testing methods for essential oils are mentioned.

6.3.1 FACTORS AFFECTING ANTIOXIDANT ACTIVITY

The antioxidant activity of essential oils and their components shows the marked difference between the reported results (Table 6.4), which depend on many factors. The magnitudes of antioxidant activities, therefore, could only be compared for given process conditions. In this review, studies on the antioxidant activity of *Cymbopogon* species and their major components published in the literature are reported, and only factors affecting their antioxidant activity, using the three methods of testing mentioned earlier, are discussed.

6.3.1.1 Choice of Oxidizable Substrate and End-Product Evaluation

The choice of an oxidizable substrate depends on whether the antioxidant is for nutritional, therapeutic, or preservative purposes. It is important to select a substrate that is representative of in situ conditions. Since the aim of assessing antioxidant efficacy of essential oils is to preserve foods, the oxidizable substrate normally used to evaluate antioxidant activity is either linoleic acid or homogenized egg yolk (Laguerre et al. 2007). Therefore, the methods often used for this assessment are β-carotene bleaching assay and TBARS assay. However, a limitation of evaluating antioxidant activity by TBARS assay is that thiobarbituric acid not only reacts with MDA, which is a secondary oxidation product of lipid peroxidation but also can react with other aldehydes (Janero 1990). Thus, essential oils containing aldehyde compounds such as citral, citronellal, heptanal, octanal, etc., may cause an underestimation of their antioxidant activity. This evidence can be seen in the study of Ruberto and Baratta (2000), which found that antioxidant activity of aldehyde compounds can be assessed by determination of conjugate diene hydroperoxides, which are primary oxidation products of linoleic acid peroxidation but undetectable with TBARS assay.

6.3.1.2 Media

There are two major types of media: homogeneous medium and heterogeneous medium. In a heterogeneous medium, the type of solvent could affect the antioxidant mechanism. Antioxidants behave differently in media with different polarities. For example, DPPH is a stable-free radical, but it is sensitive to some Lewis bases, solvent types, light, and oxygen. Polar solvents may decrease the odd electron density of the nitrogen atom in DPPH and increase the reactivity of DPPH. In general, DPPH in the methanol buffer system shows better stability than in the acetone buffer system (pH 10) (Ozcelik et al. 2003). Porter (1993) described the "polar paradox": lipophilic antioxidants are

TABLE 6.4
In Vitro Antioxidant Activity of *Cymbopogon* Species and Their Major Components

Sample	Antioxidant Activity	Method of Testing	References
Citronellol	AI at 1000 ppm = 27.5%	TBARS method	Ruberto et al. (2000)
	AI at 0.001 M = 13.3%	Conjugated diene formation	Ruberto et al. (2000)
	IC_{50} => 1000 µg/mL	DPPH test	Hierro et al. (2004)
	AI at 10 µL/mL = 15.9%	DPPH test	Unpublished data
Geraniol	AI at 1000 ppm = 34.9%	TBARS method	Ruberto et al. (2000)
	AI at .001 M = 26.5%	Conjugated diene formation	Ruberto et al. (2000)
	IC_{50} => 1000 µg/mL	DPPH test	Hierro et al. (2004)
	AI at 10 µL/mL = 19.7%	DPPH test	Unpublished data
Limonene	AI at 1000 ppm = 27.4%	TBARS method	Ruberto et al. (2000)
	AI at 10 µL/mL = 22.2%	DPPH test	Unpublished data
	AI at 0.001 M = 21.0%	Conjugated diene formation	Ruberto et al. (2000)
Alpha tocopherol	AI at 1000 ppm = 93.5%	Conjugated diene formation	Ruberto et al. (2000)
	AI at 0.001 M = 94.8%	TBARS method	Ruberto et al. (2000)
C. citratus	AI at 5 µL/mL = 26.0%	TBARS method	Unpublished data
	AI at 10 µL/mL = 31.6%	TBARS method	Unpublished data
	IC_{50} = 27.0 µL/mL	DPPH test	Lertsatittanakorn et al. (2006)
	AI at 10 µL/mL = 63.8%	DPPH test	Sacchetti et al. (2005)
	AI at 2 µL/mL = approx 50%	β-carotene bleaching assay	Sacchetti et al. (2005)
C. nardus	IC_{50} = 2.0 µL/mL	DPPH test	Lertsatittanakorn et al. (2006)
	AI at 5 µL/mL = 89.0%	DPPH test	Unpublished data
	AI at 10 µL/mL = 45.5%	TBARS method	Unpublished data
Citronellal	AI at 1000 ppm = not detectable	TBARS method	Ruberto et al. (2000)
	AI at 0.001 M = 21.9%	Conjugated diene formation	Ruberto et al. (2000)
	AI at 10 µL/mL = 16.4%	DPPH test	Unpublished data
Citral	AI at 1000 ppm = not detectable	TBARS method	Ruberto et al. (2000)
	AI at 0.0001 M = 18.3%	Conjugated diene formation	Ruberto et al. (2000)
	IC_{50} => 1000 µg/mL	DPPH test	Hierro et al. (2004)
	AI at 10 µL/mL = 21.1%	DPPH test	Unpublished data

more active in a polar medium, whereas polar antioxidants are more active in a lipophilic medium. In a heterogeneous medium, either in oil in water emulsion or aqueous suspensions of liposome or LDLs, oxidation is affected by the type of interface and its viscosity, the size distribution of oil and liposome droplets, the partition and diffusion of oxygen toward reaction centers, and the antioxidant location. According to Porter's polar paradox, in emulsion media, oxidation occurs at the oil–water interface, where lipophilic antioxidants are more efficient than hydrophilic antioxidants. In addition, lipid peroxidation is dependent on the pH of water in oil emulsion and in liposomes (Laguerre et al. 2007). Frankel (1998) reported that lipid oxidation is generally lower at high pH and, consequently, the oxidation rate increases as pH decreases.

6.3.1.3   Oxidation Conditions

Heat: Since essential oils are very volatile and undergo thermal degradation at high temperature, temperature is an important factor when assessing the extent of oxidation. It is, therefore, important to carry out tests at temperatures close to natural conditions, such as ambient or physiological temperatures (Laguerre et al. 2007).

ROO peroxy radicals: ROO˙ peroxy radicals are a prime target for assessing antiradical activity. Azoinitiators are commonly used for ROO˙ generation. In β-carotene and TBARS assays, a water-soluble 2,2′-Azobis(2-amidopropane)dihydrochloride (AAPH or ABAP) is most commonly used for ROO generation. Hanlon and Seybert (1997) found that ABAP dramatically increased the rate of lipid peroxidation as pH increased from 5 to 7, and began to plateau at a pH of about 8. In addition, the rate of lipid peroxidation depends on temperature and the viscosity of the medium. It is likely that as the rate of ROO diffusion increases, the rate of lipid peroxidation increases (Laguerre et al. 2007).

6.3.1.4 Methods Commonly Used for Testing Antioxidant Activity of Essential Oils

Three common methods used to evaluate antioxidant activity of essential oils are the β-carotene bleaching method, DPPH free radicals scavenging method, and thiobarbituric acid reactive species assay (TBARS). The principles of the assays are described in the following subsections.

6.3.1.4.1 β-Carotene Bleaching Assay

This method is based on the result of β-carotene oxidation by linoleic acid degradation products, which are catalyzed by heat. Tween is used for dispersion of linoleic acid and β-carotene in the aqueous phase. The addition of an antioxidant results in retarding β-carotene bleaching. Quantitative analysis of β-carotene is measured by UV spectrophotometry at 470 nm (Laguerre et al. 2007). Results are expressed as the percentage inhibition of β-carotene bleaching (Gachkar et al. 2007; Kulisic et al. 2004; Sacchetti et al. 2005; Wang et al. 2008), the relative antioxidant activity (RAA%) of the sample and butylated hydroxyltoluene (BHT) (Obame et al. 2007), or IC_{50}, which is a sample concentration providing 50% inhibition (Khadri et al. 2008).

6.3.1.4.2 2,2-Diphenyl-1-Perylhydrazyl Free Radicals Scavenging Test (DPPH Test)

The DPPH test is based on the reduction of 2,2-diphenyl-1-perylhydrazyl free radicals (DPPH) in the presence of hydrogen-donating antioxidant or free radical scavenging agent (Kulisic et al. 2004). The DPPH radical absorbs at 517 nm, and antioxidant activity can be determined by monitoring the decrease of DPPH radical. Results are expressed as EC_{50}, that is, the amount of antioxidant necessary to decrease the initial DPPH concentration by 50% (Antolovich et al. 2002; Lertsatitthanakorn et al. 2006; Demirci et al. 2007), or the percentage inhibition of DPPH radical (Gachkar et al. 2007; Khadri et al. 2008; Kulisic et al. 2004; Obame et al. 2007; Sacchetti et al. 2005; Singh et al. 2005; Wang et al. 2008).

6.3.1.4.3 Thiobarbituric Acid Reactive Substances (TBARS) Assay

This method is commonly used to detect lipid oxidation. This procedure measures malonaldehyde (MDA) formation, which is the split product and endoperoxide of unsaturated fatty acids resulting from oxidation of a lipid substrate. The MDA is reacted with thiobarbituric acid to form a pink pigment that is measured spectrophotometrically at 532–535 nm (Antolovich et al. 2002). Results may be expressed as the percentage inhibition of oxidation (AI%) (Ruberto and Baratta 2000; Kulisic et al. 2004) or thiobarbituric acid value (meq of malonaldehyde/g) (Singh et al. 2005).

6.4 SUMMARY

The results given in the literature suggest that essential oils from some *Cymbopogon* species show antimicrobial activity against a wide range of bacteria and fungal species, and also have a weak-to-moderate antioxidant activity. Although the oils are used as fragrance in perfumery and in the food and beverage industry, they may also have great potential for protection of food and cosmetics from microbial spoilage and as a topical antiseptic. However, because it has weak-to-moderate antioxidant activity, using the oils to prevent oxidative deterioration of lipids in food might be helpful for a certain period of time.

REFERENCES

Antolovich, M., Prenzler, P., Patsalides, E., McDonald, S., Robards, K., 2002. Methods for testing antioxidant activity. *Analyst* 127, 183–198.

Bakkali, F., Averbeck, A., Averbeck, V.D., Idaomar, M., 2008. Biological effects of essential oils—A review. *Food and Chemistry Toxicology* 46, 446–475.

de Billerbeck, V., Roque, C., Bessiere, J., Fonvieille, J., Dargent, R., 2001. Effects of *Cymbopogon nardus* (L.) W. Watson essential oil on the growth and morphogenesis of *Aspergillus niger*. *Canadian Journal of Microbiology* 47, 9–17.

Demirci, B., Kosar, M., Demirci, F., Dinc, M., Baser, K.H.C., 2007. Antimicrobial and antioxidant activities of the essential oil of *Chaerophyllum libanoticum* Boiss. et Kotschy. *Food Chemistry* 105, 1512–1517.

Devkatte, A.N., Zore, G.B., Karuppayil, S.M., 2005. Potential of plant oils as inhibitors of *Candida albicans* growth. *FEMS Yeast Research* 5, 867–873.

Duarte, M.C., Figueira, G.M., Sartoratta, A., Rehder, V.L., Delarmeliva, C., 2005. Anti-candida activity of Brazilian medicinal plants. *Journal of Ethnopharmacology* 97, 305–311.

Dutta, B.K., Karmakar, S., Naglot, A., Aich, J.C., Begam, M., 2006. Anticandidial activity of some essential oils of a mega biodiversity hotspot in India. *Mycoses* 50, 121–124.

Farag, R.S., Daw, Z.Y., Abo-Raya, S.H., 1989. Influence of some spice essential oils on *Aspergillus parasiticus* growth and production of aflatoxins in a synthetic medium. *Journal of Food Science* 54, 74–76.

Frankel, E., 1998. *Lipid Oxidation*. The Oily Press, Dundee, U.K.

Frankel, E., Meyer, A., 2000. The problems of using one-dimensional methods to evaluate multifunctional food and biological antioxidants. *Journal of Science Food Agriculture* 80, 1925–1941.

Gachkar, L., Yadegari, D., Rezaei, M.B., Taghizadeh, M., Astaneh, S.A., Rasooli, I., 2007. Chemical and biological characteristics of *Cuminum cyminum* and *Rosmarinus officinalis* essential oils. *Food Chemistry* 102, 898–904.

Hammer, K.A., Carson, C.F., Riley, T.V., 1999. Antimicrobial activity of essential oils and other plant extracts. *Journal of Applied Microbiology* 86, 985–990.

Hanlon, M., Seybert, D., 1997. The pH dependence of lipid peroxidation using water-soluble azo initiators. *Free Radical Biological Medicine* 23, 712–719.

Helander, I.M., Alakomi, H., Latva-Kala, K., Mattila-Sandholm, T., Pol, I., Smid, E., Gorris, L.G.M., von Wright, A., 1998. Characterization of the action of selected essential oil components on Gram-negative bacteria. *Journal of Agricultural and Food Chemistry* 46, 3590–3595.

Hierro, I., Valero, A., Perez, P., Gonzalez, P., Cabo, M.M., Montilla, M.P., Navarro, M.C., 2004. Action of different monoterpenic compounds against *Anisakis simplex* s.l. L3 larvae. *Phytomedicine* 11, 77–82.

Hili, P., Evans, C.S., Veness, R.G., 1997. Antimicrobial action of essential oils: The effect of dimethylsulphoxide on the activity of cinnamon oil. *Letters in Applied Microbiology* 24, 269–275.

Himejima, M., Kubo, I., 1993. Fungicidal activity of polygodial in combination with anethole and indole against *Candida albicans*. *Journal of Agricultural and Food Chemistry* 41, 1776–1779.

Inouye, S., Takizawa , T., Yamaguchi, H., 2001. Antibacterial activity of essential oils and their major constituents against respiratory tract pathogens by gaseous contact. *Journal of Antimicrobial Chemotherapy* 47, 565–573.

Inouye, S., Uchida, K., Abe, S., 2006. Vapor activity of 72 essential oils against a *Trichophyton mentagrophytes*. *Journal of Infection and Chemotherapy* 12, 210–216.

Janero, D.R., 1990. Malondialdehyde and thiobarbituric acid-reactivity as diagnostic indices of lipid peroxidation and peroxidative tissue injury. *Free Radical Biology and Medicine* 9, 515–540.

Janssen, A.M., Scheffer, J.J.C., Svendsen, A.B., 1987. Antimicrobial activity of essential oils: A 1976–1986 literature review. Aspects of the test methods. *Planta Medica* 53, 395–398.

Jirovetz, L., Buchbauer, G., Gernot, N., Benoit, M.M., Pierre, M., 2007. Composition and antimicrobial activity of *Cymbopogon giganteus* (Hochst.) Chiov. Essential flower, leaf and stem oils from Cameroon. *Journal of Essential Oil Research* 19, 485–489.

Kalemba, D., Kunicka, A., 2003. Antibacterial and antifungal properties of essential oils. *Current Medicinal Chemistry* 10, 813–829.

Khadri, A., Serralheiro, M.L.M., Nogueira, J.M.F., Neffati, M., Smiti, S., Araujo, M.E.M., 2008. Antioxidant and antiacetylcholinesterase activities of essential oils from *Cymbopogon schoenanthus* L. Spreng. Determination of chemical composition by GC-mass spectrometry and 13C NMR. *Food Chemistry* 109, 630–637.

Kulisic, T., Radonic, A., Katalinic, V., Milos, M., 2004. Use of different methods for testing antioxidative activity of oregano essential oil. *Food Chemistry* 85, 633–640.

Kurita, N., Miyaji, M., Kurane, R., 1981. Antifungal activity of components of essential oils. *Agricultural and Biological Chemistry* 45, 945–952.

Laguerre, M., Lecomte, J., Villeneuve, P., 2007. Evaluation of the ability of antioxidants to counteract lipid oxidation: Existing methods, new trends and challenges. *Progress in Lipid Research* 46, 244–282.

Lertsatitthanakorn, P., Taweechaisupapong, S., Aromdee, C., Khunkitti, W., 2006. In vitro bioactivities of essential oils used for acne control. *International Journal of Aromatherapy* 16, 43–49.

Lopez, P., Sanchez, C., Batlle, R., Nerin, C., 2005. Solid- and vapor-phase antimicrobial activities of six essential oils: Susceptibility of selected foodborne bacterial and fungal strains. *Journal of Agricultural and Food Chemistry* 53, 6939–6946.

Mahmoud, A.L., 1994. Antifungal action and antiaflatoxigenic properties of some essential oil constituents. *Letters in Applied Microbiology* 19, 110–113.

Mann, C.M., Markham, J.L., 1998. A new method for determining the minimum inhibitory concentration of essential oils. *Journal of Applied Microbiology* 84, 538–544.

Manou, I., Bouillard, L., Devleeschouwer, M.J., Barel, A.O., 1998. Evaluation of the preservative properties of *Thymus vulgaris* essential oil in topically applied formulations under a challenge test. *Journal of Applied Microbiology* 84, 368–376.

Milos, M., Mastelic, J., Jerkovic, I., 2000. Chemical composition and antioxidant effect of glycosidically bound volatile compounds from oregano (*Origanum vulgare* L. ssp. *hirtum*). *Food Chemistry* 71, 79–83.

Mishra, A.K., Dubey, N.K., 1994. Evaluation of some essential oils for their toxicity against fungi causing deterioration of stored food commodities. *Applied and Environmental Microbiology* 60, 1101–1105.

Nakahara, K., Alzoreky, N., Yoshihashi, T., Nguyen, H., Trakoontivakorn, G., 2003. Chemical composition and antifungal activity of essential oil from *Cymbopogon nardus* (Citronella grass). *Japan Agricultural Research Quarterly* 37, 249–252.

Nikaido, H., 1994. Prevention of drug access to bacterial targets: Permeability barriers and active efflux. *Science* 264, 382–388.

Obame, L., Koudou, J., Chalchat, J., Boassole, I., Edou, P., Ouattara, A., Traore, A., 2007. Volatile components, antioxidant and antibacterial activities of *Dacryodes buettneri* H.J. Lam. essential oil from Gabon. *Scientific Research and Essay* 2, 491–495.

Ohno, T., Kita, M., Yamaoka, Y., Imamura, S., Yamamoto, T., Mitsufuji, S., Kodama, T., Kashima, K., Imanishi, J., 2003. Antimicrobial activity of essential oils against *Helicobacter pylori*. *Helicobacter* 8, 207–215.

Oussalah, M., Caillet, S., Saucier, L., Lacroix, M., 2006. Antimicrobial effects of selected plant essential oils on the growth of a *Pseudomonas putida* strain isolated from meat. *Meat Science* 73, 236–244.

Ozcelik, B., Lee, J.H., Min, D.B., 2003. Effects of light, oxygen, and pH on the absorbance of 2,2-diphenyl-1-picrylhydrazyl. *Journal of Food Science* 68, 487–490.

Paranagama, P.A., Abeysekera, K.H.T., Abeywickrama, K., Nugaliyadde, L., 2003. Fungicidal and anti-aflatoxigenic effects of the essential oil of *Cymbopogon citratus* (DC.) Stapf. (lemongrass) against *Aspergillus flavus* Link. isolated from stored rice. *Letters in Applied Microbiology* 37, 86–90.

Pattnaik, S., Subramanyam, V.R., Kole, C., 1996. Antibacterial antifungal activity of ten essential oils in vitro. *Microbios* 86, 237–246.

Pattnaik, S., Subramanyam, V.R., Kole, C.R., Sahoo, S., 1995a. Antibacterial activity of essential oils from *Cymbopogon*: Inter- and intra-specific differences. *Microbios* 84, 239–245.

Pattnaik, S., Subramanyam, V.R., Rath, C.C., 1995b. Effect of essential oils on the viability and morphology of *Escherichia coli*. *Microbios* 84, 195–199.

Pawar, V.C., Thaker, V.S., 2006. In vitro efficacy of 75 essential oils against *Aspergillus niger*. *Mycoses* 49, 316–323.

Porter, W., 1993. Paradoxical behavior of antioxidants in food and biological systems. In: Williams, G. (Ed.), *Antioxidants: Chemical, Physiological, Nutritional and Toxicological Aspects*, Princeton Scientific, Princeton, pp. 93–122.

Prabuseenivasan, S., Jayakumar, M., Ignacimuthu, S. 2006. In vitro antibacterial activity of some plant essential oils. *BMC Complementary and Alternative Medicine*, 6, 39, doi:10.1186/1472-6882-1186-1139.

Prashar, A., Hili, P., Veness, R.G., Evans, C.S., 2003. Antimicrobial action of palmarosa oil (*Cymbopogon martinii*) on *Saccharomyces cerevisiae*. *Phytochemistry* 63, 569–575.

Remmal, A., Bouchikhi, T., Rhayour, K., Ettayebi, M., Tantaoui-Elaraki, A., 1993. Improved method for the determination of antimicrobial activity of essential oils in agar medium. *Journal of Essential Oil Research* 5, 179–184.

Rosato, A., Vitali, C., De Laurentis, N., Armenise, D., Antonietta Milillo, M., 2007. Antibacterial effect of some essential oils administered alone or in combination with Norfloxacin. *Phytomedicine* 14, 727–732.

Ruberto, G., Baratta, M.T., 2000. Antioxidant activity of selected essential oil components in two lipid model systems. *Food Chemistry* 69, 167–174.

Russell, A.D., Hugo, W.B., Ayliffe, G., 1992. *Principles and Practice of Disinfection, Preservation and Sterilization*. 2nd ed. Blackwell Science, Oxford.

Sacchetti, G., Maietti, S., Muzzoli, M., Scaglianti, M., Manfredini, S., Radice, M., Bruni, R., 2005. Comparative evaluation of 11 essential oils of different origin as functional antioxidants, antiradicals and antimicrobials in foods. *Food Chemistry* 91, 621–632.

Shahi, A.K., Tava, A., 1993. Essential oil composition of three cymbopogon species of Indian Thar Desert. *Journal of Essential Oil Research* 5, 639–643.

Si, W., Gong, J., Tsao, R., Zhou, T., Yu, H., Poppe, C., Johnson, R., Du, Z., 2006. Antimicrobial activity of essential oils and structurally related synthetic food additives towards selected pathogenic and beneficial gut bacteria. *Journal of Applied Microbiology* 100, 296–305.

Singh, G., Paimuthu, P., Murali, H., Bawa, A., 2005. Antioxidative and antibacterial potentials of essential oils and extracts isolated from various spice materials. *Journal of Food Safety* 25, 130–145.

Southwell, I.A., Hayes, A.J., Marham, J., Leach, D.N., 1993. The search for optimally bioactive Australian tea tree oil. *Acta Horticulturae* 334, 256–265.

Tampieri, M.P., Galuppi, R., Macchioni, F., Carelle, M., Falcioni, L., Cioni, P., Morelli, I., 2005. The inhibition of *Candida albicans* by selected essential oils and their major components. *Mycopathologia* 159, 339–345.

Tullio, V., Nostro, A., Mandras, N., Dugo, P., Banche, G., Cannatelli, M.A., Cuffini, A.M., Alonzo, V., Carlone, N.A., 2007. Antifungal activity of essential oils against filamentous fungi determined by broth microdilution and vapour contact methods. *Journal of Applied Microbiology* 102, 1544–1550.

van Zyl, R., Seatlholo, S., van Vuuren, S., Viljoen, A.M., 2006. The biological activities of 20 nature identical essential oil constituents. *Journal of Essential Oil Research* 19–133.

Wang, W., Wu, N., Zu, Y.G., Fu, Y.J., 2008. Antioxidative activity of *Rosmarinus officinalis* L. essential oil compared to its main components. *Food Chemistry* 108, 1019–1022.

Wannissorn, B., Jarikasem, S., Siriwangchai, T., Thubthimthed, S., 2005. Antibacterial properties of essential oils from Thai medicinal plants. *Fitoterapia* 76, 233–236.

Zani, F., Massimo, G., Benvenuti, S., Bianchi, A., Albasini, A., Melegari, M., Vampa, G., Bellotti, A., Mazza, P., 1991. Studies on the genotoxic properties of essential oils with *Bacillus subtilis* rec-assay and *Salmonella/* microsome reversion assay. *Planta medica* 57, 237–241.

7 Thrombolysis-Accelerating Activity of Essential Oils

Hiroyuki Sumi and Chieko Yatagai

CONTENTS

7.1 INTRODUCTION

We have presented reports that ingestion of various types of alcoholic drinks (Sumi et al. 1988, 1998) and coffee (Sumi 1997) results in changes in blood coagulation–fibrinolysis systems; in particular, a rather lengthy period of promotion of fibrinolysis in the blood was observed with drinking of *Oturui* shochu liquors (distilled only once, retaining the character of the original ingredients and known as "authentic" shochu). Studies have been made on properties of the components bringing about such effects (Sumi 2003). It has also been generally recognized that essential oils have natural healing effects, and the psychological effect of inducing a feeling of relaxation. However, regarding the physiological effects of essential oils on blood circulation, especially their direct effects on blood coagulation–fibrinolysis systems, no report has yet been presented to our knowledge.

Lemongrass being a well-known herb, we used the CLT method to check its effects on the blood coagulation system and the fibrinolysis system. Lemongrass has the effects of facilitating blood coagulation and inhibiting fibrinolysis, but it has been reported that the substantial citral and geraniol content in lemongrass contributes to strong inhibitory activity against platelet aggregation, producing fibrinolytic effects in vivo as a result. Imai et al. discovered that lemongrass oil, the essential oil component (*Cymbopogon citrus*) of lemongrass (oil yield: 0.2%), shows strong platelet-aggregation inhibitory activity (Imai et al. 1986). They reported that, when using platelet-rich plasma of humans, rabbits, and rats, aggregation through the application of ADP, collagen, and arachidonic acid was inhibited by adding lemongrass oil at an amount 1/12,000 of the amount of plasma. It was reported that the component of the oil was citral and that *trans*-citral was more effective than *cis*-citral.

In the current study, we conducted tests regarding the effects of more than 100 types of essential oils (herbs) on the blood coagulation–fibrinolysis systems, using lemongrass as the control.

7.2   TEST I

Thirty types of essential oils purchased from Holistic Origin Pvt. Ltd. (Singapore) were used. The CLT test for essential oils was conducted in the following way:

Solutions were mixed in a glass test tube (1.0×120 mm^2). They included 50 µL of a solution of each essential oil diluted with dimethyl sulfoxide (DMSO), 250 µL of 0.17 M borate-saline buffer with pH 7.8, 500 µL of bovine plasma fibrinogen manufactured by Sigma Co., Ltd. that was dissolved in 250 µL of 0.17 borate-saline buffer with pH 7.8 with a concentration of 0.6% and filtered with filter paper, and 100 µL of urokinase, a fibrinolytic enzyme in 20 IU/mL of saline. After incubation at 37°C for 1 min, 100 µL of bovine thrombin (2.25 IU/mL saline) was added and stirred, followed by another incubation. Measurements were then conducted for the time until coagulation and until complete lysis of the artificial thrombus (fibrin) were generated, in order to test the blood coagulation–fibrinolysis activity of the essential oils (Sumi 2000).

7.2.1   RESULTS

As shown in the left part of Figure 7.1, lysis time was found to be longer for many samples in comparison with the control with no additions (broken line), which means that these samples had inhibitory effects on coagulation. Each value in the figure is the average value of five repetitions of the test. As shown on the right in Figure 7.1, lysis time was seen to be longer for many samples in comparison with the control with no additions (broken line), meaning these samples had inhibitory effects on fibrinolysis. Overall, there were considerable differences among the aroma essences with regard to the blood coagulation–fibrinolysis systems. As shown by the arrows, the results for lemongrass here were that it promoted coagulation (index $6/30 = 0.20$) and reduced fibrinolysis ($24/30 = 0.80$).

7.3   TEST II

Essential oils (plant, animal, and synthetic aroma essences) supplied by Kanebo Cosmetics, Inc., were used for the test.

7.3.1   STANDARD FIBRIN PLATE METHOD

In a round, 90-mm-diameter petri dish, 30 µL of each of the samples, 30 µL of a solution with the sample diluted by an equal amount of 10 IU/mL urokinase, and 30 µL of 1% ethanol as the control were applied to the artificial thrombus (fibrin) plates, created with 10 mL of fibrinogen solution with a final concentration of 0.5% and 0.5 mL of 50 U/mL thrombin. After incubation at 37°C for 4 h, the area (mm^2) of the lysis area was measured.

7.3.2   EUGLOBULIN: FIBRIN PLATE METHOD

Citrate blood was drawn from the tail veins of etherized Wistar male rats, and the blood plasma obtained through centrifugal separation (3000 rpm, 10 min, room temperature) was diluted 20 times

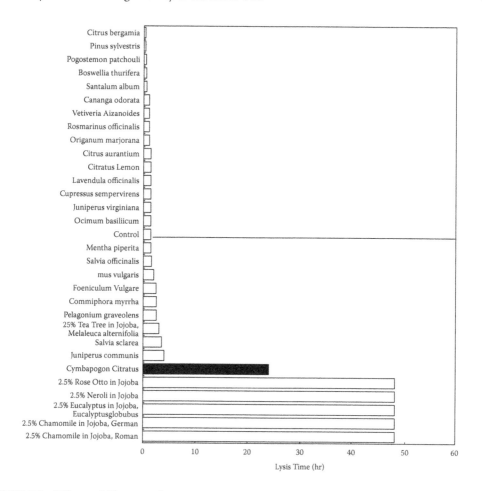

FIGURE 7.1 Effects of 30 types of aroma essences on the blood coagulation–fibrinolysis systems. Each sample dissolved to a concentration of 20% with DMSO, along with urokinase solution, was added to the fibrinogen solution. After adding thrombin solution, measurements were conducted for the time until coagulation (left, in minutes) and the time until the lysis of the coagulated fibrin (artificial thrombus).

with 0.016% acetic acid. After being left standing for 30 min at 4°C, another centrifugal separation (3000 rpm, 5 min, room temperature) resulted in euglobulin fractions, which were dissolved with 1/15 M phosphate buffer (pH 6.8) containing an amount of 0.9% sodium chloride equivalent to the amount of plasma. The euglobulin solution diluted with the same amount of each of the samples, 30 µL of the mixture was applied to the fibrin plate and after incubation at 37°C for 4 h, the area (mm²) of the lysis area was measured.

7.3.3 RESULTS

Lemongrass resulted in an index of 20/84 = 0.24 for the standard fibrin plate, 23/84 = 0.27 for ELT and for CLT, clotting promotion of 65/84 = 0.77 and a fibrinolytic reduction of 40/84 = 0.48. In comparison, results of the standard fibrin plate, ELT and CLT tests with the addition of 84 types of diluted essential oils showed that elder, cashew, and grapefruit have strong fibrinolytic activity, whereas celery, fir tree, baca, olive, and rosemary have strong inhibitory effects against fibrinolysis.

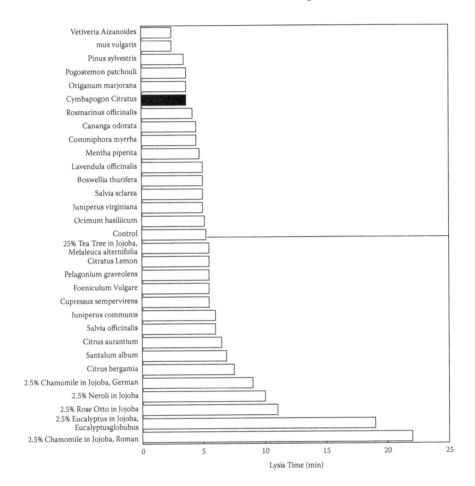

FIGURE 7.1 (continued).

7.4 TEST III

Similarly, 84 types of essential oils were used for this test. Pyrazin, its derivatives of 2-ethyl pyrazine, 2,5-dimethyl pyrazine, 2,3,5-trimethyl pyrazine, and 2,6-dimethyl pyrazine and aspirin were purchased from Sigma–Aldrich, Inc.

7.4.1 EFFECTS ON TISSUE PLASMINOGEN ACTIVATOR (tPA)-PRODUCING CELLS

Tissue plasminogen activator-free human cells were supplied by physiology Assistant Professor Yoshida of Miyazaki Medical University. After multiplying these cells using a 24-well microtest plate (Falcon) to confluence, the cultured solution inside the wells was removed using an aspirator. The cells were then washed twice with PBS(−), 450 μL of a new culture solution and 50 μL of each sample were added and incubated for 24 h, following which the culture solution was recovered (first medium). After again washing the cells twice with PBS(−), 500 μL of a new culture solution was added and incubated again for 24 h, and the culture solution was recovered (second medium). For the fibrinolytic activity, the area (mm^2) of the lysis area was measured through the standard fibrin plate method.

7.4.2 PLATELET AGGREGATION TEST

As the aggregation-inducing substance, 22 µL of 300 µM ADP or of 1.2 mg/mL collagen (final concentration of 30 µM or 1.2 mg/mL) was added, and the platelet aggregation rate was measured for 5 min at 37°C using an aggregometer (PAT-4A: Mebanix). With 100% light transmittance set for PPP, platelet aggregation activity was measured.

7.4.3 RESULTS

The effects of lemongrass on tPA-producing cells and platelet aggregation are shown with arrows. Regarding the comparative effects of the addition of other essential oils, promotion of tPA was observed for orange, basil, and clove (see Figure 7.2). In the case of orange in particular, adding 450 µL of the cultured solution and 50 µL of the added sample and incubating for 24 h inside a 5% CO_2 incubator at 37°C (first medium) showed that, compared with the control, approximately eight times the fibrinolytic activity was seen.

Regarding the effects on platelet aggregation, using ADP as the aggregation-inducing substance showed that the inhibitory activity of basil (77%) and tolu (64%) is fairly strong (see Figure 7.3). Platelet-aggregation inhibitory activity was also detected in coffee and white peach. The pyradine compounds contained in coffee had strong inhibitory effects on platelet aggregation, with strength equivalent to that of aspirin in many cases (Table 7.1).

7.5 TEST IV

As in the previous tests, 84 types of essential oils were used for this test.

7.5.1 IN VIVO FIBRINOLYTIC ACTIVITY OF EUGLOBULIN

The aroma essences were diluted 200 times with 0.5% ethanol–0.9% sodium chloride solution, and 5 mL per kg of body weight (aroma essence volume: 25 µL/kg) was orally administered to Wistar male rats using a feeding tube. Blood was drawn from the rat tail veins 1 h prior to oral administration and 1 and 2 h after administration.

For the ELT, 0.1 mL of the blood plasma obtained from citrate blood through centrifugal separation (3000 rpm, 10 min, room temperature) was diluted 20 times with 0.016% acetic acid, and left standing for 30 min at 4°C. Centrifugal separation (3000 rpm, 5 min, room temperature) resulted in euglobulin fractions, which were dissolved with 0.1 mL of 0.1 M *tris*-hydrochloric acid buffer (pH 7.4) to obtain the sample. 90 µL of the euglobulin solution and 10 µL of 100 U/mL thrombin were mixed inside a 96-well microtest plate and incubated at 37°C, following which turbidity of the dissolving thrombi was measured every 10 min at 405 nm using a Well reader (SK601: Seikagaku Corp.).

7.5.2 EFFECTS ON BLOOD COAGULATION ACTIVITY

For the measurements, Data-Fi aPTT (Dade Behring) was used for the activated partial thromboplastin time (aPTT), and thromboplastin C plus (Dade Behring) was used for the prothrombin time (PT).

All the animals used in the tests were healthy and all provisions of the Declaration of Helsinki (1964) were met.

7.5.3 RESULTS

Finally, several essential oils were orally administered to rats and changes in their blood were tested for through ELT. For mainly coffee and lemongrass, the reduction of ELT was observed,

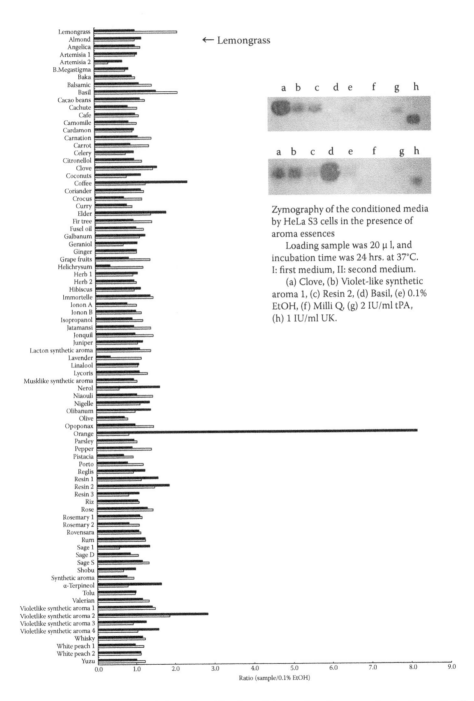

← Lemongrass

Zymography of the conditioned media by HeLa S3 cells in the presence of aroma essences

Loading sample was 20 μ l, and incubation time was 24 hrs. at 37°C. I: first medium, II: second medium.

(a) Clove, (b) Violet-like synthetic aroma 1, (c) Resin 2, (d) Basil, (e) 0.1% EtOH, (f) Milli Q, (g) 2 IU/ml tPA, (h) 1 IU/ml UK.

Ratio (sample/0.1% EtOH)

FIGURE 7.2 tPA activity generated through cultivation of tissue plasminogen activator-free cells. Further isolation of each activity through zymography possible (partial results shown). Average values (n = 3) shown; ■: first medium, □: second medium.

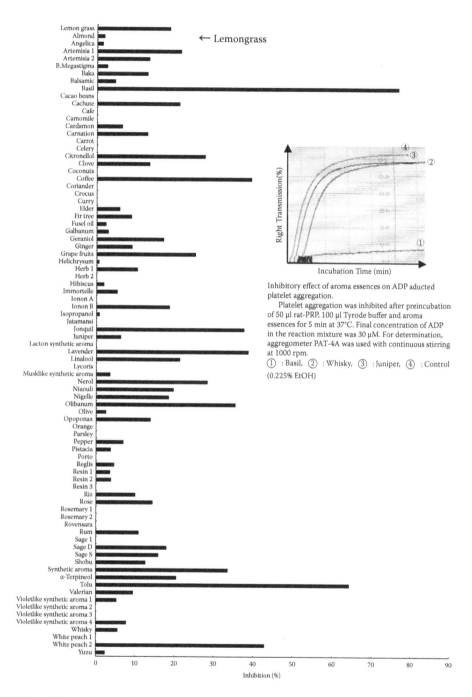

FIGURE 7.3 Effects of essential oils on platelet aggregation activity. For rat blood platelets, 30 μM ADP used as aggregation-inducing substance. Aggregation patterns from aggregometer shown. Average values (n = 3) shown.

TABLE 7.1
Antiplatelet Aggregation Effects of Pyradine Compounds

Components		Aggregation	50 mg/mL	5 mg/mL	1 mg/mL	0.5 mg/mL
Pyridine		Collagen	+++	++	+	—
		ADP	+++	++	+	—
2-ethyl pyridine		Collagen	+++	+++	++	—
		ADP	+++	+++	+++	+
2,5-dimethyl pyridine		Collagen	+++	+++	+	—
		ADP	+++	+++	—	—
2,3,5-trimethyl pyridine		Collagen	+++	+++	—	—
		ADP	+++	+++	+++	+
2,6-dimethyl-3-methyl pyridine		Collagen	+++	+++	—	—
		ADP	+++	++	+	—
Aspirin		Collagen	+++	+++	+	—
		ADP	+++	+++	++	—

Note: In the order of the strength of platelet aggregation, reactions as viewed with the naked eye are shown.

meaning a tendency to promote fibrinolysis (Figure 7.4). The aPTT (38 ± 7 s) and PT (12 ± 3 s) values as a result of the effects of coffee and lemongrass were investigated, but the changes were not significant.

7.5.4 DISCUSSION

Various tests related to the blood coagulation–fibrinolysis systems were conducted using essential oils I and II, but the results were not consistent. However, it seemed that there were effects on the promotion of tPA from vascular endothelial cells and on platelets, and a fibrinolytic tendency was observed in the test using lemongrass cells. In the case of coffee, it seemed that there were effects of the activity of pyridine compounds in promoting fibrinolysis. Yamamoto (2003) selected 6 out of

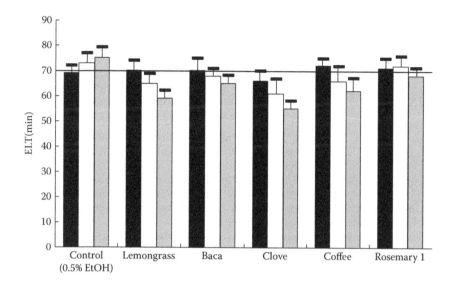

FIGURE 7.4 Oral administration. Changes in the blood caused by oral administration of several essential oils as measured by plasma ELT. Values are means ± SE (n = 4); ■, 1 h before administration; □, 1 h after administration; ▨, 2 h after administration.

26 types of herbs for their blood fluidity, through a test in which each herb (aroma essence) was added to blood and fluidity of the blood was measured. The six types selected were echinacea, thyme, hip, marigold, lavender, and lemongrass.

The primary purpose of aroma therapy using essential oils is deemed to be bringing about relaxation, but many essential oils have antibacterial activity and are effective in the sterilization of skin. Takarada et al. (2004) used manuka oil and rosemary oil to compare and study their antibacterial activities against bacteria in the mouth, which cause dental caries and periodontitis. Many applications have recently been filed for patents related to lemongrass. Examples include a stress reliever (Suntory Ltd. 2003) containing aromatic components of alcoholic drinks as effective components, bath tablets (Cosme Park 2005) containing herb ingredients, fragrances (Nitto Pharmaceutical 2005) containing 0.1%–5% lemongrass oil per weight, and an aroma essence for increasing memory (Pola Chemical Industries 2001).

If practical effects can be achieved merely by smelling aroma essences rather than ingesting them, it is possible that aromatic components may constitute a wholly new category of functional materials. In tests adding aromatic components using sweet potato-based shochu liquor, we have already shown that the amount of fibrinolytic enzymes produced from human vascular endothelial cells increases (Sumi et al. 2001). It has also been reported that many aroma essences are generated at the time of heating and distilling unprocessed shochu liquor for *Otsurui* shochu liquors, particularly sweet potato-based shochu liquors (Ota 1991). A more detailed analysis of each essential oil and identification of the in vivo effects of each essential oil would be issues to address from now on.

7.6 SUMMARY

Tests conducted on more than 100 types of essential oils showed that some of them had fibrinolytic activity similar to lemongrass. However, no correlation was found among the results of the standard fibrin plate, ELT, and CLT methods. It is thought that promoting the release of tissue plasminogen activators from cells and inhibiting platelet aggregation are significant reactions, which were revealed to be the result of complex configurations.

ACKNOWLEDGMENTS

We are grateful to Mr. S. Murakami for excellent technical assistance. Essential oils were kindly supplied by Kanebo Cosmetics, Inc. (Kanagawa).

REFERENCES

Cosme Park. 2005. Bath tablets, Published Patent Application 112794 (in Japanese).

Imai H, Tamada T, Sawai Y, Yoshida M, Takao K, Endo E, Ohshiba S, Kojisawa Y. 1986. Blood platelet aggregation inhibitory activity of lemongrass oil. 28th Convention of the Japanese Society of Clinical Hematology, Extracts, 279 (in Japanese).

Nitto Pharmaceutical Industries Ltd. 2005. Fragrances, Published Patent Application 023044 (in Japanese).

Ota T. 1991. Aroma of sweet potato-based shochu liquor, *Tokyo* **86**: 250–254 (in Japanese).

Pola Chemical Industries Inc. 2001. Aroma essences and components containing aroma essences for increasing memory, Published Patent Application 288493 (in Japanese).

Sumi H, Hamada H, Tsushima H, Mihara H. 1988. Urokinase-like plasminogen activator increased in plasma after alcohol drinking. *Alcohol Alcoholism* **23**: 33–43.

Sumi H, Iijima Y, Komine S, Sasahira T. 2001. Kagoshima Industry Support Center, Report on Commissioned Research and Development Project 2000: 1–6.

Sumi H. 1997. Abnormalities in the blood coagulation and fibrinolysis systems. *Nippon Rinsho* **712**: 233–239 (in Japanese).

Sumi H. 2000. Blood circulation: Analysis of foods and aroma essences related to blood coagulation–fibrinolysis systems. *Food Technology* **20**: 64–70 (in Japanese).

Sumi H. 2003. Improvements in blood circulation and disease prevention by food components, *Food Style* **21**(7): 47–53 (in Japanese).

Sumi H, Kozaki Y, Yatagai C, Hamada H. 1998. Effects of wine on plasma fibrinolytic and coagulation systems. *Japanese Journal of Alcohol Studies and Drug Dependence* **33**: 263–272.

Suntory Ltd. 2003. Stress reliever, Published Patent Application 171286 (in Japanese).

Takarada K, Kimizuka R, Takahashi N, Honma K, Okuda K, Kato T. 2004. A comparison of the antibacterial efficacies of essential oils against oral pathogens. *Oral Microbiology and Immunology* **19**(1): 61–64.

Yamamoto N. 2003. Effects of herbs on blood fluidity. *Food Style* **21**: 61–63 (in Japanese).

8 Analytical Methods for *Cymbopogon* Oils

Ange Bighelli and Joseph Casanova

CONTENTS

8.1 INTRODUCTION

The *Cymbopogon* genus comprises 56 species, and most of them produce essential oil (EO) by hydrodistillation or steam distillation of aerial parts. Various *Cymbopogon* species, such as *C. nardus* (L.) Rendle, *C. winterianus* Jowitt, *C. citratus* Stapf, *C. flexuosus* Stapf and *C. martinii* (Roxb.)

Wats., are well known as sources of commercially valuable compounds such as geraniol, geranyl acetate, citral (neral and geranial), and citronellal (Surburg and Panten 2006). The other species contribute to the conservation of the biodiversity through the world.

EOs, which are the raw materials used in the perfume, fragrance, and flavor industries, are, in general, complex mixtures of more than 100 individual components, containing mainly monoterpenes and sesquiterpenes that present various frameworks and functions. The identification of these compounds would be carried out using various techniques. Obviously, the choice of the analytical methods depends on whether the final objective is structural identification of an unknown substance or recognition of a substance that has already been identified.

In this chapter, we first summarize the analytical methods developed for identification of individual components of EOs; then we apply these techniques and methods to the analysis of *Cymbopogon* EOs, including studies carried out in our laboratories using a combination of various analytical techniques.

8.2 METHODS FOR EO ANALYSIS: A SUMMARY

Identification of the individual components of an EO is a difficult task that first requires an adequate choice of analytical technique and then a careful utilization of the instruments, software, spectral data libraries and, sometimes, statistical tools.

Depending on the objective of analysis (routine, quality control, detailed analysis of an EO investigated for the first time, etc.) and the complexity of the mixture (number and structure of the components and heat sensitivity), various schemes can be drawn up.

In routine analysis, identification of a component was achieved by combination of "on-line" chromatographic separation and spectroscopic identification. For instance, a fast-scanning mass spectrometer directly coupled with a gas chromatograph is the basic equipment of control in the field of EO analysis. Nevertheless, misuse—or abuse—of modern instrumentation is not unknown. Nowadays, a conventional analysis of an EO is carried out by matching against computerized mass spectral libraries and comparison of the retention indices with those of authentic samples (GC[RI] and GC-MS). However, more sophisticated techniques are needed for the analysis of EOs, and a wide variety of combinations of analytical techniques are being developed, including complex hybrids such as GC-MS-FTIR, HPLC-GC-MS, or HPLC-^{1}H NMR. In a similar fashion, individual components of an EO may be identified by comparison of the chemical shift values in the ^{13}C NMR spectrum of the mixture with those of reference compounds.

Despite the improvement of these on-line analytical techniques, the misidentification of some compounds, especially sesquiterpenes, still occurs. For this reason, several research groups developed a two-step procedure in which a small quantity of a substance is separated by a chromatographic technique and then identified by comparison of its spectral data, including ^{1}H NMR and sometimes ^{13}C NMR, with those of reference compounds. This "off-line" sequence is obviously more accurate but very time consuming.

Nowadays, in numerous research laboratories, the chemical composition of complex EOs is analyzed by combination of various techniques, the oil being prefractionated by fractional distillation and/or column chromatography and, eventually, by HPLC or preparative gas chromatography (PGC). Prefractionation of the EO considerably reduces the number of coeluted compounds. Then, analysis of the fractions is achieved by combination of spectroscopic techniques, such as MS, IR, and/or NMR.

8.2.1 CHEMICAL METHODS AND FRACTIONAL DISTILLATION

The procedures that were mostly used during the last century for the isolation of EO constituents were chemical methods and fractional distillation. Concerning chemical methods, a preparative separation of an EO into different groups of components was achieved. In this procedure, the oil was successively treated by aqueous sodium carbonate (5%), aqueous sodium hydroxide (5%), methanolic sodium hydroxide (0.5N), sodium hydrogen sulfite or Girard reagent and phthalic anhydride to separate various families of compounds: acids, phenols, esters and lactones, aldehydes and ketones, and alcohols (primary and secondary). Esters and lactones can be hydrolyzed, and the resulting acids separated as salts. The final residue, consisting of tertiary alcohols, ethers, and hydrocarbons, must be separated by physical methods such as fractional distillation, crystallization, or liquid–liquid extraction (on the basis of the difference of solubility of the components in various solvents) (Kubeczka 1985).

Fractional distillation had an important role in the preparative isolation of EO constituents, although it is seldom possible to obtain chemically pure components by this technique. The development of efficient distillation columns (spinning-band columns) enables one to fractionate milliliter amounts of EOs under mild conditions. For instance, by using a so-called Spaltrohr column, amounts of a few milliliters may be distilled at reduced pressure, which allows the distillation of thermally unstable components. With this type of column, up to 100 theoretical plates may be achieved. The efficiency of this device is clearly demonstrated by the separation of citronellyl acetate and geranyl acetate (boiling points at 1 mbar: 73.5°C and 77.0°C, respectively). However, pure compounds from EOs are rarely obtained by fractional distillation since isomerization and decomposition of labile components occasionally take place (Kubeczka, 1985).

8.2.2 CHROMATOGRAPHIC TECHNIQUES

8.2.2.1 Thin-Layer Chromatography (TLC)

Thin-layer chromatography is a type of liquid chromatography that can separate different chemical compounds based on the rate at which they move through a support under defined conditions. The support, known as the *plate*, is a layer of support material (silica gel or alumina) that has been spread out and dried on a sheet of material such as glass. The mobile phase is a solvent or a mixture of solvents. Individual components of a mixture are identified after separation by migration, by comparison of their Rf values with those of reference compounds obtained under the same experimental conditions. Due to its simplicity and speed, TLC was an important pioneering method in the analysis of EOs (Kubeczka 1985; Rouessac and Rouessac 1994).

TLC can also be used for preparative separation. By means of plates up to 1 m long and a layer thickness up to 2 mm, several grams of a mixture of compounds can be applied as a line. The various bands are scraped off, collected, and the components eluted with appropriate solvents.

8.2.2.2 Gas Chromatography

Gas chromatography (GC), which allows the separation of volatile compounds from complex mixtures as well as their quantification, has become one of the most important tools in the analysis of EOs (Kubeczka, 1985). Nowadays, the development of efficient capillary columns allows the individualization of several hundreds of components. For instance, more than 200 compounds have been distinguished in the chromatogram of labdanum oil (*Cistus ladanifer* oil) or vetiver oil (Weyerstahl et al. 1998, 2000). Concerning the identification of individual components, the comparison of their retention times with those of reference compounds is generally not sufficient, even using two columns of different polarity. For that reason, utilization of retention indices is generally preferred. Retention indices are determined relative to the retention times of a series of *n*-alkanes with constant temperature (Kováts indices [KIs]) (Kováts 1965) or with linear temperature programmed (retention indices [RIs]) (Van Den Dool and Kratz 1963). Identification is then carried out by comparison of retention indices with those of authentic compounds or literature data (Figure 8.1).

However, identifying sesquiterpenes (or diterpenes) by comparison of their RIs remains a difficult task. Indeed, as mentioned by Joulain (1994), more than 230 natural sesquiterpenes, which possess a molecular mass of 204, have been reported, and the values of their GC RIs differ by less than 30 points. Moreover, RI values reported in the literature for the same compound can vary by more

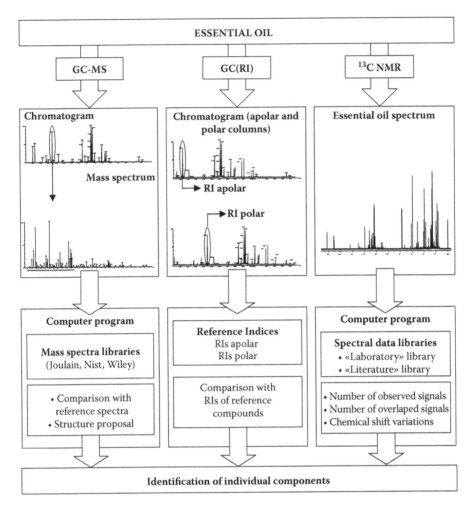

FIGURE 8.1   Combined analysis of essential oils by GC-MS, GC(RI) and ^{13}C NMR.

than 10 points, particularly those concerning the polar column (Grundschober 1991). Consequently, identification of all the components of a complex EO using only GC appeared very difficult, if not impossible, for sesquiterpenes, even by taking into account their RIs on polar and apolar columns.

8.2.3   HYPHENATED TECHNIQUES

8.2.3.1   GC-MS and GC-MS Combined with GC(RI)

A fast-scanning mass spectrometer using electronic impact (EI) mode, directly coupled with a gas chromatograph (GC-MS), is the basic equipment of control in the field of EO analysis. The two most-used analyzers in this field are the quadrupole and ion-trap types. Identification of the components in the oil is achieved by computer matching with laboratory-made or commercial mass spectral libraries containing several thousands spectra (Figure 8.1). Among the best-known commercial computerized libraries, we could mention the Wiley Registry of Mass Spectral Data (McLafferty and Stauffer 1994), the National Institute of Standards and Technology EPA/NIH Mass Spectral Library (NIST 1999), and the Terpenoids and Related Constituents of Essential Oils library (König et al. 2001). Comparison of experimental spectra may be done with those of literature data compiled in various paper libraries published, among others, by McLafferty and Stauffer (1988), Joulain and König (1998), and Adams (2007). Better reliability is generally obtained when comparing the experimental spectra with those of reference compounds recorded in the laboratory with the same experimental conditions.

Analysis of EOs by GC-MS began in the 1960s and was applied to common EOs of commercial interest. However, it is quite difficult to avoid misidentification of some compounds, especially those that possess insufficiently differentiated mass spectra. This problem concerns not only epimers such as α-cedrene and α-funebrene, stereoisomers such as (Z)-β-farnesene and (E)-β-farnesene, or diastereoisomers such as α-bisabolol and α-epi-bisabolol, but also compounds with different skeletons, such as (E,Z)-α-farnesene (linear sesquiterpene) and *cis*-α-bergamotene (bicyclic sesquiterpene) (Schultze et al. 1992), or 1-endo-bourbanol (tricyclic sesquiterpene) and 1,6-germacradien-5-ol (monocyclic sesquiterpene) (Joulain and Laurent 1989).

For that reason, and although analyses by GC-MS are unfortunately still reported in the literature, identification of a component is much more accurate when GC-MS is used in combination with GC(RI) (Figure 8.1). Today, it is considered by specialized reviews that analysis of an EO by at least two techniques, for instance, MS in combination with RIs measured on two columns of different polarity, should be done to make the work publishable.

Vernin et al. (1986, 1990) developed a computerized procedure based on the combination of mass spectral data and retention index on polar and apolar columns to identify the individual components of several EOs. Cavaleiro (2001) used the same procedure for the characterization of several *Juniperus* EOs.

In some cases, positive or negative chemical ionization mass spectrometry [(PCI)MS or (NCI) MS] allows the identification of compounds that possess similar mass spectra. Chemical ionization produces mass spectra that exhibit "quasi-molecular" ions from which the molecular mass of the compounds can be easily deducted; that is not always possible using EI. As in the case of EI mode, the quadrupole or ion-trap analyzer can be used in the chemical ionization mode. Utilization of (CI)MS allowed the differentiation of alcohols, such as the four isomers of isopulegol (Lange and Schultze 1988a), or esters (Lange and Schultze 1988b), whereas the mass spectra of these compounds are generally similar when EI mode is used. In the same way, several isomers bearing the pinane skeleton (α-pinene, β-pinene, and *cis*-δ-pinene; pinocamphone and isopinocamphone; nopinone and isonopinone) were differentiated by (PCI)MS by using ammoniac as reactive gas (Badjah Hadj et al. 1988). The choice of the reactive gas (methane, isobutene, or ammoniac for PCI and a mixture of N_2O and methane or ammoniac for NCI) to differentiate some isomers depends on the structure and the functionalization of the molecules under consideration. For instance, GC-(CI) MS with isobutane and ammonia chemical ionization allowed the distinguishing of sesquiterpene hydrocarbons (Schultze et al. 1992). (PCI)MS and (NCI)MS proved to be useful for the analysis of complex EOs (Paolini et al. 2005). However, due to the difficulty of obtaining reproducible spectra, chemical ionization mass spectrometry is not usable as unique technique, but it is recommended in combination with GC-(EI)MS for the analysis of complex EOs.

8.2.3.2 GC×GC-MS

In some very complex mixtures in which the low peak resolution in GC prevents good characterization, the use of two-dimensional GC (GC×GC) can be useful to improve the resolution (Dugo et al. 2000). For instance, using this two-dimensional technique, Mondello et al. (2005) eluted separately β-bisabolene, neryl acetate, and bicyclogermacrene in an EO of citrus, these three compounds being coeluted by conventional GC.

8.2.3.3 GC-MS-MS

Combination of GC and 2D mass spectrometry (GC-MS-MS) has been used for the analysis of complex EOs. For instance, Decouzon et al. (1990) clearly identified the four stereoisomers of dihydrocarveol using MS-MS (NCI) by observation of some characteristic fragments, which is not possible by monodimensional mass spectrometry [(NCI)MS]. Cazaussus et al. (1988) identified khusimone, a particular sesquiterpene, by GC-MS-MS in EI mode, in the oxygenated fraction of a complex vetiver EO that contained more than 100 components.

8.2.3.4   GC-FTIR, GC-MS-FTIR

The improvement of the threshold of detection of FTIR in the last decade allowed the development of this on-line technique for the analysis of various families of organic compounds (Coleman et al. 1989). Concerning EO components, GC-FTIR is particularly well suited to identifying compounds that possess insufficiently differentiated mass spectra. For instance, this technique has been used to identify two sesquiterpene alcohols, 1-endo-bourbanol (tricyclic) and 1,6-germacradien-5-ol (monocyclic) (Joulain and Laurent 1989), on the basis of their infrared mass spectra combined with their RIs. CG-FTIR is also well suited for analysis of compounds that suffer rearrangement in MS, such as germacrene-B and bicyclogermacrene, correctly identified by this method in citrus EOs (Chamblee et al. 1997). Conversely, it has been also reported that GC-FTIR was not appropriate to distinguish β-gurjunene and *cis*-1,3-dimethylcyclohexane, two compounds that possess quite different structure (Hedges and Wilkins 1991).

It seems that GC-FTIR is not really appropriate for the analysis of EOs, but it provides useful information when utilized in combination with GC-MS. For that reason, hyphenated techniques such as GC-FTIR-MS have been developed and successfully used in the field of EOs. For example, Hedges and Wilkins (1991) identified, by GC-FTIR-MS, 26 minor components together with the main compound, 1,8-cineole, in an EO from *Eucalyptus australiana*. A better separation of the components was obtained when multidimensional GC is directly combined with FTIR and MS for analysis of eucalyptus EOs (Krock et al. 1994).

8.2.3.5   HPLC-MS, HPLC-¹H NMR, and HPLC-GC-MS

Some EOs contain a variable amount of poor volatile compounds. Their analysis can be carried out by HPLC directly coupled with MS. Utilization of these combined techniques allowed the identification of oxygen heterocyclic compounds, coumarins, psoralenes, and flavones in citrus EOs (Dugo et al. 2000).

More recently, HPLC has also been combined with ¹H NMR, which provides important structural information. Two methods can be used: (1) *stopped flow*, in which the elution is stopped to record the spectra of a component, and (2) *continuous flow*, in which the spectra of a component is recorded during the elution of the liquid. Today, the use of high-field spectrometers (up to 21 T) allows one to record spectra of pure compounds with a quantity near to 1 ng (Korhammer and Bernreuther 1996).

Even though HPLC-¹H NMR is mainly used in the pharmaceutical industry, it was also regularly used to identify some terpenes such as three sesquiterpene lactones present in a solvent extract of *Zaluzania grayana* (Spring et al. 1995).

With the aim of limiting the number of coeluted compounds that can occur in complex EOs, HPLC was combined with another chromatographic technique such as GC before spectroscopic identification. This method (HPLC-GC-MS) has been used to identify very minor compounds in citrus EOs (Mondelo et al. 1995, 1996).

8.2.3.6   Essential Oil Analysis by ¹³C NMR

Since its discovery, NMR has been a powerful tool for structure elucidation of organic molecules. Even ¹H NMR spectra recorded with low field spectrometers provided structural information not available with other techniques at that time. The introduction of high-field spectrometers, combined with the discovery of Fourier Transform NMR and two- and three-dimensional NMR (2D and 3D NMR), made the use of that technique essential in structure elucidation of natural compounds isolated from plants (Derome 1987).

Concerning the analysis of natural mixtures, the chief idea was to take benefit of the information provided by ^{13}C NMR, avoiding, or at least limiting as far as possible, the separation steps.

Carbon-13 is preferred to proton, despite its low natural isotopic abundance (1.1%), for several reasons:

1. Carbon constitutes the backbone of organic molecules, and the slightest structural modification induces measurable chemical shift variations.
2. The sweep width, around 240 ppm, is much wider than that of protons, leading to a better spectral dispersion.
3. ^{13}C NMR spectra can easily be simplified by complete decoupling of protons, so that each signal appears as a singlet.
4. Transverse relaxation time, T_2, is higher than that of the proton, leading a better resolution.
5. To avoid any thermal degradation, spectra were recorded at room temperature. Furthermore, the sample may be recovered and then submitted to other spectroscopic analyses.

In their pioneering work, Formácek and Kubeczka (1982a,b) used ^{13}C NMR, in combination with GC, to confirm the occurrence of a compound previously identified (or suspected) by GC. Since that time, ^{13}C NMR has been regularly employed, in various laboratories, to contribute to the analysis of EOs (De Medici et al. 1992; Alencar et al. 1997; Ramidi et al. 1998; Ahmad and Jassbi 1999; Núñez and Roque 1999; Ferreira et al. 2001a, 2001b; Al-Burtamani et al. 2005). A computerized procedure has been proposed by Ferreira et al. (2001a, 2001b).

In 1992, our group developed a computerized method, based on the analysis of the ^{13}C NMR spectrum, that allows the direct identification of the (major) components of a mixture (Corticchiato and Casanova 1992; Tomi et al. 1995).

In that procedure, there is neither separation nor individualization of the components before their identification. The computer program, made up in our laboratories from Microsoft Access, compared the chemical shift of each carbon in the experimental spectrum with the spectra of pure compounds listed in our spectral libraries (Figure 8.1). Each compound is then identified, taking into account three parameters that are directly available from the computer program:

1. The number of observed carbons with respect to the number of expected signals
2. The number of overlapped signals of carbons that possess the same chemical shift
3. The difference of the chemical shift of each signal in the mixture spectrum and in the reference spectra

Two complementary spectral data libraries were created, the first one containing spectra of mono-, sesqui-, and diterpenes, and phenylpropanoids recorded in our laboratories under the same experimental conditions (solvent, concentration, pulse sequence). The reference compounds were commercially available or isolated from essential oils and extracts. The second library was constructed with literature data.

Quantitative determination of some components may be carried out by NMR when necessary (nonvolatile compounds, thermolabile compounds) (Rezzi et al. 2002; Baldovini et al. 2001). Analysis of complex EOs may be achieved by a combination of CC and ^{13}C NMR (Gonny et al. 2004; Blanc et al. 2006).

Our results have been reviewed (Bradesi et al. 1996; Tomi and Casanova 2000) with a special insight into the application of NMR to the analysis of EOs from *Labiatae* (Tomi and Casanova 2006).

8.2.3.7 Enantiomeric Differentiation

EOs contain mainly terpenes that are present in pure enantiomeric form or as a nonracemic mixture of both enantiomers. This point is quite important since olfactory characteristics of a terpene are generally different, depending on the considered enantiomer. For instance, (R)-(+)-limonene smells like orange, while the (S)-(–)-enantiomer smells like turpentine. (–)-Menthol smells and tastes sweetish-minty and is fresh and strongly cooling, in contrast to the (+)-enantiomer, which smells and tastes herby-minty and is weakly cooling. (S)-(+)-Carvone possess the typical odor of caraway, and its (R)-(–)-enantiomer smells more like peppermint (Bretmaier 2006).

Moreover, in an aromatic plant, the two enantiomers of a compound usually have different origin. For instance, (+)-limonene is the major component of orange zest oil, whereas (–)-limonene is characteristic of eucalyptus, mint, or pine oils (Mosandl 1988). Coriander oil contains mainly (+)-linalool, whereas (–)-linalool is present in basil oil (Boelens et al. 1993).

Therefore, measurement of enantiomeric excess of selected terpenes in EOs is an important parameter for the characterization of a plant as well as for quality control. For example, the presence of pure (+)-fenchone in fennel EO indicates the natural origin of this oil. Conversely, (–)-fenchone is present only in *Artemisia* and cedar oils (Ravid et al. 1992).

The enantiomeric differentiation of monoterpenes and monoterpenoids, and the quantitative determination of the enantiomers, is typically carried out using a chiral GC column. Permethylated cyclodextrins are the most commonly used chiral phases as they are efficient for many terpenoids present in various EOs (Bicchi et al. 1997, 1999).

Using a single chiral column is not recommended since the enantiomeric differentiation of most components of the oil induces an increase of the number of peaks in the chromatogram and, consequently, the number of coelution. So, it is not easy to attribute the right peak to one enantiomer of a compound. In such a case, GC needs to be coupled with MS.

Two-dimensional GC is preferred. The components of the oil are first separated on a nonchiral column contained in the first oven of the chromatograph and, then, only some compounds (selected with respect to their retention time and order of elution) are injected into the second chiral column where they are enantiomerically differentiated.

Enantiomeric differentiation of oxygenated terpenes could be carried out using (^{1}H and ^{13}C) NMR spectroscopy and a chiral lanthanide shift reagent (CLSR). Although that technique was hardly used, some examples have been reported. The enantiomeric composition of linalool isolated from the oil of coriander was established by ^{1}H NMR, and the result was in agreement with that obtained by conventional polarimetry (Ravid et al. 1985). ^{13}C NMR and CLSR were successfully employed by Fraser et al. (1973) for the determination of the enantiomers of alcohols and acetates. In our laboratories, enantiomeric differentiation was performed for oxygenated monoterpenes in the pure form and in EOs: camphor and fenchone in *Lavandula stoechas* EO (Ristorcelli et al. 1998) and bornyl acetate in the EO of *Inula graveolens* (Baldovini et al. 2003).

8.2.3.8 Two-Step Procedure

In this procedure, a small quantity of a substance is separated by fractionated distillation or a chromatographic technique—liquid chromatography (LC), HPLC, preparative GC—and then identified by comparison of its spectral data—MS, FTIR, UV, ^{1}H NMR, and sometimes ^{13}C NMR—with those of reference compounds.

Weyerstahl et al. (2000) have used this procedure for several years to fully characterize very complex EOs. For instance, they partitioned Haitian vetiver oil by fractionated distillation, column chromatography, and transformation of alcohols into their methyl esters. In total, they identified by GC-MS and NMR more than 150 components. Among them, several tens are new sesquiterpenes; they were fully characterized by ^{1}H and ^{13}C NMR. In the same way, these authors analyzed an EO of *Cistus ladaniferus* gum (Weyerstahl et al. 1998). They fractionated 400 g of a commercial sample by various methods (treatment by alkaline solution, fractionated distillation, CC, and TLC). Compounds present as mixtures were identified by GC-MS, whereas pure components were fully characterized by usual spectroscopic techniques: MS and ^{1}H, and ^{13}C NMR. The authors identified, in total, 186 components; among them, 26 are new compounds.

Bicchi et al. (1998) used the same techniques to fractionate an essential oil from *Artemisia roxburghiana*. They identified 108 components, mainly by GC(RI) and GC-MS. However, 23 of these components were sesquiterpene isomers, and their identification required the use of ¹H NMR and, sometimes, ¹³C NMR.

This two-step procedure is obviously more accurate, but it is very time consuming and requires a large quantity of EO.

8.3 ANALYSIS OF *CYMBOPOGON* OILS

The composition of *Cymbopogon* oils was investigated a long time ago, in connection with the properties of the oils. Most studies concerned oils of commercial importance: citronella, lemongrass, and palmarosa oils. Citronella oils are usually divided into two types: Ceylon type and Java type obtained from *C. nardus* (L.) Rendle and *C. winterianus* Jowitt. Lemongrass oil is found either in *Cymbopogon citratus* Stapf (sometimes known as West Indian lemongrass) or in *C. flexuosus* Stapf. Palmarosa oil is produced from *C. martinii* (Roxb.) Wats. var *martinii*. The compositions of these oils have been reviewed by Lawrence (1979, 1989, 1993, 1995, 2003, 2006). However, the compositions of various oils obtained from other species of *Cymbopogon* are reported in the literature, among others: *C. afronardus, C. distans, C. giganteus, C. jawarancusa, C. microstachys, C. olivieri, C. parkeri, C. travancorensis,* and *C. validus.*

Cymbopogon oils contain mainly oxygenated monoterpenes: geranial, neral (citral), citronellol, geraniol, etc. (Figure 8.2). However, a great variety of compounds was reported as minor components: monoterpene hydrocarbons, sesquiterpene hydrocarbons and oxygenated sesquiterpenes, phenylpropanoids, and nonterpenic acyclic compounds.

Depending on the various techniques (chemical, chromatographic, spectroscopic) used to identify and quantify the individual components of *Cymbopogon* EOs, we first remember the analysis by chemical methods. Then, we shall develop the analysis by chromatographic techniques, analysis by mass spectrometry coupled with gas chromatography, and analysis by hyphenated techniques. We shall summarize the enantiomeric differentiation of terpenes by chiral GC and by NMR. Finally, the combined analysis of EOs by various techniques will be exemplified on *C. giganteus* oil.

FIGURE 8.2   **1** = geranial, **2** = citronellal, **3** = neral, **4** = geraniol, **5** = citronellol, **6** = myrcene.

8.3.1 EARLIEST STUDIES: ANALYSIS BY CHEMICAL METHODS (ISOLATION OF COMPOUNDS AND MISCELLANEOUS METHODS)

According to Lawrence (1989c), the first studies date back to 1948 when Dasgupta and Narain used fractional distillation and chemical derivatization in combination with physicochemical data to identify geraniol and a few derivatives in a sample of palmarosa oil from India. Later, the content of citral in lemongrass oil was determined by condensation with barbituric acid (Lawrence 1995b).

Several sesquiterpene hydrocarbons were isolated from citronella oil from Java: β-bourbonene, δ-cadinene, α-cubebene, β-elemene, and α-selinene (Lawrence 1989a).

Various physical and chemical methods were developed to isolate the major constituents of *Cymbopogon* oils. For instance, the sodium method was used to isolate the alcoholic constituents (mostly geraniol and citronellol) of the EO of Java citronella, and it was suggested that this method was more efficient than removal of alcohols by fractionated distillation (Lawrence 1979a). Various chemical and chromatographic methods were compared to evaluate the total citral content in lemongrass oil: bisulfite method, oxime method, barbituric acid method, GC with internal standard, and the GC method after NaBH$_4$ reduction. The authors recommended that the GC method using phenylethyl alcohol as internal standard was the most accurate method for determination of citral (Lawrence 1995b).

Only a few new compounds have been isolated from *Cymbopogon* oils and their structure elucidated.

C. flexuosus oil contained *iso*-intermedeol, a sesquiterpene alcohol (Thappa et al. 1979; Huffman and Pinder 1980), and later, both the oil and the alcohol were investigated for their ability to induce apoptosis in human leukemia HL-60 cells because dysregulation of apoptosis is the hallmark of cancer cells (Kumar et al. 2008).

Evans et al. (1982) reported that the sesquiterpene diol proximadiol (Figure 8.3), with antispasmodic properties, earlier isolated from *Cymbopogon proximus*, is in fact identical with cryptomeridiol.

An eudesmandiol, [2R-(2α,4aβ,8α,8aα)]-decahydro-8a-hydroxy-α,α4a,8-tetramethyl-2-naphthalenemethanol, was isolated from the EO of *Cymbopogon distans*. It was studied spectroscopically, and the absolute configuration was determined by means of x-ray diffraction (Mathela et al. 1989).

FIGURE 8.3 **7** = proximadiol, **8** = 5α–hydroperoxy-β-eudesmol, **9** = 3-eudesmen-1β,11-diol, **10** = selin-4(15)-en-1β,11-diol, **11** = 7α,11-dihydroxycadin-10(14)ene, **12** = α-oxobisabolene.

Cymbodiacetal, a novel dihemiacetal bis-monoterpenoid, isolated from the EO of *C. martinii*, was identified by means of x-ray diffraction (Bottini et al. 1987).

Recently, various sequiterpene alcohols, including two new compounds, 5α-hydroperoxy-β-eudesmol and 7α,11-dihydroxycadin-10(14)-ene (Figure 8.3), have been isolated from a pentane extract of *C. proximus* (El-Askari et al. 2003).

Finally, it should be noted that the presence of α-oxobisabolene (bisabolone) (12%) (Figure 8.3) was reported in a sample of Ethiopian *C. citratus* oil (Abegaz et al. 1983). Melkani et al. (1985) also reported the occurrence of that compound in concentrations ranging from 18% to 68% in one of the oils of two varieties of *C. distans*. Isolation of (+)-1-bisabolone from *C. flexuosus* and its utilization as antibacterial agent has been patented.

However, the chemical techniques were supplanted by the introduction, first of chromatographic techniques (TLC and, obviously, GC) and later of spectroscopic techniques (MS, IR, NMR).

8.3.2 ANALYSIS BY CHROMATOGRAPHIC TECHNIQUES [TLC, HPLC, GC(RT) OR GC(RI)]

Due to the complexity of the EOs (number of constituents, diversity of the structures and functionalities, and also similarities resulting from the biosynthesis of terpenes from the same building block, the isoprene unit), TLC has not been considered a practical tool for analysis of *Cymbopogon* oils. Nevertheless, TLC has been used to characterize the quality of lemongrass oil from Bangladesh (Lawrence 1995). The authors concluded that the oil could be competitive on the market of citral-rich oils.

Following an improvement of the column for gas chromatography that allowed a good individualization of the components of essential oil, this technique became more and more useful for analysis of *Cymbopogon* EOs.

The first experiments concerned the (tentative) identification of individual components, by comparison of their retention times with those of reference compounds:

- As early as 1962, retention times were used to identify ten monoterpene hydrocarbon components in Ceylon citronella oil (Lawrence 1989a).
- A few years later, eugenol and methyl eugenol were identified in the same way, besides monoterpenes, in a sample of Java citronella oil dominated by citronellal and geraniol (Lawrence 1989a);
- In the meantime, Peyron (1973) was able to distinguish, among other monoterpenes, (Z) and (E) isomers of ocimene in palmarosa oil from Brazil, largely dominated by geraniol;
- The same year, Wijesekera et al. (1973) reported the utilization of GLC profiles to distinguish both varieties of citronella oil, although both types contained comparable amounts of geraniol: Ceylon variety (Lenabatu) contained large amounts of monoterpene hydrocarbons, while the Java variety (Mahapengiri) contained only small amounts. In addition, the Ceylon type contained tricyclene, methyl eugenol, methyl isoeugenol, eugenol, and borneol;
- Mohammad et al. (1981) determined the composition of two oil samples of lemongrass (*C. flexuosus*) and detected trace constituents using the peak enrichment technique (addition of a known compound in the EO and comparison of the new chromatogram with that of the pure oil). Besides neral and geranial, by far the main components, the authors identified various monoterpenes as well as decanal. The two samples also contained appreciable contents of farnesol and farnesal. However, the correct isomer of the two acyclic sesquiterpenes was not identified.

More than 20 years later, Rauber et al. (2005) investigated a completely different approach to the quantitative determination of citral in *C. citratus* volatile oil. An HPLC method was developed (Spherisorb® CN column, *n*-hexane: ethanol mobile phase, and UV detector) and validated (precision, accuracy, linearity, specificity, quantification, and detection limits). The concentration of citral in *C. citratus* volatile oil obtained with this assay was 75%.

8.3.3 ANALYSIS BY MASS SPECTROMETRY COUPLED WITH GAS CHROMATOGRAPHY

In parallel, during the 1970s, owing to the improved sensitivity of mass spectrometers, identification of components became possible by comparison of their mass spectra with those of reference compounds. A gas chromatograph coupled with a mass spectrometer became the basic instrument for EO analysis. Using this technique, any volatile known compound, even present at the trace level, could be theoretically identified by comparison of its mass spectral data with those of reference compounds compiled in either paper or "computerized" libraries.

A GC-MS study of the hydrocarbon fraction and the fraction containing oxygenated compounds showed the presence of 12 monoterpene hydrocarbons (28.4%), 13 sesquiterpene hydrocarbons (32.8%), 3 sesquiterpene alcohols (27.2%), 2 esters (7.2%), and 3 carbonyl compounds (4.4%) in the EO of *C. distans*. Among these, 27 compounds have been identified (Mathela and Joshi 1981).

Analysis of an oil sample of *C. flexuosus* of Indian origin, by GC-MS, allowed the identification of monoterpenes (including perillene), various sesquiterpene hydrocarbons, as well as methyl eugenol and elemicin (Taskinen et al. 1983).

In practice, the main drawback of that technique, particularly for the analysis of EOs, consists in the occurrence of insufficiently differentiated spectra for compounds built up with the same isoprene unit. Therefore, it was determined that sesquiterpenes bearing quite different skeletons exhibit similar (if not identical) mass spectra. Moreover, superimposable mass spectra are commonly observed for isomers (stereoisomers, diastereoisomers). Consequently, it is not surprising that the correct isomer of various monoterpenes and overall sesquiterpenes could not be identified by MS. In his excellent reviews, Lawrence pointed out the lack of specification of the correct isomer. For instance:

- The correct isomer of various monoterpenes (menthone, isopulegols, and rose oxide) was not specified in the composition of citronella oils from Java and Sri Lanka investigated by GC-MS (Lawrence 1989a);
- The isomer of verbenol contained in lemongrass (*C. citratus*) grown in Zambia was not reported (Chisowa et al. 1998). Conversely, both isomers were found, as minor components, in oil samples from the Ivory Coast (Chalchat et al. 1998).
- The correct isomer of elemene remained frequently unassigned in old papers and more surprisingly, in recent papers, as well as the stereoisomer of farnesol (Lawrence 2006d).

Moreover, misidentification frequently occurred when the analysis was carried out only by MS. Concerning *Cymbopogon* oils, Lawrence, in his reviews, pointed out misidentified common compounds simply by the observation of the order of elution on apolar (or polar) column. Misidentifications concerned, in general, minor compounds but, sometimes, compounds present at appreciable contents.

Anyway, year after year, MS became essential in EO analysis. Reliable results are obtained by combination of GC and MS. The combined use of both techniques will be developed in the next subsection.

8.3.4 COMBINED ANALYSIS BY CHROMATOGRAPHIC AND SPECTROSCOPIC TECHNIQUES: TLC, CC, GC(RT) OR GC(RI) AND GC-MS

In the previous text, we have seen the importance of both GC and MS for identification of individual components of EOs. Combination of the two techniques appeared really useful for analysis of volatiles. Identification of individual components resulted from comparison of mass spectral data in the case of MS and retention times or retention indices in the case of GC with those of reference compounds. More accurate and reproducible results are obtained by using retention indices (Van Den Dool and Kratz 1963; Kováts 1965) instead of retention times (see Part I).

Reliable results are obtained when individual components of EO are identified by MS in combination with RIs measured on two columns of different polarity (apolar and polar). Nowadays, it is considered by specialized reviews that analysis of an EO by at least two techniques should be done, to keep the work publishable. To facilitate identification of components, prefractionation of the oil by liquid chromatography (LC) could be useful.

8.3.4.1 *Cymbopogon* Oils of Commercial Interest

The first combined analyses, by GC and GC-MS, of *Cymbopogon* oils of commercial interest (citronella, lemongrass, and palmarosa oils), were published in the 1970s and have been reviewed by Lawrence (1979, 1989, 1993, 1995, 2003, 2006). Although the oil composition was always dominated by oxygenated monoterpenes (geranial, neral, and geraniol), the number and the structure of identified compounds varied substantially from paper to paper.

The composition of lemongrass oil (*C. flexuosus*) from Guatemala was reported in 1976. Among unusual monoterpenes, the nonterpenic aldehyde decanal, as well as germacrene D and γ-cadinene were identified. The correct isomer of ocimene and allo-ocimene was not specified (Lawrence 1979b).

A few years later, 2-undecanone was detected, by GC and GC-MS, in an Indian lemongrass oil sample containing more than 72% of citral (Lawrence 1989b).

In 1983, two samples of Turkish lemongrass oil (*C. citratus*) were investigated by GC and GC-MS, after prefractionation of the oil by liquid chromatography. Both oils were dominated by citral, and they contained appreciable amounts of myrcene (up to 19%) as well as nonanal and 2-undecanone (Lawrence 1989b).

Retention times and GC-MS data were taken into account to investigate an oil sample of *C. winterianus* from Bangladesh. Usual monoterpene hydrocarbons and monoterpenols were identified, as well as elemol and α-eudesmol. The authors also found 2,3-dimethyl-5-heptenal probably misidentified, according to Lawrence (1995a), instead of methylheptenone.

Twelve samples of palmarosa oil produced in Madagascar were investigated by GC(KI) and GC-MS. Besides geraniol (major component) and usual monoterpenes, numerous sesquiterpenes were identified, although the coelution of various sesquiterpene hydrocarbons was observed. Three isomers of farnesol, (Z,E), (E,Z), and (E,E), were found and differentiated (Figure 8.4) (Randriamiharisoa and Gaydou 1987). The same authors reported the detailed analysis of the hydrocarbon fraction of that oil (4.75%), separated by CC (SiO$_2$) and analyzed by GC-MS in combination

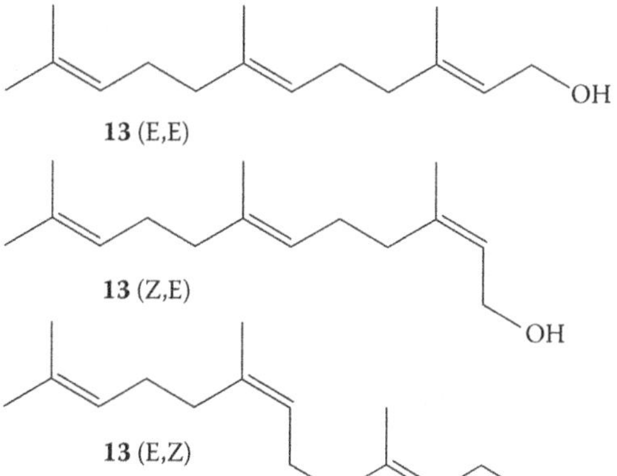

13 (E,E)

13 (Z,E)

13 (E,Z)

FIGURE 8.4 Structure of farnesol stereoisomers **13**.

with KI. Twenty-eight sesquiterpene hydrocarbons were identified. In addition, long-chain alkanes were characterized. We note also the occurrence of the three isomers of cymene (Gaydou and Randriamiharisoa 1987).

Ntezurubanza et al. (1992) identified the individual components of lemongrass oil from Rwanda, using MS in combination with RIs measured on two columns of different polarity (apolar and polar). They found 2-tridecanone and lavandulyl acetate, besides the usually seen monoterpenes.

The chemical composition of commercial citronella oils of various geographical origins has been examined, using MS data and RIs (ethyl esters). Fifty-five compounds were identified, including various sesquiterpenes and phenyl propanoids (eugenol, methyl eugenol, E and Z isomers of methyl isoeugenol) (Lawrence 2003a).

A Brazilian citronella oil sample, examined using GC(KI) and GC-MS, was found to contain citronellal and neral as major components while the content of geranial was very low. The phenyl propanoid elemicin accounted for 7% of the oil, which also contained β-elemene (0.5%) and elemol (3.7%) (Lawrence 1995a).

Despite the improvement of analytical techniques, the analysis of complex mixtures, such as EOs, remains a difficult task and needs careful work. Unfortunately, despite using two analytical techniques (GC and GC-MS), misidentification of individual components still occurs as illustrated by the analysis of *C. nardus* oil from Zimbabwe, where numerous sesquiterpenes appeared misidentified on the basis of their relative elution order (Lawrence 2003a).

The EO from seeds of Indian *Cymbopogon martinii* (Roxb.) Wats. var. *motia* Burk. collected from three different geographical locations was analyzed by capillary GC (KI measured on apolar and polar columns and peak enrichment by co-injection of authentic samples) and GC-MS (Mallavarapu et al., 1998). The composition of the oil samples was compared with that of the oil of flowering palmarosa herb. Besides the main constituent, geraniol (74.5%–81.8%), 55 other constituents were identified in the seed EOs. Although the composition of the seed oils is similar to that of the herb oil, quantitative differences in the concentration of some constituents were observed. The seed oil was found to contain lower amounts of geranyl acetate and higher amounts of (E,Z)-farnesol than the herb oil (two isomers of farnesol were differentiated in the oils).

The composition of the leaf oil of *C. citratus* (DC) Stapf, growing on the campus of Lagos State University (Nigeria), was determined by the use of GC (KI on polar column) and GC-MS (Kasali et al. 2001). Twenty-three (97.3%) constituents were identified. The main components were geranial (33.7%), neral (26.5%), and myrcene (25.3%). An unusual saturated monoterpene hydrocarbon (0.1%) was also detected and probably misidentified as 2,6-dimethyloctane.

The steam-distilled volatile oil obtained from partially dried grass (citronella grass) *C. nardus* (Linn.) Rendle, cultivated in the Nilgiri Hills, India, was analyzed by capillary GC and GC-MS (Mahalwal and Ali 2003). The prominent monoterpenes were citronellal (29.7%), geraniol (24.2%), γ-terpineol (9.2%), and *cis*-sabinene hydrate (3.8%). The predominant sesquiterpenes were (E)-nerolidol (4.8%), β-caryophyllene (2.2%), and germacren-4-ol (1.5%). An irregular monoterpene compound eluted among sesquiterpenes was identified as 3,3,6-trimethylhepta-1,5-diene, and two 1,2-benzenze dicarboxylic acid derivatives were tentatively identified (although they are probably not naturally occurring substances).

The EOs of palmarosa (*C. martinii* (Roxb.) Wats. var. *motia* Burk) flowering herbs from three different geographical locations in India (Hyderabad, Lucknow, and Amarawati) were analyzed by high-resolution GC (apolar column) and GC–MS (Raina et al. 2003). In all three samples, geraniol (67.6%–83.6%) was the major constituent. However, quantitative differences in the concentration of some constituents were observed. Among the three oils analyzed, Amarawati oil was of the highest quality due to higher geraniol (83.6%) and lower geranyl acetate (2.3%) and geranial (1.0%) content. Isopropyl propionate and butyrate (as well as cyclohexanone!) were detected in that oil.

The supercritical fluid extraction (SFE) of *C. citratus* was performed in sequential and dynamic extraction modes (Sargenti and Lanças 1997). Principal compounds in the SFE extract were analyzed by GC and GC-MS and identified as neral, geraniol, and geranial. Nerolic acid and geranic acid were identified as minor components in the essential oil. Different chromatographic profiles were obtained depending of the experimental conditions.

The same technique was applied by Marongiu et al. (2006) who analyzed the influence of pressure on the supercritical extraction, in order to maximize citral content in the extract oil. Dried and ground leaves of lemongrass (*C. citratus* Stapf) were used as a matrix. The collected extracts were analyzed by GC-MS, and their composition was compared with that of the EO isolated by hydrodistillation and by steam distillation. At optimum conditions (90 bar and 50°C) citral represented more than 68% of the extract. At higher solvent density, the extract aspect changes because of the extraction of high-molecular-mass compounds.

8.3.5 OTHER *CYMBOPOGON* OILS

The composition of EO of other species has been also investigated:

- The composition of the essential oil of *C. jawarancusa* (Khavi grass), was investigated by GC in combination with GC-MS (Saeed et al. 1978), and 64 compounds were identified. The oil exhibited a high content of piperitone (Figure 8.5; 60%–70%), which is mainly responsible for the smell of Khavi grass. The chemical variability of that oil was evidenced, and races rich in piperitone, phellandrene, and other chemical constituents have been identified (Dhar et al. 1981).
- Analysis of the volatile oil of *C. parkeri* has been carried out by GC-MS (Rizk et al. 1983). To facilitate identification, prefractionation of the oil by adsorption high-performance liquid chromatography was performed. The major compounds are geraniol (33.5%), nerol (22.2%), geranyl acetate (8.9%), neryl acetate (3.8%), and farnesol (3.7%). A sesquiterpene alcohol that accounted for 4.8% remained unidentified. A quite different composition, determined by the use of GC and GC-MS, was reported for an oil sample obtained from aerial parts of *C. parkeri* Stapf, collected at flowering stage from Kerman province of Iran. The main constituents were piperitone (80.8%) and germacrene D (5.1%) (Bagheri et al. 2007).
- The chemical composition of the essential oil of *C. travancorensis* Bor (Poaceae) was investigated by capillary GC and GC-MS. Thirty-five compounds were identified. The oil contains terpenes and phenyl propanoids (22.04%). The main constituents of the oil are limonene (18%), elemicin (17%), camphene (12%), elemol (11%), and borneol (10%) (Mallavarapu et al. 1992).
- The steam-distilled oils from wild and cultivated *C. validus* (Stapf) Stapf ex Burtt Davy (Gramineae) were analyzed by GC and GC-MS (Chagonda et al. 2000). The major components from wild *C. validus* were the following: myrcene (23.1%–35.6%), (E)-β-ocimene (10.3%–11.5%), geraniol (3.4%–8.3%), linalool (3.2%–3.7%), and camphene (5.2%–6.0%).

FIGURE 8.5   **14** = piperitone, **15** = eugenol, **16** = (E)-methyl isoeugenol, **17** = elemicin.

Cultivated mature plants contained myrcene (11.6%–20.2%), (E)-β-ocimene (6.0%–12.2%), borneol (3.9%–9.5%), geraniol (1.7%–5.0%), and camphene (3.3%–8.3%) as the major components. Young nursery crop/seedlings (20–30 cm high) contained oil with myrcene (20.6%), geraniol (17.1%), and germacrene-D-4-ol (8.3%) as main components. Geranyl acetate (4.5%), linalol (4.5%), and borneol (2.9%) were notable minor components.

- The constituents of the essential oil obtained by hydrodistillation of the aerial parts of *C. olivieri* (Boiss.) Bor, growing wild in Iran, was investigated by GC (RI on apolar column) and GC-MS (Norouzi-Arasi et al. 2002). Forty-two components, representing 97.6% of the oil, were identified, of which piperitone (53.3%), α-terpinene (13.6%), elemol (7.7%), β-eudesmol (4.4%), torreyol (3.3%), limonene (2.9%), and α-cadinol (2.1%) were the major components.

C. microstachys (Hook.f.) Soenarko is reported to be found wild in some pockets of the Himalayan foothills of Assam, Uttar Pradesh, and West Bengal. Mathela et al. (1990) have studied the composition of its oil and identified 31 compounds. The oil was found to contain about 60% of phenyl propanoids with methyl eugenol (19.5%), methyl isoeugenol (4.2%), elemicin (25.3%), and isoelemicin (11.0%) (Figure 8.5). Another sample of *C. microstachys* had quite a different composition, with (E)-methyl isoeugenol being the main constituent (56.4%–60.7%) and myrcene (7.8%–12.2%) (Rout et al. 2005).

C. afronardus oils, analyzed by GC and GC-MS, contained myrcene and intermedeol as major constituents. The oil is used in Uganda in the formulation of locally manufactured toothpaste (Baser et al. 2005).

8.3.6 ENANTIOMERIC DIFFERENTIATION BY CHIRAL GC

Ravid et al. (1992) carried out the enantioselective analysis of citronellol (10.2% in an oil sample of citronella) through a trifluoroacetyl derivative of the isolated compound; the (R)-(+)/(S)-(−) ratio was 74/26.

Mosandl et al. (1990) used a combination of heart-cutting 2D GC and chiral GC to determine the enantiomeric distribution of monoterpene hydrocarbons in citronella oil: α-pinene [(+)/(−) = 77/23] and limonene [(+)/(−) = 4/96] (Figure 8.6), and in lemongrass oil: α-pinene [(+)/(−) = 4/96], limonene [(+)/(−) = 0/100], and β-pinene [(+)/(−) = 0/100]. The enantiomeric composition of oxygenated monoterpenes was investigated: citronellol, linalool (Figure 8.6), terpinen-4-ol, *cis* and *trans* rose-oxides (Kreis and Mosandl 1994), linalool (Wang et al. 1995) lemongrass, and borneol (Ravid et al. 1996).

The EO from aerial parts of *C. winterianus* Jowitt, cultivated in Southern Brazil, was analysed by GC-MS. Enantiomeric ratios of limonene [(R)-(+)/(S)-(−) = 13/87], linalool [(R)-(−)/(S)-(+) = 39/61]), citronellal [(R)-(+)/(S)-(−) = 91/09]), and β-citronellol [(R)-(+)/(S)-(−) = 85/15] were obtained by

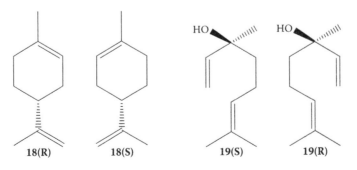

FIGURE 8.6   18(R) = (R)-(+)-limonene, **18(S)** = (S)-(−)-limonene, **19(R)** = (R)-(−)-linalool, **19(S)** = (S)-(+)-linalool.

multidimensional gas chromatography, using a developmental model setup with two GC ovens. The enantiomeric distributions are discussed as indicators of origin authenticity and quality of this oil (Lorenzo et al. 2000).

8.3.7 COMBINED ANALYSIS OF *CYMBOPOGON* OILS BY VARIOUS TECHNIQUES: TLC, CC, IR, GC(RI), GC-MS, AND ¹³C NMR

In various examples, the analysis was carried out by combination of chromatographic (TLC and GC) and spectroscopic (IR) techniques. This type of work is illustrated by the analysis of lemongrass oil (*C. citratus*) from Egypt. About 28 compounds were clearly detected by GC, from which 17 were identified, citral being mentioned as a major component. The components belonged mostly to the acyclic-oxygenated monoterpenes. δ-Cadinene and (E)-β-caryophyllene were also identified (Abdallah et al. 1975). The oil was also analyzed by TLC, with a gradient of solvents as eluent. Citral, citronellal, and various alcohols were detected (Zaki et al. 1975). Finally, the infrared spectrum of the lemongrass oil proved that the changes in the samples during storage could be detected, especially the changes of the carbonyl groups (Foda et al. 1975).

A combined analysis of citronella oil (*C. winterianus*) by fractional distillation, TLC, GC, IR, and NMR was described: besides citronellal and geraniol, the main components, four acyclic oxygenated monoterpenes, a cyclic one, a sesquiterpenol (elemol), and a phenylpropanoid (eugenol) were identified (Lawrence 1989a).

An elegant and unusual technique was used to analyze the nitrogenous fraction of palmarosa oil. The compounds were separated by preparative GC and characterized using a combination of GC (nitrogen and sulfur detectors), GC-MS, IR, and NMR. Eighteen pyrazines, as many pyridines, four thiazoles, and a few other nitrogen-containing compounds were identified (Lawrence 1993).

Very recently, the composition of the sample of palmarosa oil that has proven antimicrobial properties against cells of *Saccharomyces cerevisiae* was determined as 65% geraniol and 20% geranyl acetate, as confirmed by GC–FTIR (Prashar et al. 2003).

In the early 1980s, NMR began to be employed to analyze EO. In 1981, Chiang et al. used proton NMR to quantify citronellal in Taiwanese citronella oil (average: 38.4%) (Lawrence 1989a). One year later, Formácek and Kubeczka (1982a) used a combination of GC and ¹³C NMR to identify various monoterpenes, as well as 2-nonanone and (E)-β-caryophyllene in East Indian lemongrass. More recently, the same authors again used a combination of techniques, capillary GC, and ¹³C NMR to compare the composition of citronella oil (*C. nardus*) and Java-type oil (*C. winterianus*) (Kubeczka and Formácek 2002). Both oils differed overall by the content of monoterpene hydrocarbons (camphene, limonene), oxygenated monoterpenes (citronellal, citronellol, borneol), and (E)-methyl isoeugenol. Forty-eight compounds were identified, including various sesquiterpenols and phenylpropanoids.

Seven oil samples of *C. schoenanthus* were recently analyzed by GC(RI), GC-MS, and ¹³C NMR, without fractionation (Khadri et al. 2008). The composition of samples was dominated by monoterpene hydrocarbons, limonene, β–phellandrene, and δ-terpinene. The last samples contained higher amounts of sesquiterpenes, including β-eudesmol, valencene, and δ-cadinene. All the components were identified by GC-MS and GC(RI). The identification of 17 components was confirmed by ¹³C NMR.

In our laboratories, we developed a computerized procedure allowing the identification of individual components of EOs, as illustrated in the first part of this chapter. We applied this method to the analysis of two samples of *C. winterianus* and a sample of *C. tortilis*, all cultivated in Vietnam (Ottavioli et al. 2008a). In all, 47 components were identified by a combination of GC(RI) on two columns of different polarity, GC-MS and ¹³C NMR, in *C. winterianus* leaf oil. The composition of the first sample was largely dominated by geraniol (43.3%), citronellal (15.0%), citronellol (9.2%), and geranial (8.9%). It is noteworthy that all the 12 compounds, accounting for 91.6% of

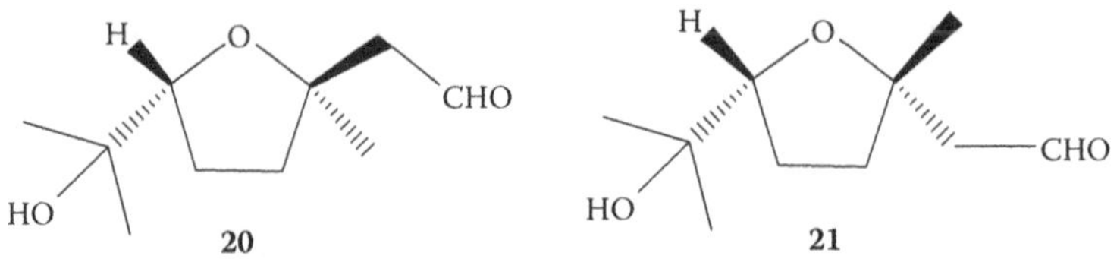

FIGURE 8.7 Structure of **20** = trans-7-hydroxy-3,7-dimethyl-3,6-oxyoctanal, **21** = *cis*-7-hydroxy-3,7-dimethyl-3,6-oxyoctanal.

the composition of the oil, were identified by GC-MS and by NMR. The composition of the second sample was also dominated by geraniol (50.2%) and geranial (8.3%), but citronellal and citronellol were present at very low levels. In that sample, 46 components were identified, 13 of these (accounting for 0.6%–50.2% each) by NMR. The same procedure was applied to the analysis of an oil sample from *C. tortilis,* without fractionation. The composition of that sample was dominated by geranial (32.0%) and neral (19.1%). The oil contained epoxyneral and epoxygeranial as well as caryophyllene oxide at appreciable contents. It differed drastically from the methyl-eugenol-rich oil from China (Liu et al. 1981).

A leaf oil sample of *C. flexuosus* was analyzed by GC(RI), GC-MS, and ^{13}C NMR (Ottavioli et al. 2008b). The composition of the investigated sample was dominated by geranial (39.2%) and neral (24.1%), and was somewhat similar to that of various samples reported in the literature (Surburg and Panten 2006). Most of the identified compounds were classical components found in EOs, and they were identified by GC-MS by computer matching against commercial mass spectral data libraries and by comparison of their retention indices with those of reference compounds on two columns of different polarity. All the compounds, present at appreciable contents, were also identified by ^{13}C NMR, following the computerized method described earlier. However, two compounds, accounting for 4.5% and 4.4%, respectively, remained unassigned. Their retention indices on apolar and polar columns (compound A: RI = 1252 and 1957; compound B: RI = 1257 and 1940) led us to suspect two oxygenated monoterpenes. Computer matching of the ^{13}C NMR data against a home-made library constructed with literature data suggested the occurrence of *cis-* and *trans*-7-hydroxy-3,7-dimethyl-3,6-oxyoctanal (Figure 8.7).

To ensure the occurrence of both isomers, the oil from *C. flexuosus* was partitioned by flash chromatography on silica gel. Indeed, in the most polar fraction, eluted with diethyl oxide, both compounds accounted for 21.5% and 21.3%, respectively. Their NMR data were in agreement with those reported by Yarovaya et al. (2002). In parallel, analysis of the other fractions of chromatography by ^{13}C NMR confirmed the presence of minor monoterpenes, borneol, nerol, as well as sesquiterpenes, *trans*-α-bergamotene, 1,5-*diepi*-aristolochene, δ-cadinene, cuparene, β-elemol, and (2E,6E)-farnesol.

To conclude this chapter, we illustrate the various techniques or combination of techniques and methods developed for analyzing EOs on a unique *Cymbopogon* species. To do that, we chose *C. giganteus.*

8.3.8 ANALYSIS OF *C. GIGANTEUS* OIL: A SUMMARY OF ANALYTICAL METHODS INVOLVED IN THE ANALYSIS OF *CYMBOPOGON* OILS

C. giganteus (Hochst.) Chiovenda is a perennial and sweet-smelling grass that grows spontaneously in the savannahs of Asian and African tropical regions. Aerial parts (leaves, flowers, stems) produce by vapor distillation or water distillation an EO whose composition has been studied by various techniques.

The physiochemical properties of an oil sample (leaf and flower oil) from Côte d'Ivoire have been measured: density and refraction index as well as viscosity. The authors concluded that the oil is not Newtonian (Kanko et al. 2004).

The composition of essential oils, obtained from aerial parts of plants growing in different countries, has been investigated: Benin (Ayedoun et al. 1997; Alitonou et al. 2006), Burkina Faso (Menut

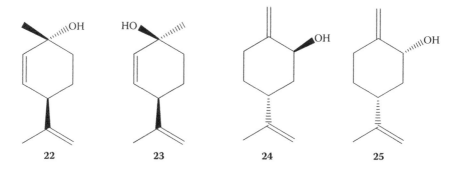

22   23   24   25

FIGURE 8.8   Structure of **22** = trans-p-mentha-2,8-dien-1-ol, **23** = *cis-p*-mentha-2,8-dien-1-ol, **24** = *trans-p*-mentha-1(7),8-dien-2-ol, **25** = *cis-p*-mentha-1(7),8-dien-2-ol.

et al. 2000), Cameroon (Ouamba 1991; Jirovetz et al. 2007), Côte d'Ivoire (Kanko et al. 2004; Boti et al. 2006), and Mali (Popielas et al. 1991; Keita 1993; Sidibe et al. (2001).

Most analyses were carried out by a combination of GC(RI) and GC-MS, and the number of identified components varied substantially from 6 to 55. Whatever the origin, the composition of *C. giganteus* oils was dominated by monoterpenes, mostly oxygenated p-menthane derivatives, such as *cis-* and *trans-p*-mentha-2,8-dien-1-ols, *cis-* and *trans-p*-mentha-1(7),8-dien-2-ols, *cis-* and *trans*-isopiperitenols, limonene oxides, carvone, and limonene, the only monoterpene hydrocarbon found at appreciable content (Figure 8.8).

Conversely, the oils differed from sample to sample by the occurrence of different minor components. Indeed, although several tens of compounds were identified in total in all the studies, only the six main compounds were identified in all the oil samples: *cis-* and *trans-p*-mentha-2,8-dien-1-ols, *cis-* and *trans-p*-mentha-1(7),8-dien-2-ols, carvone, and limonene. Conversely, numerous compounds were identified only in one paper, including monoterpenes, acyclic nonterpenic compounds, and benzenoid derivatives.

The first study on the chemical composition of *C. giganteus* essential oil was reported in 1991, for an oil sample from Bamako, Mali (flowering stage), investigated by GC(RT) and GC-MS (apolar column) (Popielas et al. 1991). Eighteen compounds were identified, the major ones being p-mentha-2,8-dien-1-ol (10.7%, isomer nonspecified), *cis* and *trans-p*-mentha-1(7),8-dien-2-ols (17.2% and 20.4%, respectively). 1,2-Limonene oxide (13.7%, isomer nonspecified) was also among the major components. Isopiperitenol and trans-carveol coeluted.

The next year, Keita (1993) published a short analysis of another oil sample of the plant of the same geographical origin, also carried out by GC(RT) and GC(MS). The menthadienols were not identified, although they were probably the main components (quoted as unidentified). Isopiperitenone accounted for 10.2% and carveol (isomer unspecified) for 8.0%.

Finally, *C. giganteus* oil of Mali was also investigated by Sibidie et al. (2001) by GC (polar column) and GC-MS. Beside p-menthadienols, by far the main components, two p-menthatrienes and (Z)-2-phenylbut-2-ene were also identified.

Ayedoun et al. (1997) investigated three oil samples hydrodistilled from plants harvested in different provinces of Benin. The three samples were characterized by high amounts of limonene (18%–24%) beside the four p-menthadienols. The essential oil of *C. giganteus* of Benin was also analyzed by GC and GC-MS. Once again, the major constituents were as follows: *trans*-p-1(7),8-menthadien-2-ol (22.3%), *cis*-p-1(7),8-menthadien-2-ol (19.9%), *trans*-p-2,8-menthadien-1-ol (14.3%), and *cis*-p-2,8-menthadien-1-ol (10.1%). This study reports the inhibitory effect produced by the chemical constituents of the essential oil, in vitro on 5-lipoxygenase (Alitonou et al. 2006).

In the course of their studies on aromatic plants of tropical West Africa, Menut et al. (2000) reported the composition of an oil sample of *C. giganteus* from Burkina Faso. The analysis was carried out by GC (two columns, apolar and polar) and GC-MS. The oil contained mainly limonene and a set of *p*-menthadienols. Two unusual esters, 3-methylbutyl hexanoate and octanoate, were also identified as minor components. The oil's antioxidant and antiradical activities, which are low, were studied.

The composition of *C. giganteus* oil from Cameroon has been first investigated by Ouemba (1991). A more detailed study was published very recently (Jirovetz et al. 2007). The EOs of fresh flowers, leaves, and stems of *C. giganteus* were investigated by GC and GC-MS. More than 55 components have been identified in the three oil samples, the major ones being, as usual, the four p-menthadienol isomers. Additional components in higher concentrations, responsible for the characteristic aroma of these samples, are especially limonene, *trans*-verbenol, and carvone, as well as some other mono- and sesquiterpenes. Antimicrobial activities of the four oils were found against Gram-(+)- and Gram-(−)-bacteria, as well as the yeast *Candida albicans*, and these results were discussed with the compositions of each sample.

The *Cymbopogon giganteus* oil from Côte d'Ivoire was only briefly investigated some time ago (nine components identified; Ouemba 1991). A more recent paper was no more informative (six compounds identified, method of analysis unspecified; Kanko et al. 2004).

In order to get better insight into the composition of *C. giganteus* leaf oil from Côte d'Ivoire, a sample was analyzed using a combination of CC, GC(RI), GC-MS, and ^{13}C NMR (Boti et al. 2006). The bulk sample was analyzed by GC(RI), GC-MS, and ^{13}C NMR, and the fractions of chromatography were investigated by GC(RI) and ^{13}C NMR. In all, 46 constituents, which represented 91.0% of the oil, were identified. As expected, *cis*- and *trans*-p-mentha-2,8-dien-1-ols (8.7% and 18.4%), *cis*- and *trans*-p-mentha-1(7),8-dien-2-ols (16.0% and 15.7%), and limonene (12.5%) dominated largely the composition. Beside the main components, various compounds were present at appreciable contents: 3,9-oxy-mentha-1,8(10)-diene (2.2%), *cis*- and *trans*-isopiperitenols (2.2% and 3.1%), carveol (2.2%), and carvone (2.7%).

It should be pointed out that 22 compounds were identified by the three techniques. Obviously, all the major components were included in this group, which also contained some unusual compounds in *C. giganteus* oil, such as *p*-cymenene, 3-methylbutyl hexanaote and octanoate, safrole, precocene I, and caryophyllene oxide. A few other sesquiterpenes were identified for the first time in *C. giganteus* oil by GC-MS (β-elemene, (E)-β-caryophyllene, β-selinene, and δ-cadinene), besides nonterpenic esters (2-phenylethyl hexanoate and 2-phenylethyl octanoate).

It is noteworthy that various compounds that were not identified by GC-MS of the whole oil sample were identified by ^{13}C NMR in the fractions of chromatography, in combination with RIs on apolar and polar columns: *trans*-dihydroperillaldehyde, geraniol, geranial, perillaldehyde, thymol, and ascaridole. The same association was useful for the identification of compounds that coeluted on apolar column, and consequently, although they were suggested by MS, they could not be accurately identified by this technique. For instance, *trans*-dihydrocarvone and *cis*-dihydroperillaldehyde (RI = 1171) were identified by ^{13}C NMR in the fractions of chromatography, and then quantified by GC(FID) in the EO by comparison of their RIs on apolar and polar columns.

A last interesting point is the identification of 3,9-oxy-p-mentha-1,8(10)-diene, not present in our MS and ^{13}C NMR data libraries. Its structure was determined by extensive NMR studies carried out on a fraction of chromatography where the oxide accounted for 80.0%, and then the NMR data were compared with those reported in the literature.

In conclusion, the detailed analysis carried out by a combination of chromatographic and spectroscopic techniques, after fractionation of the bulk sample, gives better insight into the composition of the *C. giganteus* leaf oil from Côte d'Ivoire. Indeed, 25 compounds were reported for the first time in *C. giganteus* leaf oil, including oxygenated acyclic monoterpenes (citronellal, geraniol, geranial), oxygenated menthane derivatives (*cis*- and *trans*-dihydroperylaldehydes, thymol), phenylpropanoids (safrole), phenylethyl derivatives (hexanoate, octanoate), and oxides (ascaridole, precocene, caryophyllene oxyde), as well as sesquiterpene hydrocarbons (β-elemene, (E)-β-caryophyllene, β-selinene, δ-cadinene) and linear aldehydes (nonanal, decanal) and alkanes.

8.4 CONCLUSION

Nowadays, identification and quantitative determination of the individual components of EOs isolated from various species of *Cymbopogon* is mostly achieved by a combination of "on-line" chromatographic separation and spectroscopic identification. For that reason, a fast-scanning mass spectrometer directly coupled with a gas chromatograph is the basic equipment for such analyses. However, MS data should be considered in combination with retention indices (RIs). In summary, a component is identified (1) by comparison of its GC retention indices on polar and apolar columns, determined relative to the retention times of a series of *n*-alkanes with linear interpolation with those of authentic compounds or literature data; (2) on computer matching against laboratory-made and commercial mass spectral libraries and comparison of spectra with those of laboratory-made library or literature data. Identification by only one of the two techniques should be avoided, even for routine analysis.

^{13}C NMR spectroscopy is an alternative method that has been used, at the beginning, to confirm the occurrence of a component, previously identified (or suggested) by MS or by RIs. Actually, the computerized comparison of the chemical shifts of signals in the ^{13}C NMR spectrum of the essential oil (EO) with those of reference spectra compiled in a library allows the identification of its main components, without previous separation.

When the composition of an EO is investigated for the first time, more sophisticated techniques could be needed and a wide variety of combinations of analytical techniques are being developed, including complex hybrids such as GC-MS-MS, GC-MS-FTIR, HPLC-GC-MS, and HPLC-^{1}H NMR-MS. Concerning *Cymbopogon* oils, to the best of our knowledge, none of these techniques have been used. Conversely, analyses of *Cymbopogon* oils have been carried out by a combination of various techniques, the oil being prefractionated by fractional distillation and/or column chromatography. Then, the fractions of chromatography have been analyzed by complementary techniques, GC(RI), GC-MS, ^{1}H, and ^{13}C NMR. Such types of analyses, which obviously are time consuming, provide unequivocal identification of the individual components present in EOs. Use of NMR spectroscopy for the analysis of natural mixtures is strongly encouraged by perfume, fragrance, and flavor manufacturers and quality control organizations.

ACKNOWLEDGMENTS

The authors are indebted to their coworkers and PhD students, who contributed substantially to this research. They appreciated the cooperation of colleagues of the universities of Corsica, Abidjan (Ivory Coast), and Hanoi (Vietnam). They acknowledge the Collectivité Territoriale de Corse and the European Community for partial financial support.

REFERENCES

Abd Allah M. A., Foda Y. H., Saleh M., Zaki M. S. A., and Mostafa M. M. (1975) Identification of the volatile constituents of the Egyptian lemongrass oil. Part I. Gas chromatographic analysis. *Nahrung*, 19, 195–200.

Abegaz B., Yohannes P. G., and Dieter R. K. (1983) Constituents of the essential oil of Ethiopian *Cymbopogon citratus* Stapf. *J. Nat. Prod.*, 46, 424–426.

Adams R. P. (2007) *Identification of Essential Oil Components by Gas Chromatography/Mass Spectrometry*, 4th ed., Allured Publishing, Carol Stream, IL.

Ahmad V. U. and Jassbi A. R. (1999) Analysis of the essential oil of *Echinophora sibthorpiana* Guss. by means of GC, GC/MS and 13C-NMR techniques. *J. Essent. Oil Res.*, 11, 107–108.

Al-Burtamani S. K. S., Fatope M. O., Marwah R. G., Onifade A. K., and Al-Saidi S. H. (2005) Chemical composition, antibacterial and antifungal activities of the essential oil of *Haplophyllum tuberculatum* from Oman. *J. Ethnopharmacol.*, 96, 107–112.

Alencar J. W., Vasconcelos Silva M. G., Lacerda Machado M. I., Craveiro A. A., Abreu Matos F. J., and de Abreu Magalhães R. (1997) Use of ¹³C-NMR as complementary identification tool in essential oil analysis. *Spectroscopy*, 13, 265–273.

Alitonou G. A., Avlessi F., Sohounhloue D. K., Agnaniet H., Bessiere J. -M., and Menut C. (2006) Investigations on the essential oil of *Cymbopogon giganteus* from Benin for its potential use as an anti-inflammatory agent. *Internat. J. Aromatherapy*, 16, 37–41.

Ayedoun M. A., Moudachirou M., and Lamaty G. (1997) Composition chimique des huiles essentielles de deux espèces de *Cymbopogon* du Bénin exploitable industriellement. *Bioressources, Energie, Développement, Environnement*, 8, 4–6.

Badjah Hadj Ahmed A. Y., Meklati B. Y., and Fraisse D. (1988) Mass spectrometry with positive chemical ionization of some α-pinene and β-pinene derivatives. *Flavour Fragr. J.*, 3, 13–18.

Bagheri R., Mohamadi S., Abkar A., and Fazlollahi A. (2007) Essential oil compositions of *Cymbopogon parkeri* STAPF from Iran. *Pakistan J Biol. Sci.*, 10, 3485–3486.

Baldovini N., Tomi F., and Casanova J. (2003) Enantiomeric differentiation of bornyl acetate by carbon-13 NMR using a chiral lanthanide shift reagent. *Phytochem. Anal.*, 14, 241–244.

Baldovini N., Tomi F., and Casanova J. (2001) Identification and quantitative determination NMR of furanodiene, a heat-sensitive compound, in essential oils by ¹³C-NMR. *Phytochem. Anal.*, 12, 58–63.

Baser K. H. C., Özek T., and Demirci B. (2005) Composition of the essential oil of *Cymbopogon afronardus* Stapf from Uganda. *J. Essent. Oil Res.*, 17, 139–140.

Bicchi C., D'Amato A., and Rubiolo P. (1999) Cyclodextrin derivatives as chiral selectors for direct gas chromatographic separation of enantiomers in the essential oil, aroma and flavour fields. *J. Chromatogr. A*, 843, 99–121.

Bicchi C., D'Amato A., Manzin V., and Rubiolo P. (1997) Cyclodextrin derivatives in GC separation of racemic mixtures of volatiles. Part XI. Some applications of cyclodextrin derivatives in GC enantioseparations of essential oil components. *Flavour Fragr. J.*, 12, 55–61.

Bicchi C., Rubiolo P., Marschall H., Weyerstahl P., and Laurent R. (1998) Constituents of *Artemisia roxburghiana* Besser essential oil. *Flavour Fragr. J.*, 13, 40–46.

Blanc M. C., Bradesi P., Gonçalves M. J., Salgueiro L., and Casanova J. (2006) Essential oil of *Dittrichia viscosa* ssp. *viscosa*: Analysis by ¹³C-NMR and antimicrobial activity. *Flavour Fragr. J.*, 21, 324–332.

Boelens M. H., Boelens H., and Van Germert L. J. (1993) Sensory properties of optical isomers, *Perf. Flav.*, 18, 1–16.

Boti J. B., Muselli A., Tomi F., Koukoua G., N'Guessan T. Y., Costa J., and Casanova J. (2006) Combined analysis of *Cymbopogon giganteus* Chiov. Leaf oil from Côte d'Ivoire by GC/RI, GC-MS and 13C-NMR. *C.R. Chimie*, 9, 164–168.

Bottini A. T., Dev V., Garfagnoli D. J., Hope H., Joshi P., Lohani H., Mathela C. S., and Nelson T. E. (1987) Isolation and crystal structure of a novel dihemiacetal bis-monoterpenoid from *Cymbopogon martini*. *Phytochemistry*, 26, 2301–2302.

Bradesi P., Bighelli A., Tomi F., and Casanova J. (1996) L'analyse des mélanges complexes par RMN du carbone-13. *Can. J. Appl. Spectrosc.*, 41, 15–24 and 41, 41–50.

Bretmaier E. (2006) *Terpenes, Flavors, Fragrances, Pharmaca, Pheromones*. Wiley-VCH-Verlag GmbH and Co., Weinheim.

Cavaleiro C. (2001) Óleos essenciais de *Juniperus* de Portugal. Thèse de doctorat, Université de Coïmbra.

Cazaussus A., Pes R., Sellier N., and Tabet J. C. (1988) GC-MS and GC-MS-MS analysis of a complex essential oil. *Chromatographia*, 25, 865–869.

Chagonda L. S., Makanda C., and Chalchat J. C. (2000) The essential oils of wild and cultivated *Cymbopogon validus* (Stapf) Stapf ex Burtt Davy and *elionurus muticus* (Spreng.) Kunth from Zimbabwe. *Flavour Fragr. J.*, 15, 100–104.

Chalchat J. C., Garry R. P., Harama M., and Sidibé L. (1998) A study of aromatic plants from Mali. Chemical composition of essential oils of two *Cymbopogon* varieties. *Cymbopogon citratus* and *Cymbopogon giganteus*. *Riv. Ital. EPPOS, Spec. Num.*, 741–752.

Chamblee T. S., Karelitz R. L., Radford T., and Clark B. C. (1997) Identification of sesquiterpenes in *Citrus* essential oils by cryofocusing GC/FT-IR. *J. Essent. Oil Res.*, 9, 127–132.

Chisowa E. H., Hall D. R., and Farman D. I. (1998) Volatile constituents of the essential oil of *Cymbopogon citratus* Stapf frown in Zambia. *Flavour Fragr. J.*, 13, 29–30.

Coleman III W. M., Gordon B. M., and Lawrence B. M. (1989) Examinations of the matrix isolation Fourier transform infrared spectra of organic compounds: part XII, *Appl. Spectrosc.*, 43, 298–304.

Corticchiato M. and Casanova J. (1992) Analyse des mélanges complexes par RMN du Carbone-13. Application aux huiles essentielles. *Analusis*, 20, M51–57.

De Medeci D., Pieretti S., Salvatore G., Nicoletti M., and Rasoanaivo P. (1992) Chemical analysis of essential oils of Malagasy medicinal plants by gas chromatography and NMR spectroscopy. *Flavour Fragr. J.*, 7, 275–281.

Decouzon M., Géribaldi S., Rouillard M., and Sturla J. -M. (1990) A new look at the spectroscopic properties of dihydrocarveol stereoisomers. *Flavour Fragr. J.*, 5, 147–152.

Derome A. E. (1987) *Modern NMR Techniques for Chemistry Research*, Vol. 6, Organic chemistry series, J. E. Baldwin Ed., Pergamon Press, Oxford.

Dhar A. K., Thappa R. K., and Atal C. K. (1981) Variability in yield and composition of essential oil in *Cymbopogon jawarancus*. *Planta Med.*, 41, 386–388.

Dugo P., Mondello L., Dugo L., Stancanelli R., and Dugo G. (2000) LC-MS for the identification of oxygen heterocyclic compounds in citrus essential oils. *J. Pharm. Biomed. Anal.*, 24, 147–154.

El-Askari H. I., Meselhy M. R., and Galal A. M. (2003) Sesquiterpenes from *Cymbopogon proximus*. *Molecules*, 8, 670–677.

Evans F. E., Miller D. W., Cairns T., Vernon Baddeley G., and Wenkert E. (1982) Structure analysis of proximadiol (cryptomeridiol) by ^{13}C NMR spectroscopy. *Phytochemistry*, 21, 937–938.

Ferreira M. J. P., Costantin M. B., Sartorelli P., Rodrigues G. V., Limberger R., Henriques A. T., Kato M. J., and Emerenciano V. P. (2001a) Computer-aided method for identification of components in essential oils by ^{13}C NMR spectroscopy. *Anal. Chim. Acta*, 447, 125–134.

Ferreira M. J. P., Brant A. J. C., Rodrigues G. V., and Emerenciano V. P. (2001b) Automatic identification of terpenoids skeletons through ^{13}C nuclear magnetic resonance data disfunctionalization. *Anal. Chim. Acta*, 429, 151–170.

Foda Y. H., Abdallah M. A., Zaki M. S., and Mostafa M. M. (1975) Identification of the volatile constituents of the Egyptian lemongrass oil. Part III. Infrared spectroscopy. *Nahrung*, 19, 395–400.

Formácek V. and Kubeczka K. H. (1982a) *Essential Oils Analysis by Capillary Gas Chromatography and Carbon-13 NMR Spectroscopy*, John Wiley & Sons, Chichester.

Formácek V. and Kubeczka K. H. (1982b) *13C NMR Analysis of Essential Oils*, in *Aromatic Plants: Basic and Applied Aspects*, pp. 177–18, Margaris N., Koedam A., Vokou D., Eds., Martinus Nijhoff Publishers, La Haye.

Fraser R. R., Stothers J. B., and Tan C. T. (1973) Determination of optical purity with chiral shift reagents and ^{13}C magnetic resonance. *J. Magn. Reson.*, 10, 95–97.

Gaydou E. M. and Randriamiharisoa R. P. (1987) Gas chromatographic diastereomer separation of linalool derivatives. Application to the determination of the enantiomeric purity of linalool in essential oils. *J. Chromatogr.*, 396, 378–381.

Gonny M., Bradesi P., and Casanova J. (2004) Identification of the components of the essential oil from wild Corsican *Daucus carota* L. using ^{13}C-NMR spectroscopy. *Flav. Fragr. J.*, 19, 424–433.

Grundschober F. (1991) The identification of individual components in flavorings and flavoured foods. *Z. Lebensm Unters Forsch.*, 192, 530–534.

Hedges L. M. and Wilkins C. L. (1991) Components analysis of eucalyptus oil by gas chromatography–Fourier transform–infrared spectrometry–mass spectrometry. *J. Chromatogr. Sci.*, 29, 345–350.

Huffman J. W. and Pinder A. R. (1980) Comments on the structure of isointermedeol, a sesquiterpene reported from *Cymbopogon flexuosus*. *Phytochemistry*, 19, 2468–2469.

Jirovetz L., Buchbauer G., Eller G., Ngassoum M. B., and Maponmetsem P. M. (2007) Composition and antimicrobial activity of *Cymbopogon giganteus* (Hochst.) Chiov. Essential flower, leaf and stem oils from Cameroon. *J. Essent. Oil Res.*, 19, 485–489.

Joulain D. (1994) Methods for analyzing essential oils. Modern analysis methodologies: use and abuse. *Perfumer and Flavorist*, 19, 5–17.

Joulain D. and Laurent R. (1989) Two closely related sesquiterpenols: 1-endo-bourbonanol and 1,6-germacradien-5-ol. *J. Essent. Oil Res.*, 1, 299–301.

Joulain D. and König W. A. (1998) The atlas of spectral data of sesquiterpene hydrocarbons, E.B. Verlag, Hamburg.

Kanko C., Koukoua G., N'Guessan Y. T., Fournier J., Pradère J. P., and Toupet L. (2004) Etude des propriétés physico-chimiques des huiles essentielles de *Lippia multiflora*, *Cymbopogon citratus*, *Cymbopogon nardus*, *Cymbopogon giganteus*. C.R. Chimie, 7, 1039–1042.

Kasali A. A., Oyedeji A. O., and Ashilokun A. O. (2001) Volatile leaf oil constituents of *Cymbopogon citratus* (DC) Stapf. *Flavour Fragr. J.*, 16, 377–378.

Keita A. (1993) Terpènes de l'huile essentielle des inflorescences de « ce kala » *Cymbopogon anteus* (Chiov.). *Médecine Afr. Noire,* 40, 229–233.

Khadri A., Serralheiro M. L. M., Nogueira J. M. R., Neffati M., Smiti S., and Araujo M. E. M. (2008) Antioxidant and antiacetylcholinesterase activities of essential oils from *Cymbopogon schoenanthus* L. Spreng. Determination of chemical composition by GC–mass spectrometry and ^{13}C NMR. *Food Chemistry,* 109, 630–637.

König W. A., Hochmuth D. H., and Joulain D. (2001) Terpenoids and related constituents of essential oils. Library of MassFinder 2.1, Institute of Organic Chemistry, Hamburg.

Korhammer S. A. and Bernreuther A. (1996) Hyphenation of high-performance liquid chromatography (HPLC) and other chromatographic techniques (SFC, GPC, GC, CE) with nuclear magnetic resonance (NMR): A review. *Fresenius J. Anal. Chem.,* 354, 131–135.

Kováts E. (1965) Gas chromatographic characterization of organic substances in the retention index system, in *Advances in Chromatography,* vol. 1, pp. 229–247, Giddings J.C., and Keller R.A., Eds., Marcel Dekker, New York.

Kreis P. and Mosandl A. (1994) Chiral compounds of essential oils. VVVII: Simultaneous stereoanalysis of *Cymbopogon* oil constituents. *Flavour Fragr. J.,* 9, 257–260.

Krock K. A., Ragunathan N., and Wilkins C. L. (1994) Multidimensional gas chromatography coupled with infrared and mass spectrometry for analysis of *Eucalyptus* essential oils, *Anal. Chem.,* 66, 425–430.

Kubeczka K. H. (1985) Progress in isolation techniques for essential oil constituents, in *Essential Oils and Aromatic Plants,* pp. 107–126, Baerheim Svendsen and Scheffer Eds., Nijhoff Junk pub, Dordrecht.

Kubeczka K. H. and Formácek V. (2002) Essential oils analysis by capillary gas chromatography and carbon-13 NMR spectroscopy. Second ed. John Wiley & Sons, NY, pp. 67–80.

Kumar A., Malik F., Bhushan S., Sethi V. K., Shahi A. K., Kaur J., Taneja S. C., Qazi G. N., and Singh J. (2008) An essential oil and its major constituent isointermedeol induce apoptosis by increased expression of mitochondrial cytochrome C and apical death receptors in human leukaemia HL-60 cells. *Chem.-Biol. Interactions,* 171, 332–347.

Lange G. and Schultze W. (1988a) Differentiation of isopulegol isomers by chemical ionization mass spectrometry, in *Bioflavour* '87, p. 115–122, P. Schreier Ed., W. de Gruyter and Co., Berlin.

Lange G. and Schultze W. (1988b), Studies on terpenoid and non-terpenoid esters using chemical ionization mass spectrometry in GC/MS coupling, in *Bioflavour'87,* pp. 105–114, Schreier P. Ed., de Gruyter W. and Co., Berlin.

Lawrence B. M. (1979a) Citronella oil, in *Essential Oils* 1976–1978, p. 22, Allured Publishing Co., Carol Stream, IL.

Lawrence B. M. (1979b) Lemongrass oil, in *Essential Oils* 1976–1978, p. 20 and pp. 30–31, Allured Publishing, Carol Stream, IL.

Lawrence B. M. (1989a) Citronella oil, in *Essential Oils* 1981–1987, pp. 29–31 and pp. 147–149, Allured Publishing, Carol Stream, IL.

Lawrence B. M. (1989b) Lemongrass oil, in *Essential Oils* 1981–1987, p. 111, Allured Publishing, Carol Stream, IL.

Lawrence B. M. (1989c) Palmarosa oil, in *Essential Oils* 1981–1987, pp. 237–239, Allured Publishing, Carol Stream, IL.

Lawrence B. M. (1993) Palmarosa oil, in *Essential Oils* 1988–1991, pp. 81–82, Allured Publishing, Carol Stream, IL.

Lawrence B. M. (1995a) Citronella oil, in *Essential Oils* 1992–1994, pp. 82–83 and p. 196, Allured Publishing, Carol Stream, IL.

Lawrence B. M. (1995b) Lemongrass oil, in *Essential Oils* 1992–1994, p. 60, Allured Publishing, Carol Stream, IL.

Lawrence B.M. (2003a) Citronella oil, in *Essential Oils* 1995–2000, pp. 52–53 and pp. 201–203, Allured Publishing, Carol Stream, IL.

Lawrence B. M. (2003b) Ginger grass oil, in *Essential Oils* 1995–2000, pp. 345–346, Allured publishing, Carol Stream, IL.

Lawrence B. M. (2003c) Lemongrass oil, in *Essential Oils* 1995–2000, pp. 206–210, Allured Publishing, Carol Stream, IL.

Lawrence B. M. (2003d) Palmarosa oil, in *Essential Oils* 1995–2000, pp. 334–335, Allured Publishing, Carol Stream, IL.

Lawrence B. M. (2006a) Citronella oil, in *Essential Oils* 2001–2004, pp. 142–145, Allured Publishing, Carol Stream, IL.

Lawrence B. M. (2006b) Lemongrass oil, in *Essential Oils* 2001–2004, pp. 128–132, Allured Publishing, Carol Stream, IL.

Lawrence B. M. (2006c) Palmarosa oil, in *Essential Oils* 2001–2004, pp. 77–78 and pp. 313–314, Allured Publishing, Carol Stream, IL.

Lawrence B. M. (2006d) Citronella oil. *Perf. Flav.*, 31, 10, 46.

Liu C., Zhang J., Yiao R., and Gan L. (1981) Chemical studies on the essential oils of *Cymbopogon* genus. *Huaxue Kuebao* 241–247.

Lorenzo D., Dellacassa E., Atti-Serafini L., Santos A. C., Frizzo C., Paroul N., Moyna P., Mondello L., and Dugo G. (2000) Composition and stereoanalysis of *Cymbopogon winterianus* Jowitt oil from Southern Brazil. *Flavour Fragr. J.*, 15, 177–181.

Mahalwal V. S. and Ali M. (2003) Volatile constituents of *Cymbopogon nardus* (Linn.) Rendle. *Flavour Fragr. J.*, 18, 73–76.

Mallavarapu G. R., Rajeswara Rao B. R., Kaul P. N., Ramesh S., and Bhattacharya A. K. (1998) Volatile constituents of the essential oils of the seeds and the herb of palmarosa (*Cymbopogon martinii* (Roxb.) Wats. var. *motia* Burk. *Flavour Fragr. J.*, 13, 167–169.

Mallavarapu G. R., Ramesh S., Kulkarni R. N., and Syamasundar K. V. (1992) Composition of the essential oil of *Cymbopogon travancorensis*. *Planta Med.*, 58, 219–220.

Marongiu B., Piras A., Porcedda S., and Tuveri E. (2006) Comparative analysis of the oil and supercritical CO^2 extract of *Cymbopogon citratus* Stapf. *Nat. Prod. Res.*, 20, 455–459.

Mathela C. S., Melkani A. B., Pant A., Dev V., Nelson T. E., Hope H., and Bottini A. T. (1989) A eudesmanediol from *Cymbopogon distans*. *Phytochemistry*, 28, 936–938.

Mathela C. S. and Joshi P. (1981) Terpenes from the essential oil of *Cymbopogon distans*. *Phytochemistry*, 20, 2770–2771.

Mathela C. S., Pandey C., Pant A. K., and Singh A. K. (1990) Phenylpropanoid constituents of *Cymbopogon microstachys*. *J. Ind. Chem. Soc.*, 67, 526–528.

McLafferty F. W. and Stauffer D. B. (1988) *Wiley/NBS Registry of Mass Spectral Data*, 4th ed., Wiley-Interscience, New York.

McLafferty F. W. and Stauffer D. B. (1994) *Wiley Registry of Mass Spectral Data*. 6th ed. Mass Spectrometry Library Search System Bench–Top/PBM, version 3.10d, Palisade, Newfield, UK.

Melkani A. B., Joshi P., Pant A. K., Mathela C. S., and Dev V. (1985) Constituents of the essential oils from two varieties of *Cymbopogon distans*. *J. Nat. Prod.*, 48, 995–997.

Menut C., Bessiere J. M., Samate D., Djibo A. K., Buchbauer G., and Schopper B. (2000) Aromatic plants of tropical west Africa. XI. Chemical composition, antioxidant and antiradical properties of the essential oils of three *Cymbopogon* species from Burkina Faso. *J. Essent. Oil Res.*, 12, 207–212.

Mohammad F., Nigam M. C., and Rahman W. (1981) Detection of new trace constituents in the essential oils of *Cymbopogon flexuosus*. *Perf. Flav.*, 6, 29–31.

Mondello L., Casilli A., Tranchida P. Q., Dugo P., and Dugo G. (2005) Comprehensive two-dimensional GC for the analysis of citrus essential oils. *Flavour Fragr. J.*, 20, 136–140.

Mondello L., Dugo G., Dugo P., and Bartle K. D. (1996) On-line HPLC-HRGC in the analytical chemistry of *Citrus* essential oils. *Perf. Flav.*, 21, 25–49.

Mondello L., Dugo P., and Bartle K. D. (1995) Automated HPLC-HRGC: A powerful method for essential oils analysis. Part V. Identification of terpene hydrocarbons of bergamot, lemon, mandarin, sweet orange, bitter orange, grapefruit, clementine and Mexican lime oils by coupled HPLC-HRGC-MS (ITD). *Flavour Fragr. J.*, 10, 33–42.

Mosandl A. (1988) Chirality in flavour chemistry—recent developments in synthesis and analysis. *Food Rev. Intern.*, 4, 1–43.

Mosandl A., Hener O., Kreis P., and Schmorr H. G. (1990) Enantiomeric distribution of a-pinene, b-pinene and limonene in essential oils and extracts. Part I. Rutaceae and gramineae. *Flavour Fragr. J.*, 5, 193–199.

NIST (1999) National Institute of Standards and Technology, PC Version 1.7 of The NIST/EPA/NIH mass spectral library. Perkin–Elmer Corp., Norwalk, CT.

Norouzi-Arasi H., Yavari I., Ghaffarzade F., and Mortazavi M. S. (2002) Volatile constituents of *Cymbopogon olivieri* (Boiss.) Bor. from Iran. *Flavour Fragr. J.*, 17, 272–274.

Ntezurubanza L., Collin G., Deslauriers G., and Nizeyimana J. B. (1992) Huiles essentielles de Geranium et de lemongrass cultivées au Rwanda. Rivista Ital. *EPPOS (Num. Spec.)*, 631–639.

Núñez C. V. and Roque N. F. (1999) Sesquiterpenes from the stem bark of *Guarea* guidonia (L.) Sleumer (Meliaceae). *J. Essent. Oil Res.*, 11, 439–440.

Ottavioli J., Bighelli A., Casanova J., Bui Thi Bang, and Pham Van Y. (2008a) unpublished results.

Ottavioli J., Bighelli A., Casanova J., Bui Thi Bang, and Pham Van Y. (2008b) submitted.

Ouamba J.-M. (1991) Valorisation chimique des plantes aromatiques du Congo. Extraction et analyse des huiles essentielles, Oximation des aldéhydes naturels, Thèse de Doctorat d'Etat, Université Montpellier II.

Paolini J., Costa J., and Bernardini A. F. (2005) Analysis of the essential oil from aerial parts of *Eupatorium cannabinum* subsp. *corsicum* (L.) by gas chromatography with electron impact and chemical ionisation mass spectrometry, *J. Chromatogr. A.*, 1076, 170–178.

Peyron L. (1973) Sur quelques essences en provenance du Mato Grosso. *Parfum. Cosmet. Savon. France*, 3, 371–378.

Popielas L., Moulis C., Kéita A., Fouraste I., and Bessière J. M. (1991) The essential oil of *Cymbopogon giganteus*, *Planta Med.*, 57, 586–587.

Prashar A., Hili P., Veness R. G., and Evans C. S. (2003) Antimicrobial action of palmarosa oil (*Cymbopogon martinii*) on *Saccharomyces cerevisiae*. *Phytochemistry*, 63, 569–575.

Raina V. K., Srivastava S. K., Aggarwal K. K., Syamasundar K. V., and Khanuja S. P. S., (2003) Essential oil composition of *Cymbopogon martinii* from different places in India, *Flavour Fragr. J.*, 18, 312–315.

Ramidi R., Ali M., Velasco-Negueruela A., and Pérez-Alonso M. J. (1998) Chemical composition of the seed oil of *Zanthoxylum alatum* Roxb. *J. Essent. Oil Res.*, 10, 127–130.

Randriamiharisoa R. P. and Gaydou E. M. (1987) Composition of palmarosa (*Cymbopogon martini*) essential oil from Madagascar. *J. Agr. Food Chem.*, 35, 62–66.

Rauber C. S., da Guterres S. S., and Schapoval E. E. S. (2005) LC determination of citral in *Cymbopogon citratus* volatile oil. *J. Pharm. Biomed. Anal.*, 37, 597–601.

Ravid U., Putievsky E., and Katzir I. (1992) Chiral GC analysis of enantimerically pure fenchone in essential oils. *Flavour Fragr. J.*, 7, 169–172.

Ravid U., Putievsky E., and Katzir I. (1996) Stereochemical analysis of borneol in essential oils using permethylated beta-cyclodextrin as a chiral stationary phase. *Flavour Fragr. J.*, 11, 191–195.

Ravid U., Putievsky E., Weinstein V., and Ikan R. (1985) Determination of the enantiomeric composition of natural flavouring agents by 1H-NMR spectroscopy, pp. 135–138 in *Essential Oils and Aromatic Plants*, Baerheim Svendsen and Scheffer Eds., Nijhoff Junk, Dordrecht.

Rezzi S., Bighelli A., Castola V., and Casanova J. (2002) Direct identification and quantitative determination of acidic and neutral diterpenes using ^{13}C-NMR spectroscopy. Application to the analysis of oleoresin of *Pinus Nigra*. *Appl. Spectrosc.*, 56, 312–317.

Ristorcelli D., Tomi F., and Casanova J. (1998) Carbon-13 NMR as a tool for identification and enantiomeric differentiation of major terpenes exemplified by the essential oil of *Lavandula Stoechas*. *Flavour Fragr. J.*, 13, 154–158.

Rizk A. M., Heiba H. I., Sandra P., Mashaly M., and Bicchi C. (1983) Constituents of plants growing in qarar, constituents of the volatile oil of *Cymbopogon Parkeri*. *J. Chromatogr.*, 279, 145–150.

Rouessac F. and Rouessac A. (1994) Analyse chimique. Méthodes et Techniques Instrumentales Modernes. Masson, Ed., Paris.

Rout P. K., Sahoo S., Rao Y. R. (2005) Essential oil composition of *Cymbopogon* microstachys (Hook.) Soenarke occurring in Manipur. *J. Essent. Oil Res.*, 17, 358–360.

Saeed T., Sandra P., and Verzele M. (1978) Constituents of essential oil of *Cymbopogon jawarancusa*. *Phytochemistry*, 17, 1433–1434.

Sargenti S. R. and Lanças F. M. (1997) Supercritical fluid extraction of *Cymbopogon citratus*. *Chromatographia*, 46, 285–290.

Schultze W., Lange G., and Schmaus G. (1992) Isobutane and ammonia chemical ionization mass spectrometry of sesquiterpene hydrocarbons. *Flavour Fragr. J.*, 7, 55–64.

Sidibé L., Chalchat J. C., Garry R. P., and Lacombe L. (2001) Aromatic plants of Mali (IV): Chemical composition of essential oils of *Cymbopogon citratus* (DC) Stapf and *Cymbopogon giganteus* (Hochst) Chiov. *J. Essent. Oil Res.*, 13, 110–112.

Spring O., Buschmann H., Vogler B., Schilling E. E., Spraul M., and Hofmann M. (1995) Sesquiterpene lactone chemistry of *Zaluzania grayana* from on-line LC-NMR measurements. *Phytochemistry*, 39, 609–612.

Surburg H. and Panten J. (2006) *Common Fragrance and Flavor Materials*, 5th ed., Wiley-VCH Verlag GmbH and Co., Weinheim.

Taskinen Jyrki, Mathela D. K., and Mathela C. S. (1983) Composition of the essential oil of *Cymbopogon flexuosus*. *J. Chromatogr. A*, 262, 364–366.

Thappa R. K., Dhar K. L., and Atal C. K. (1979) Isointermedeol, a new sesquiterpene alcohol from *Cymbopogon flexuosus*. *Phytochemistry*, 18, 671–672.

Tomi F. and Casanova J. (2000) Contribution de la RMN du carbone-13 à l'analyse des huiles essentielles. *Ann. Fals Expert. Chim.*, 93, 313–330.

Tomi F. and Casanova J. (2006) ^{13}C NMR as a tool for identification of individual components of essential oils from Labiatae—A review. *Acta Horticult.*, 723, 185–192.

Tomi F., Bradesi P., Bighelli A., and Casanova J. (1995) Computer-aided identification of individual components of essential oil using carbon-13 NMR spectroscopy. *J. Magn. Reson. Anal.*, 1, 25–34.

Van Den Dool H. and Kratz P. D. (1963) A generalization of the retention index system including linear temperature programmed gas-liquid partition chromatography, *J. Chromatogr.*, 11, 463–471.

Vernin G., Metzger J., Suon K.N., Fraisse D., Ghiglione C., Hamoud A., and Párkányi C. (1990) GC–MS–SPECMA bank analysis of essential oils and aromas. GC-MS (EI-PCI) data bank analysis of sesquiterpenic compounds in juniper needle oil—Application of the mass fragmentometry SIM technique, *Lebensmittel-Wissenschaft Technol.*, 23, 25–33.

Vernin G., Petitjean M., Poite J. C., Metzger J., Fraisse D., and Suon K. N. (1986) *Computer Aids to Chemistry*, Chap. VII, pp. 294–333, Vernin G., Chanon M., Horwood E., Eds., Mass Spectra and Kováts' Indices Databank of Volatile Aroma Compounds, Chichester.

Wang X. H., Jia C. R., and Wan H. (1995) The direct chiral separation of some optically active compounds in essential oils by multidimensional gas chromatography. *J. Chromatogr. Sci.*, 33, 22–25.

Weyerstahl P., Marschall H., Splittgerber U., and Wolf D. (2000) 1,7-Cyclogermacra-1(10),4-dien-15-al, a sesquiterpene with a novel skeleton, and other sesquiterpenes from Haitian vetiver oil. *Flavour Fragr. J.*, 15, 61–83.

Weyerstahl P., Marschall H., Weirauch M., Thefeld K., and Surburg H. (1998) Constituents of commercial Labdanum oil. *Flavour Fragr. J.*, 13, 295–318.

Wijesekera R. O. B., Jayewardene A. L., and Fonseka B. D. (1973) Varietal differences in the constituents of citronella oil, *Phytochemistry*, 12, 2697–2704.

Yarovaya O. I., Salomatina O. V., Korchagina D. V., Polovinka M. P., and Barkhash V. A. (2002) Transformations of 6,7-epoxy derivatives of citral and citronellal in various acidic media. *Russian J. Org. Chem.*, 38, 1649–1660.

Zaki M. S. A., Foda Y. H., Mostafa M. M., and Abd Allah M. A. (1975) Identification of the volatile constituents of the Egyptian lemongrass oil. Part II. Thin Layer Chromatography. *Food Nahrung*, 19, 201–205.

9 Citral from Lemongrass and Other Natural Sources

Its Toxicology and Legislation

David A. Moyler

CONTENTS

9.1 INTRODUCTION

Natural essential oils containing citral have been used in traditional and oriental folk medicines for millennia. The industrial manufacturing and chemical synthesis of citral has been known for more than a century, and even in the early stages of development, its reactivity and readily oxidizable characteristics were well known. It has long been considered best practice in laboratories and small-scale manufacturing units that spillages collected in absorbent paper or rags should be soaked in water before disposal to prevent combustion of citral and organic matter in the air. Citral is also active on human skin and can cause irritation, and even sensitization in some cases, although under "good manufacturing practice" and using personal protection equipment current in the industry, cases of reported problems are fortunately few.

This chapter contains many acronyms used throughout this chapter and the trade, and these are listed in Table 9.1. More complete acronym lists are provided by some trade associations for their members, for example, BEOA, IFRA. Legislatively, citral is in the EU Annex I register of regulated substances as 605-019-00-3, in the ATP 19 (EFFA 2006). Citral consists of two isomers, neral [CAS 106-26-3] and geranial [CAS 141-27-5], usually occurring naturally in the ratio of 40:60.

TABLE 9.1
Acronyms Listed in Alphabetical Order

AISE	Association Internationale de la Savonnerie de la Detergence & des Produits d'Entreti d'Entretien (European Soap and Detergent Assn.)
ATP	Adaptation of Technical Progress (EU Annex I), followed by revision no.
BCF	Bio Concentration Factor (REACH)
BEMA	British Essence Manufacturer's Association
BEOA	British Essential Oils Association
BFA	British Fragrance Association
CA	Competent Authority (REACH)
CAS	Chemical Abstracts Service
C&L	Classification and Labeling
CHIP	Chemical Hazard Information Packaging
CMR	Carcinogen, Mutagen or Reproductive toxin (REACH)
COE	Council of Europe
COLIPA	Cote de Liaison European de Industrie de la Parfumerie de Produits Cosmetiques et de Toilette (European Cosmetic, Toiletry, and Perfumery Association)
CREOD	Center for Research Expertise in Occupational Disease (U.S.)
CSA	Chemical Safety Assessment (REACH)
CSR	Chemical Safety Report (REACH)
CWG	Commission Working Group (of the EU)
DIN	Deutsches Institut fur Normung (German Institute for Standardization)
DNEL	Derived No Effect Level — human health (REACH)
DPD	Dangerous Preparations Directive (88/379/EEC as amended)
DSD	Dangerous Substances Directive (67/548/EEC as amended)
DU	Downstream User (REACH)
EC	European Community (now European Union)
EC_0	Environmental concentration 0, the minimal quantity of a substance administered orally or dermally that does not kill any target population within a specified time
EC 3	Prioritize and refer individuals for further assessment and care (EUSC)
EC_{10}	Environmental concentration 10: quantity of a substance administered orally or dermally required to kill 10% of a target population within a specified time
EC_{50}	Environmental concentration 50: quantity of a substance administered orally or dermally required to kill 50% of a target population within a specified time
ECB	European Chemicals Bureau
E Ch A	European Chemicals Agency (REACH)
EEIII	Joint Associations of 'EFFA / EFEO / IFEAT / IFRA / IOFI'
EFEO	European Federation for Essential Oils
EFFA	European Flavour and Fragrance Association
EINECS	European Inventory Notified Existing Chemical Substances
ELINCS	European List of Notified Chemical Substances
ES	Exposure Scenario (REACH)
e-SDS	Extended Safety Data Sheet, SDS for healthcare professionals (REACH)
ESIS	European Chemical Substances Information System (REACH)
ESR	Existing Substances Regulation (793/93/EEC)
ETF	Environmental Task Force (of IFRA)
EU	European Union
EUCLID	European Chemicals Information Database (REACH)
EUROTOX	European Federation of Toxicological Societies (REACH)
EUSC	End User Support Centre

TABLE 9.1 (continued)
Acronyms Listed in Alphabetical Order

F&F	Flavour & Fragrance (industry)
FCC	Food Chemical Codex (U.S.)
FDA	Food and Drug Administration (U.S.)
FEMA	Flavor Extract Manufacturers Association (U.S. Flavor Trade)
FEXPAN	FEMA Expert Panel
FFIDS	Flavor/Fragrance Ingredient Data Sheet
FMA	Fragrance Materials Association (U.S.)
GC-MS	Gas (liquid) Chromatography (combined) Mass Spectroscopy
GHS	Globally Harmonized System (REACH)
GLC	Gas Liquid Chromatography
GLP	Good Laboratory Practice
GMP	Good Manufacturing Practice
HCWG	Hazard Communication Working Group (of EFFA/IFRA/IOFI jointly)
HPV	High Production Volume Chemicals (REACH)
HRIPT	Human Repeat Insult Patch Test
HSDS	Health and Safety Data Sheet (EU)
HSE	Health and Safety Executive (U.K.)
IC_{50}	Infectious concentration 50: quantity of a substance administered orally or dermally required to kill 50% of a target population within a specified time
IFEAT	International Federation of Essential Oil and Aroma Trades
IFRA	International Fragrance Association
IHCP	Institute for Health and Consumer Protection
INCI	International Nomenclature of Cosmetic Ingredients
In vitro	Biological testing carried out "in glass," for example, test tube, petri dish
In vivo	Biological testing carried out with humans or animals
IOFI	International Organisation of the Flavour Industry
IOFI-CE	IOFI Committee of Experts
ISO	International Standards Organization
IUCLID	International Uniform Chemicals Information Database (REACH)
IUPAC	International Union for Pure and Applied Chemistry
JAG	Joint Advisory Group of IFRA
K o/c	Coefficient of partition between Oil and Soil (log scale)
K o/w	Coefficient of partition between Oil and Water (log scale)
LC_{50}	Lethal concentration 50: quantity of a substance administered by inhalation, required to kill 50% of a target population within a specified time
LD_{50}	Lethal dose 50: quantity of a substance administered orally or dermally required to kill 50% of a target population within a specified time
LLNA	Local Lymph Node Assay
LOAEL	Lowest Observed Adverse Effect Level
LOEL	Lowest Observed Effect Level
LOEC	Lowest Observed Effect Concentration
LPV	Low Production Volume Chemicals (REACH)
MAX	Maximum Allowable Concentration
MCS	Multiple Component Substances (e.g., methyl ionone isomers)
M/I	Manufacturer/Importer (REACH)
MITI	Ministry of Trade and Industry (Japan)
MSDS	Material Safety Data Sheet (U.S.)

(continued on next page)

TABLE 9.1 (continued)
Acronyms Listed in Alphabetical Order

NCS	Natural Complex Substance (e.g., essential oils and extracts)
NESIL	No Expected Sensitization Induction Level
NI	Nature Identical
NLP	No-Longer Polymers (REACH)
NOAEL	No Observed Adverse Effect Level
NOEC	No Observed Effect Concentration
NOEL	No Observed Effect Level
NOEL-MAX	No Observed Effect Level for Maximum Allowable Concentration
NORMAN	Network Of Reference Laboratories Monitoring Emerging Environmental Pollutants
OECD	Organisation for Economic Cooperation and Development
ORATS	Online European Assessment Tracking System, EEC 793/93
PADI	Possible Average Daily Intake
PBT	Persistent Bio-accumulative Toxic Substance (REACH)
PNEC	Predicted No-Effect Concentration — environmental (REACH)
PPORD	Product and Process Oriented Research & Development (REACH)
Ph Eur	European Pharmacopoeia
PRODAROM	Syndicate National des Fabricants de Produits Aromatiques
PRODUCE	Piloting Reach On Downstream Users Compliance Exercise (REACH)
QMRF	QSAR Model Reporting Formats (ECB - for fish toxicity testing) (REACH)
QRA	Quantitative Risk Assessment (REACH)
QSAR	Quantitative Structure Activity Relationship (REACH)
REACH	Registration Evaluation Authorisation of Chemicals
REXPAN	RIFM Expert Panel
RIFM	Research Institute for Fragrance Materials
RIP's	Reach Implementation Projects (REACH)
RRIx	Relative Retention Index (GLC term)
RIPT	Repeat Insult Patch Test
RSC	Royal Society of Chemistry (London)
SAR	Structure Activity Relationship (REACH)
SARA	Structure Activity Relationship Assessment (REACH)
SCCNFP	European Scientific Committee Cosmetics and Non Food Products (now SCCP)
SCCP	Scientific Committee on Cosmetology (EC)
SHE	Committee for Occupational Safety, Health and Environment
SIEF	Substance Information Exchange Forum (REACH)
SME	Small- and Medium-Sized Enterprise (REACH)
SVHC	Substances of Very High Concern (REACH)
TEAMSPACE	A Web site for posting all of the Reach Implementation Projects (RIPs)
TGD	Technical Guidance Document (REACH)
TNO	Information and Communication Technology (Netherlands)
T/yr	Tonnes per year (REACH)
UNECE	United Nations Economic Commission for Europe
UNITIS	European Organisation for Cosmetic Ingredient Industry and Services
UVCB	Unknown Variable Complex Botanicals (REACH)
VPVB	Very Persistent Very Bioaccumulative (REACH)
WAF	Water Accommodated Fraction (for eco-toxicity testing) (REACH)
WGK	Water Pollution Class (Wassergefahrdungsklasse), Germany
WHO	World Health Organisation
WOE	Weight of Evidence

TABLE 9.2
Concentration of Citral in Final Fragrance Product

	Soap (%)	Detergent (%)	Creams, Lotions (%)	Perfume (%)
Usual	0.02	0.002	0.005	0.2
Maximum	0.2	0.02	0.02	0.8

9.2   USES AND DOSE RATE

Applications of citral in flavoring are widespread up to a reported average maximum level of 430 ppm in chewing gum, a mean average daily consumption of 137 mg in baked goods, and a PADI of 25.27 ppm (RIFM 2006). It is reported as being used in beverages, baked goods, cheese, chewing gum, condiments, frozen dairy, puddings, gravies, candy, and meat products. Its use in fragranced products is limited by the IFRA standard, which restricts the level of use of citral in consumer goods (IFRA 2007). Hence, its characteristic perfumery notes are often substituted by alternative ingredients that have the added advantage of being more stable. This is particularly true in functional products, where stability is an issue and a use level in final product is reported (Food Cosmetics 1979) in Table 9.2.

9.3   NATURAL SOURCES

Although citral occurs in many NCS (see Table 9.3), the first two on the list are the main commercial sources used for its extraction.

Other more uncommon oils available commercially containing high levels of citral:

Lindera citriodora	~65%
Backhousia citriodora	~95%
Calypranthes parriculata	~62%
Leptospermum liversidgei var. A	~75%
Ocimum gratissimum	~66%

9.3.1   *LITSEA CUBEBA* OIL

The medium-sized tree *Litsea cubeba*, a member of the Lauraceae family, is commercially grown for essential oil distillation in the Yunnan and Guangxi Provinces of China, where it is known locally as May Chang. The fresh berries and leaves are cut and allowed to wilt before they are used for steam-distilled oil production. There are some differences in the composition of the oils derived from pure berries and pure leaves, and the collection of wood is best avoided because it can introduce low levels of toxic safrole into the oil.

Litsea cubeba oil is mobile, pale yellow in color, with a powerful, fresh, intense herbal-lemon odor that is quite prominent. Virtually its only use is as a source of natural citral, which has several applications. In lemon flavors, it reinforces the main note, but it has limited stability in functional fragrance products. Citral is used for the synthesis of vitamin A and various ionone aroma chemicals.

It has been featured in ISO standard 3214, last reviewed in 2000. The citral quoted therein was 69.0% to 75.0%, measured by the ISO and FCC IV methods of GLC using a nonpolar column (Food Chemical Codex 1996). Classical wet analysis gives slightly higher results than this due to reaction of other minor components, leading to the usual commercial term of "70/75."

TABLE 9.3
Natural Citral Sources

Citral (neral + geranial)	Lemongrass oil	<90%
	Litsea cubeba oil	<78%
	Petitgrain lemon oil	<26%
	Lime oil (expressed)	<6.5%
	Verbena (*Lippia citriodora*) absolute	<5%
	Lemon oil (single fold)	<3%
	Lime oil (distilled, single fold)	<1%
	Geranium oil(s)	<1.5%
	Citronella oil Java	<1.3%
	Citronella oil Ceylon	<1.1%
	Petitgrain bergamot oil	<1%
	Petitgrain bitter orange oil (Paraguay)	<1%
	Rose oil(s)	<1%
	Ginger oil (India)	<0.8%
	Bergamot oil expressed	<0.7%
	Citronella oil China	<0.7%
	Cardamom oil	<0.6%
	Bergamot oil distilled	<0.4%
	Neroli oil	<0.3%
	Petitgrain bitter orange oil	<0.3%
	Grapefruit oil	<0.15%
	Orange sweet oil(s)	<0.1%
	Orange bitter oil	<0.1%
	Petitgrain mandarin oil	<0.1%

The physical constants recorded in ISO 3214:2000 (ISO 2000):

Relative density at 20°C	0.880 to 0.892
Refractive index at 20°C	1.480 to 1.490
Optical rotation at 20°C	+3° to +12°

The oil from Yunnan Province in China is the main commercial source, conforms to this analysis, and has the following registration and other properties:

CAS	90063-59-5 [ISO, RIFM, EFEO]
EINECS	290-018-7 [ISO, RIFM, EFEO]
TSCA	68855-99-2 [RIFM]
FEMA	3846
RIFM	910
EC	491n
FDA	182.20
INCI	*Litsea cubeba* oil; use: tonic
Flash point	65°C [EFEO], 66°C [FMA]
Log Kow calc.	>3 calculated

Assay typical: citral 72%, limonene 12%, linalol 1.5%, citronellol 0.2%, nerol 0.5%, geraniol 1%, 1,8 cineole 1%, 6-methyl hepten-2-one 2%, pinenes 2%, sabinene 1%, β-myrcene 1%, verbenols 2%, citronellal 1%, α-terpineol 0.5%, β-caryophyllene 1.5%, other monoterpene hydrocarbons 0.5%, other sesquiterpene hydrocarbons 0.2% = 99.9% (RSC EOC 2007).

9.3.2 LEMONGRASS OILS

The plants of lemongrass *Cymbopogon citratus* and *C. flexuosus* are members of the Poaceae family; they are medium-sized grasses that are commercially grown for essential oil distillation in Guatemala (*C. citratus)* and Cochin, India (*C. flexuosus)* (Arctander 1962). The fresh grass is cut and allowed to wilt before it is used for steam-distilled oil production. Oils from different regions have somewhat different compositions, Guatemalan West Indian (W.I.) having a slightly different citral content than East Indian (E.I.) Cochin origin.

Lemongrass oil is mobile, pale yellow in color, with a powerful, fresh, floral-herbal odor that is quite prominent. Similar to *Litsea cubeba* oil, it is used as source of natural citral, but because this oil is more expensive and contains a higher level of geraniol (which is difficult to remove by distillation due to the closeness of its boiling point to that of the geranial isomer of citral), it is less popular; also, its flavor profile is slightly different compared to *Litsea*. Removal of the geraniol by absorption of the citral with alkaline bisulfite is a chemical treatment that results in the citral losing its natural status.

These oils have been featured in the ISO standards 3217 (ISO 1974) and 4718 (ISO 2004); citral, quoted therein, was 70.0% to 75.0%.

The physical constants recorded in ISO were ISO 3217:1974 and ISO 4718:2004

Relative density at 20°C	0.872 to 0.897 0.885 to 0.905
Refractive Index at 20°C	1.483 to 1.489 1.483 to 1.489
Optical rotation at 20°C	+1° to −3° +1° to − 4°

The samples of oils analyzed of Guatemalan and Cochin origins conformed to these values and have the following registration and properties:

CAS	WI; 89998-14-1 [ISO, EFFA, EFEO, BEOA]
	EI; 91844-92-7 [ISO, EFEO, RIFM, BEOA]
EINECS	WI; 289-752-0 [ISO, EFEO, BEOA]
	EI; 295-161-9 [ISO, EFEO, BEOA, RIFM]
TSCA	8007-02-1 [INCI, RIFM]
FEMA	2624
RIFM	5585
EC	38n
FDA	182.20
INCI	*Cymbopogon schoenanthus* oil; uses: tonic, masking
Flash point:	71°C [EFEO, FMA]
Log Kow calc.	>3 calculated

Assay typical W.I: citral 73%, geraniol 2.5%, nerol 1%, 6-methyl hepten-2-one 1%, limonene 8%, linalol 1%, eugenol 0.1%, citronellol 0.5%, citronellal 0.3%, neryl + geranyl acetates, 3%, verbenols 3%, nonanone 1%, B-caryophyllene 2%, caryophyllene oxide 1%, other monoterpene hydrocarbons

2%, other sesquiterpene hydrocarbons 0.3% = 99.7% [7]. $_{E.I}$: citral 78%, geraniol 3.5%, nerol 1.2%, 6-methyl hepten-2-one 2.3%, limonene 0.3%, linalol 1%, eugenol 0.1%, citronellol 0.5%, citronellal 0.3%, neryl + geranyl acetates 3%, verbenols 3%, nonanone 1%, B-caryophyllene 2%, caryophyllene oxide 1%, other monoterpene hydrocarbons 2%, other sesquiterpene hydrocarbons 0.3% = 99.5% (RSC 2007).

9.4   MANUFACTURING METHODS

Ex-chemical synthesis: legislation: US—artificial, EU—nature identical. Assay 99%.
   The manufacturing methods are described in several books (Arctander 1969; Ohloff 1993). Synthesis can be based on acetylene chemistry (Kimel 1953) or isoprene (Leets et al. 1957).

Ex-NCS by absorption: legislation: US—artificial, EU—nature identical. Assay 99%.
   The essential oil is mixed with water, buffered to mildly alkaline conditions and a solution of sodium metabisulphite added, then thoroughly mixed by prolonged stirring. The citral dissolves in this aqueous phase and when stirring is stopped, the nonwater miscible phase floats to the surface and is separated. The aqueous phase is then neutralized with acid, and the citral separates as a floating layer. It is then washed with water and separated. It is then dried under vacuum or by filtering through a bed of anhydrous sodium sulfate.

Ex-NCS by distillation: legislation: US—natural, EU—natural. Assay 92, 95, and 96%.
   Depending on the NCS source, some impurities such as geraniol are codistilled because their boiling points are close to that of citral. The current commercial preference is to use *Litsea cubeba oil*, with its lower geraniol content than lemongrass oil, as a higher assay citral can be obtained. However, some flavorists still prefer the flavor of citral ex lemongrass in lemonade applications, as the flavor is said to be closer to that of lemons. As can be seen, the production method affects the legislative status of the citral.

9.5   ANALYSIS METHODS

GLC. The experienced chromatographer will often carefully examine GLC traces of the assay of citral using the FCC method (Food Chemical Codex 1996) on polymethylsiloxane phase columns, to look for the impurities that elute close to the citral peaks. The ISO standard method is to use an internal standard; it quotes a minimum 70% for the citral content of *Litsea cubeba* oil. On these nonpolar phases, the components elute in the order of their boiling points, as there is no interaction of the component with the stationary phase coating on the interior of the column.

So, peaks that elute close to the citral isomers have similar boiling points and are particularly difficult to fractionate from citral, even under high vacuum on high theoretical plate Sulzer-packed fractionation columns. If these closely eluting peaks are present, it is difficult to achieve the higher assay, premium citral grades by distillation.

GC-MS (Adams 1995) or GLC with relative retention indices on polarity-calibrated columns (RSC 1997) can be used to identify these components if necessary.

Attempts to analyze citral using polar phase columns such as polyethylene glycol result in leading peaks and difficult quantification, as well as incomplete resolution of components. This is due to interaction of the aldehyde components with the alcohol radicals of the glycol phase. These effects can be overcome by reducing the citral to the corresponding alcohols with sodium borohydride before analysis (Jones et al. 1977). However, the widespread commercial acceptance of the FCC and ISO methods on nonpolar columns has made this effective but longer method all but obsolete.

Titration. Assay of citral by titration of the aldehydes with standardized volumetric solutions gives good precision and accuracy, unless the oil under test contains citronellal or other aldehydes.

Absorption. A simple method of analysis, particularly suited to field-testing of citral-containing oils, is absorption into alkaline bisulfite solution (see manufacturing methods) in a cassia flask. This is usually a 200 mL conical flask with a long, slim, accurately graduated 10 mL by 0.1 ml neck. This flask can also be used for the determination of cinnamaldehyde in cassia oil (hence the name of the flask) and eugenol in clove, bay, and other oils containing it, by absorption into strong alkali. For citral determination, the flask is filled with alkaline bisulfite solution to the base of the flask neck zero line; 10.0 mL of the test citral oil is added, and the flask capped (usually with the technician's thumb!) and vigorously shaken for several minutes, while the technician's other hand supports the main bulb of the flask. The flask is then set aside to allow the phases to separate, and any unabsorbed oil components float to the top of the flask. The meniscus of this oil layer, which usually consists of terpene hydrocarbons, is simply read from the scale, subtracted from 10.0 mL of the test sample, and multiplied by 10 to get the percentage of absorbed citral. So, under the conditions described, if 10.0 mL of oil leaves 2.5 mL after absorption:

$$10.0 \text{ (minus) } 2.5 \text{ mL} = 7.5 \times 10 = 75\% \text{ citral in the test oil}$$

9.6 TOXICOLOGY

In terms of its PBT properties, citral has been rated as follows:

$$\text{Persistence} = \text{low; bioaccumulation} = \text{low; toxicity} = \text{medium}$$

The tests that are used for classification and labeling (C&L) are described as endpoints at which an effect is calculated. A list of these is shown in Table 9.4, with the results for citral given where available.

QSAR. As can be seen from Table 9.4, there are many data gaps for the endpoints for citral, as they are not yet in the public domain. This presents a problem for the consortia registrants under the REACH regulations. A potential solution is to use the principles of QSAR to predict or supplement insufficient test data with predictions based on the structural relationships of other molecules, if they are available.

Sometimes called SAR or SARA, this can be useful for the prediction of the level of concern regarding a material or even a mixture, with commercial computer programs available to aid in the prediction of toxicological concerns regarding a particular chemical structure (Lawrence et al. 2007).

To quote from the draft regulations—

... to facilitate the considerations of a (Q)SAR model for regulatory purposes, it should be associated with the following information:

1. A defined endpoint
2. An unambiguous algorithm
3. A defined domain of applicability
4. Appropriate measures of goodness-of-fit, robustness, and predictivity
5. A mechanistic interpretation, if possible

9.6.1 RIFM

Citral. According to RIFM, the Fragrance Structure—Activity Group for citral is "Aldehydes, branch chain, unsaturated," and they have summarized the sensitization potency of citral. These are shown in Table 9.5.

The potency classification of citral is as a weak sensitizer.

TABLE 9.4
Endpoints in the QMRF with Values for Citral

Physicochemical Effects

Melting point °C	= <–15
Boiling point °C	= 230
Flash point—DIN 51 758 c.c. °C	= 98
Water solubility—at pH 7	= mg/L = 420
Vapor pressure	mmHg 20°C = 0.07
Surface tension	mM/m at 25°C = <35
Partition octanol/water—OECD 107	log K = 2.76
Partition air/water—H	N/A

Environmental Fate

Persistence: Abiotic in water	N/A
Persistence: Abiotic in air	OECD 301 C, 28 d = >90%
Persistence: biodegradation	Lag phase: 3 d, begin plateau: 7 d
	"Readily biodegradable," modified MITI
	in water 19%, half-life 15 days
	in soil 80%, half-life 30 days
	in air 0%, half-life 0.021 days
Bioaccumulation: bioconcentration	BCF 90
Bioaccumulation: biomagnification	N/A
Adsorption/desorption in soil	N/A
Partition: vegetation/water	N/A
Partition: vegetation/air	N/A
Partition: vegetation/soil	N/A
Partition: soil/sorption—log K_{oc}	N/A

Ecotoxic Effects

Acute toxicity daphnia—*D. magna*	24 h EC_{50} = 11 mg/L
Acute toxicity algae—DIN 38 412	72 h EC_{10} = 4.9 mg/L
Acute toxicity fish—*Leuciscus idus*	96 h LC_{50} = 4.6 to 10 mg/L
Long-term daphnia—*D. magna*	48 h EC_0 = 3.13 mg/L
Long-term algae—DIN 38 412	96 hour EC_{50} = 19 mg/L
Long-term fish—DIN 38 412	NOEC = 4.6 mg/L
Daphnia reproduction	N/A
Microbial inhibition—DIN 38 412 *P. putida*	30 min EC_{50} = 2100 mg/L
Toxicity to soil microorganisms	N/A
Toxicity to earthworms	N/A
Toxicity to plants	N/A
Toxicity to sediment organisms	4 0.3 mg DOC/L
Toxicity to soil invertebrates	N/A
Toxicity to birds	N/A

Human Health Effects

Acute inhalation toxicity	>27 mg/mL (rabbit)
Acute oral toxicity	6800 mg/kg (rat)
Acute dermal toxicity	>5000 mg/kg (rabbit)
Skin irritation	at 20%, NOEL 4%
Skin corrosion	N/A
Acute photo-irritation	N/A
Skin sensitization	NESIL 1400 µg/cm²

TABLE 9.4 (continued)
Endpoints in the QMRF with Values for Citral

Respiratory sensitization	N/A
Photosensitization	N/A
Eye irritation/corrosion	Irritation at 5%
Mutagenicity	*E. coli*; 0.1 mg/plate, no effect
Photo mutagenicity	N/A
Carcinogenicity	Strong suppressing activity, $IC_{50} < 0.00625\%$
Photocarcinogenicity	N/A
Repeated dose toxicity	N/A
Developmental toxicity	22 d non-GLP, LOAEL = 60 mg/kg bw/d
Fertility	20 d non-GLP, NOAEL = 0.43 mg/L
Endocrine disruption: receptor binding	Inhibited estrogen binding
Endocrine disruption: gene expression	N/A
Toxicokinetics: skin penetration	63 min (mice)
Toxicokinetics: ocular penetration	N/A
Toxicokinetics: gastrointestinal penetration	N/A
Toxicokinetics: blood–brain penetration	N/A
Toxicokinetics: placental penetration	N/A
Toxicokinetics: blood–testis penetration	N/A
Toxicokinetics: blood–lung penetration	N/A
Toxicokinetics: metabolism	N/A
Toxicokinetics: protein binding	N/A

* N/A = Not available.

TABLE 9.5
Sensitizing Potency for Citral

LLNA weighted mean EC 3 value	In $\mu g/cm^2$ = 1414 (11 studies)
NOEL-HRIPT induction on human skin	In $\mu g/cm^2$ = 1400
NOEL-MAX induction on human skin	N/A
LOEL induction on human skin	In $\mu g/cm^2$ = 3876
WOE-NESIL	In $\mu g/cm^2$ = 1400

The REXPAN conclusion is that the review of "the critical data for citral, based on the weight of evidence, established the No Expected Sensitization Induction Level as 1400 $\mu g/cm^2$. They recommended the limits for the 11 different product categories, which derive from the application of the exposure-based quantitative risk assessment approach for fragrance ingredients, which is detailed in the QRA Expert Group technical dossier of March 15th 2006 (IFRA 2007).

The 11 product categories can be found in Table 9.6 and replace the "wash-off" and "leave on" categories for consumer products that were used for many years.

Cosmetics Directive EU. The EU 26th Cosmetics Directive, 7th amendment, lists citral as one of the 16 SCCNFP naturally occurring "alleged skin allergens" (there are 8 more aroma chemicals, plus oakmoss and treemoss for a total of 26). A paper was presented at the IFEAT conference 2001 that explained these regulations, giving an XL spreadsheet of the maximum levels of these "16" in fragrance NCS (Moyler 2001). This XL was published by EFFA and distributed to its members at the same time, to enable the levels of the 16 in fragrance compounds to be calculated from the amounts added as such, plus the maximum contribution from the NCS, without the need to

TABLE 9.6
IFRA Classes of Product Types (QRA)

Categories

1. Lip products, insect repellents, toys
2. Deodorants and antiperspirants
3. Men's facial balms and aftershaves, tampons
4. Colognes, hair styling products, body lotions, foot care, strips for "scratch and sniff"
5. Women's makeup, hand cream, face masks
6. Mouthwash, toothpaste, oral care
7. Toilet paper, intimate wipes, baby wipes
8. Make-up removers, nonaerosol hair styling products, nail care, talc
9. Shampoo and conditioners, liquid soap, body and face cleansers, shaving creams, depilatory, shower gels, soap, feminine hygiene products, bath products, aerosols
10. Hand-wash products, laundry cleaners, household cleaners, dish-wash cleaners, diaper cleaners, dry cleaning products, pet shampoos
11. Candles, air fresheners, shoe polish, carpet cleaners, insecticides, toilet blocks, incense, machine dish wash, plastics, fuels, paints, cat litter, starch sprays, odorized water for steam irons

Litsea cubeba

Toxicology acute toxicity	LD$_{50}$: oral, rat: >5000 mg/kg; dermal, rabbit: 4800 mg/kg
Local effects	LLNA, EC 3 8.4%; weak sensitizer [RIFM]
Hydrocarbon content	15% (BEOA technical comm.)
Kinematic viscosity	<7 × 10^{-6} m^2/s at 40°C
Surface tension	<35 mN/m at 25°C
Ecotoxicity data	WGK = 2 [EFEO], EC$_{50}$ = 6.0 mg/L daphnia

Lemongrass W.I.

Toxicology acute toxicity	LD$_{50}$: oral, rat: WI > 5000 mg/kg; dermal, rabbit: WI > 5000 mg/kg
Local effects	LLNA, EC 3 6.5% positive; weak sensitizer [RIFM]
Hydrocarbon content	WI 9 % (BEOA technical comm.)
Kinematic viscosity	<7 × 10^{-6} m^2/s at 40°C
Surface tension	<35 mN/m at 25°C
Ecotoxic data	WGK = 2 [EFEO], EC$_{50}$ = 27.0 mg/L daphnia

Lemongrass E.I.

Toxicology acute toxicity	LD$_{50}$: oral, rat: EI 5600 mg/kg; dermal, rabbit: EI 2000 mg/kg
Local effects	LLNA, EC 3 6.5% positive; weak sensitizer [RIFM]
Hydrocarbon content	EI 2% (BEOA technical comm)
Kinematic viscosity	<7 × 10^{-6} m^2/s at 40°C
Surface tension	<35 mN/m at 25°C
Ecotoxic data	WGK = 2 [EFEO], EC$_{50}$= 27.0 mg/L daphnia

routinely analyze a complex fragrance. This is particularly helpful for SMEs, who then only need to use the IFRA GC-MS methods (Chaintreau et al. 2003; Leijs et al. 2005) for validation of the calculated result.

9.7 CLASSIFICATION AND LABELING

Since the publication of the EFFA Code of Practice in August 2006 (EFFA 2006), the classification and labeling (C&L) of natural complex substances (NCS) in attachments II and VI is now based on the principles of the DPD rather than the DSD, which was the case in the 2005 edition.

This means that the C&L, for an essential oil that contains citral, is based on that of citral at the near maximum level (usually the 95 percentile) to be found in the oil, unless the actual test data for the oil is "robust." Robust test data are considered to be results based on modern protocols such as LLNA and, if available, they can be used to overrule the calculated data based on the DPD.

Examples of the C&L for citral, *Litsea cubeba* oil, and lemongrass oils are as follows:

Citral
Transport—Not Regulated

Evaluation: Sensitizer—Xi;

Conclusion: RISKS: R 38 irritating to skin

R 43 May cause sensitization by skin contact

SAFETY: S(2) Keep out of reach of children (when sold to general public)

S 24/25 Avoid contact with skin and eyes

S 37 Wear suitable gloves

Litsea cubeba
Evaluation: Sensitizer—Xi, Harmful by Aspiration—Xn, Toxic to Environment—N

Conclusion: Xn+N; R 38–43–51/53-65; S 24/25–37–61-62

The "Xn" label and its associated R 65—S 62 phrases apply because the >10% hydrocarbon content of this oil creates the potential for the oil to be an aspiration hazard. This means that if the oil is swallowed by accident and the patient is induced to vomit, the mobile low viscosity hydrocarbons have the potential to coat the inner surface of the lungs and prevent the entry of breathed oxygen into the bloodstream, the potential danger being asphyxiation (BEOA 2006). This "Xn label" overrides the "Xi label" of the citral when both hazards are present.

Lemongrass
Evaluation: Sensitizer—Xi, Toxic to the Environment—N

Conclusion: Xi+N; R 38-43-51/53; S 24-37-61

The "N" label and its associated R 51/53—S 61 phrases apply to lemongrass and *Litsea cubeba* oils because the terpene hydrocarbons that they contain are slow to biodegrade in the environment and represent a potential hazard.

9.8 SAFETY DATA SHEETS

The data from the C&L is incorporated into safety data sheets (ISO 2001), which also include the physical properties, safe-handling instructions, and transportation guidelines for the material. These SDS are the mechanism for the communication of data for safe handling throughout the supply chain from manufacturer to the user, and a useful start for data gathering under the REACH registration.

9.8.1   GHS

In 1992, UNECE recognized that there were different regulatory standards being applied globally (EU: HSDS; U.S.: MSDS) and started work in a concerted way to bring these together in what has become known as the Globally Harmonized System.

It is planned that an eSDS will be made available for professional users, which will contain additional toxicological and compositional data more detailed than that found on the HSDS.

With the introduction of REACH (in Europe, the GHS is planned to be adopted at the same time), this has become an urgent matter, and there are technical specialist task forces operating in the U.S. and Europe that will then come together for the harmonization.

This is to be achieved (simplistically) by enlarging the number of bands within each hazard group to accommodate the different classifications; for example, toxicity classes are expanded from the EU = 3 to GHS = 5, spanning the range up to LD_{50} of 5000 mg/kg (based on rat oral tests). Some new categories are added as well, for example, respiratory sensitizers.

There are some countries without their own system that had the option of adopting the GHS early; Brazil and New Zealand have done so, and Japan followed suit in December 2006.

The rules of the GHS system do not allow for a "pick and mix" approach; either the CHIP or GHS system can be used until the introduction of the GHS, not some of each.

One of the visible changes is the use of "diamond" pictograms for hazards, some of which are different from the "square" St. Andrews cross type, used under the CHIP regulations in the EU.

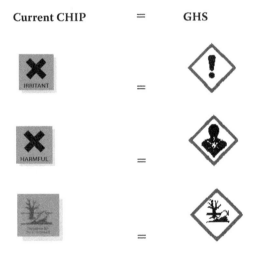

9.8.2   REACH

The European Parliament is introducing a registration system for chemicals that shifts the emphasis for the safe use of industrial materials from government to industry. All materials that are used at more than 1 tonne per annum are required to be registered, a registration that includes the modes of use. Depending on the volume of use and the hazard assessment, the amount of test data required to establish safe use increases, with the very high risk materials potentially substituted with safer alternatives.

The mechanism for registration will be online by using IUCLID 5 [24] data, the latest version of the database that was first created in 1993 to meet the EU requirements of the ESR [25–28].

9.9   CONCLUSION

The toxicologic and legislative status of citral from all sources is covered by the EU legislation of Annex I. This is a listing that is binding under EU law, in recognition of the potentially harm to human health properties.

This chapter has reviewed the status with respect to the latest guidelines and uses of citral within the flavor and fragrance industries.

The sources, production, toxicology, "classification and labeling," industrial uses, and dose rates have been covered for citral itself and its natural sources.

Future requirements of the GHS and REACH systems are also briefly explained.

Citral is an invaluable material within the F&F industries, with a long, safe history of use and an important future, when used under GMP guidelines to protect those who handle it.

ACKNOWLEDGMENTS

I gratefully acknowledge the advice given by Dr. C. Letizia on toxicology and M. Milchard on testing data.

My thanks to the directors of Fuerst Day Lawson for their support and kind permission to publish this paper.

REFERENCES

Adams R. P. 1995. Identification of EO Components by GC/MS, Allured Publishing, Carol Stream, IL.
Arctander S. 1969. *Perfume and Flavor Chemicals*, Allured Publishing, Carol Stream, IL.
Arctander S. 1962. *Perfume and Flavors of Natural Origin*, Allured Publishing, Carol Stream, IL.
BEOA 2006. CHIP List and Handbook BEOA for members (Aug.).
Chaintreau, A. et al. 2003. *J. Agric. Food Chem* 51: 6398–6403.
EFFA Code of Practice August 2008.
Food Chemical Codex 1996. National Academy Press, Washington DC ISBN 0-309-05394-3 IV July.
Food Cosmetics Toxicology 1979. 17(3): 259–266.
IFRA Standard 43rd Draft 2009.
ISO Standard 3214. 2000. *Litsea cubeba* oil.
ISO Standard 11041. 2001. Format for Safety Data Sheets.
ISO Standard 3217. 1974. Lemongrass oil West Indian.
ISO Standard 4718. 2004. Lemongrass oil East Indian.
Jones R. A., Neale M. E., Ridlington J. 1977. *J. Chromatography* 130: 368.
Kimel S. 1953. US patent 2,661,368 Hoffmann-LaRoche.
Lawrence B. M., Hayes J. R., Stavanja M. S. 2007. Biological and Toxicological Properties in *Mint—The Genus Mentha*, CRC Press, Boca Raton, FL.
Leets et al. 1957. *J. Gen. Chem. USSR* 27 1584; 1959. *Chem. Abst.* 53: 4336e.
Leijs H. et al. 2005. *J. Agric. Food Chem* 53: 5487–5491.
Moyler D. A. November 2001. On behalf of EFFA, Proceedings IFEAT Conf. Argentina.
Ohloff G. 1993. *Scent and Fragrances*, Springer-Verlag. ISBN 0-387-57108-6.
RIFM-FEMA. 2009. Database monograph 116.
RSC essential oils committee. 2009. *Perfumer & Flavorist* to be published.
RSC essential oils committee. 1997. *The Analyst (London)* 122: 1167–1174.

OTHER SOURCES

http://ecb.jrc.it/iuclid5/ 2006.
http://ec.europa.eu/enterprise/reach/prep_guidance_en.htm 2006.
http://ecb.jrc.it/REACH.htm 2006.
http://europa.eu.int/comm/environment/chemicals/index.htm 2006.
http://europa.eu.int/comm/enterprise/chemicals/index.htm 2006.
http://www.reachready.co.uk 2007.

Index

Printed and bound by CPI Group (UK) Ltd, Croydon, CR0 4YY

23/10/2024

01778246-0012